CHEMISTRY AND CYTOCHEMISTRY OF NUCLEIC ACIDS AND NUCLEAR PROTEINS

BY

C. SCHOLTISSEK
TÜBINGEN

B. M. RICHARDS
LONDON

R. VENDRELY AND C. VENDRELY
VILLEJUIF

D. P. BLOCH
AUSTIN

WITH 67 FIGURES

1966

SPRINGER-VERLAG

WIEN · NEW YORK

ISBN-13:978-3-211-80782-8   e-ISBN-13:978-3-7091-5569-1
DOI: 10.1007/978-3-7091-5569-1

Softcover reprint of the hardcover 1st edition 1966

TITEL-NR. 8795

Protoplasmatologia
  V. Karyoplasma (Nucleus)
    3. Chemistry and Cytochemistry of Nucleic Acids and (Nuclear) Proteins
    a) The Chemistry and Biological Role of Nucleic Acids

The Chemistry and Biological Role of Nucleic Acids

By

Christoph Scholtissek*

Max-Planck-Institut für Virusforschung, Tübingen, Germany

With 30 Figures

Table of Contents

*) Present address: Institut für Virologie, Gießen, Germany.

Introduction

The field of nucleic acids has grown to such a tremendeous size that it is impossible to include all publications concerning the chemistry and biological role of nucleic acids in an article of the length presented in this volume. Therefore, it is necessary to select the most important contributions and those not included in well-known reviews. In many cases reference is made only to the authors who summarized their specialized field in chapters of the three volumes of "The Nucleic Acids" (eds. E. Chargaff and J. N. Davidson, Acad. Press, New York 1955 and 1960) or to the "Nucleic Acid Outlines" (V. R. Potter, Burgess Publishing Comp. Minneapolis), where further literature and more detailed discussions may be found. Facts and theories will be dealt with, but not lists of references. Therefore it is not possible to follow in all cases the historical development of an idea and to acknowledge all publications which might be important and interesting from another point of view. Very little is mentioned about methods in the field of nucleic acids.

For the purpose of brevity many chemical or metabolic reactions will be expressed in formulae or schemes without referring to all those who elaborated the different and often very complicated pathways leading to the macromolecules or their building blocks. Only the newer literature will be discussed in more detail. Concerning the function and metabolism of the nucleic acids, emphasis is laid on reactions in the cell nucleus. Publications on bacterial or viral nucleic acids are included only if they are necessary for the understanding of processes occuring in the cell nucleus. Thus it is hoped to give a survey of the chemistry of the nucleic acids rather than a detailed description of all the published facts.

I. History of Nucleic Acids

It was in 1869 when F. Miescher, as a pupil of Hoppe-Seyler in Tübingen, isolated a material which he called "nuclein" from pus cells by digestion with pepsin in the presence of HCl. Another "nuclein," which he obtained from salmon sperm, had a phosphorus content of 9.59%, which is in good agreement with recent analytical data of nucleic acids. The term "nucleic acid" was first used by Altmann in 1889, who isolated protein-free nucleic acids from yeast and animal tissues. Piccard, in 1874, recognized guanine and hypoxanthine (derived from adenine) as constituents of nucleic acids. Schulze and Bosshard found guanosine, the first nucleoside (1885). Thymine was isolated already by Miescher, but it was not before 1894 that it was properly identified by Kossel and Neumann. Ascoli (1900 to 1901) obtained uracil from yeast nucleic acid, and Kossel and Steudel and also Levene (1902 to 1903) cytosine from thymus nucleic acid. In 1909, Levene and Jacobs characterized the carbohydrate component of yeast nucleic acid as ribose. Relatively recently (1930) Levene et al. established the structure of the very unstable sugar of thymus nucleic acid as deoxyribose.

Since the hydrolysis of nucleic acids from thymus and other animal tissues yielded d e o x y r i b o s e, but that of yeast or wheat embryo r i b o s e, it was thought that two different types of nucleic acids existed, one only in animals, the deoxyribonucleic acid (DNA), and the other one only in plants, the ribonucleic acid (RNA). However, this supposition was never free of objections. In the years from 1920 to 1924 it was shown by the work of FEULGEN, E. HAMMERSTEN, and JORPES that RNA was also present in animals. Later FEULGEN and also others found DNA in plants. These findings were further supported by the histochemical studies of BRACHET starting about 1933, and the spectrophotometric experiments of CASPERSSON beginning about 1936.

Because of the analytical data available at that time, which showed the four bases in roughly equal amounts, the theory was put forward that the nucleic acid consists of repeating subunits of tetranucleotides. This theory was maintained until about 1944, when it was disproved by better analytical data.

For more detailed information about the early history of the nucleic acids one may look in the first chapter of "The Nucleic Acids" (DAVIDSON and CHARGAFF 1955) and of the "Nucleic Acid Outlines" (POTTER 1960). What we know from this first period of nucleic acid research is that there exist two types of nucleic acids: The D N A consisting of the four bases adenine, guanine, cytosine, and thymine, D-2-deoxyribose and phosphate: and the R N A consisting of the four bases adenine, guanine, cytosine, and uracil, D-ribose and phosphate, both present in plants a n d animals.

One important step forward in the chemistry of nucleic acids was the discovery by BROWN and TODD in 1952 and 1953 that the nucleosides in the nucleic acids are joined by phosphodiester linkages between the 3'- and 5'-OH-group of the sugars. CHARGAFF had already noticed in 1950 that regularities existed in the ratios of adenine to thymine and guanine to cytosine in DNA's of different origin. These facts, together with X-ray diffraction data, led WATSON and CRICK (1953) to their theory of a double-helical structure of DNA.

The first biosynthesis of RNA i n v i t r o, starting from the nucleoside diphosphates, was obtained by GRUNBERG-MANAGO and OCHOA in 1955. One year later (1956) KORNBERG et al. isolated from bacteria an i n v i t r o system capable of synthesizing DNA from deoxynucleoside triphosphates. BOLLUM and POTTER (1957) found similar enzymes in animal cells. Later WEISS and GLADSTONE (1959) described an i n v i t r o system, in which they used the four ribonucleoside triphosphates as starting material for RNA synthesis. This RNA synthesis is DNA-dependent.

KHORANA and coworkers (TENER et al. 1958) synthesized polynucleotides of relatively low molecular weights by chemical means, which were very useful as model compounds. SCHRAMM et al. (1962), using a different method, obtained nucleic acids of a somewhat higher degree of polymerization.

Concerning the function of the nucleic acids one of the most important events in the history of nucleic acids was the discovery of AVERY et al. (1944), who succeeded in transforming an unencapsulated strain of pneumococcus into an encapsulated one by DNA extracted from encapsulated cells. This was the first direct evidence that a specific property of a cell is inherited through its DNA. Hershey and Chase (1953) labelled phages by P32 (DNA) and by S35 (protein). They found that only the DNA was injected into the bacterial cell, while 97% of the protein stayed outside. They concluded that the DNA is the only genetic material of the phages. Later Gierer and Schramm (1956) isolated an infectious RNA from tobacco mosaic virus. In this case R N A functions as the genetic material. Since that time, infectious nucleic acids have been extracted from a variety of RNA- and DNA-containing viruses.

Schuster and Schramm (1958) showed that infectious RNA can be deaminated by nitrous acid without destroying the macromolecular structure. This was the basis for the mutational studies of the genetic material in vitro by Gierer and Mundry (1958).

From the fact that the DNA is contained in the chromosomes of the cell nucleus, while protein synthesis proceeds mainly on the ribosomes in the cytoplasm, the conclusion was drawn that the DNA does not function directly as a template for protein synthesis, but needs a messenger carrying the information from the cell nucleus to the cytoplasm. This messenger was thought to be RNA. The messenger concept was put forward by Jacob and Monod (1961) working in the field of bacterial genetics. They postulated for such a messenger RNA a high rate of turnover in order to account for its regulation by genes; it should be synthesized on the DNA and function on the ribosomes, the protein factories of the cell, and it should have a structure complementary to the coding DNA.

A rapidly-turning-over RNA with a base composition complementary to the coding DNA was first found by Volkin and Astrachan in phage-infected bacteria (1956, Astrachan and Volkin 1958). But the best evidence for the messenger function of such an RNA was given by Brenner et al. (1961) and Gros et al. (1961) in bacterial systems. One important discovery which strengthened the messenger theory was the finding of a hybrid between DNA and RNA in vitro (Schildkraut et al. 1961) and also in vivo (Hayashi and Spiegelman 1961).

Concerning animal systems, much metabolic heterogeneity was found in nuclear RNA by Allfrey and Mirsky (1957). Scholtissek et al. (1958, 1962 a) demonstrated that rat liver nuclei contain a small RNA fraction having some of the particular properties one might expect for such a messenger RNA. Later, unstable RNA fractions were described in several animal cells. Recently it has been demonstrated that the messenger RNA forms aggregates with ribosomes. Only these "polysomes" are active in protein synthesis (Wettstein et al. 1963, Warner et al. 1963, Gierer 1963).

The first step in the elucidation of the amino acid code—how the base sequence directs the sequence of the amino acids within a protein—was the finding of Nirenberg and Matthaei (1961) that synthetic polyuridylic acid stimulates the incorporation of C14-phenylalanine into an acid-insoluble product, using a messenger RNA-exhausted bacterial system. Later Ochoa and his group (Lengyel et al. 1961), Wittmann (1961), and Nirenberg and his group (Matthaei et al. 1962) worked out the preliminary code for 20 different amino acids. The best evidence for a triplet code—

this means that 3 nucleotides determine one amino acid—was contributed by the work of CRICK et al. (1961).

Looking back to the time when the nucleic acids were first discovered by MIESCHER in 1869, one sees how long it took to recognize the highly important role of these macromolecules in the living cell. Most studies were devoted to their analysis, synthesis, and function. So far only a few experiments have been made with the aim of altering the genetic material without destroying it. But this might be the most interesting field for future research: To alter the genetic substances or their expression within a cell in a p r e d i c t a b l e f a s h i o n. Such experiments might lead to an understanding of cell differentiation, organ specificity, and to the cure of cancer.

II. Chemistry of Nucleic Acids

1. Purine and Pyrimidine Bases

As mentioned already in chapter I there exist five major bases in nucleic acids. Three of them, adenine, guanine, and cytosine, are found in both DNA and RNA. Thymine occurs only in DNA, and uracil only in RNA. Since little is known about the existence and the role of other trace

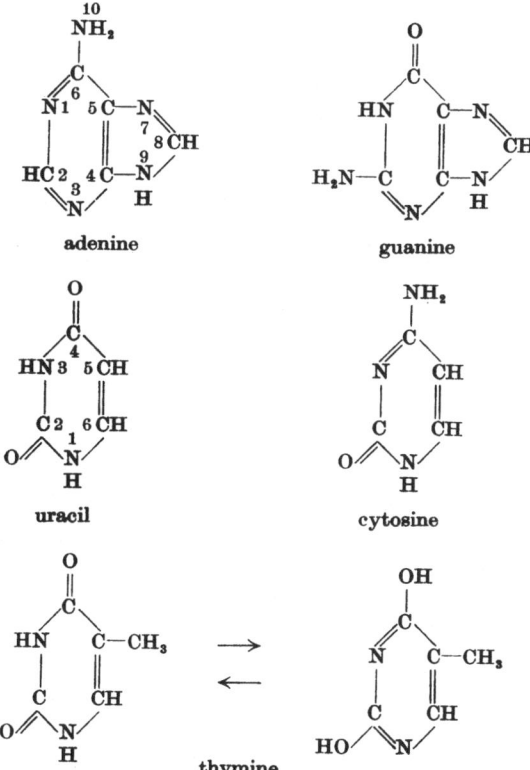

Fig. 1. Purines and pyrimidines occuring in nucleic acids.

purines and pyrimidines in nuclear DNA and RNA, such rare bases will not be treated in this article.

The structure and numbering of the five bases used in this article are shown in Fig. 1. The bases may be written also in other tautomeric forms, as shown for thymine (see also chapter III). Adenine and guanine belong to the purines (pu) while uracil, cytosine, and thymine are pyrimidines (py).

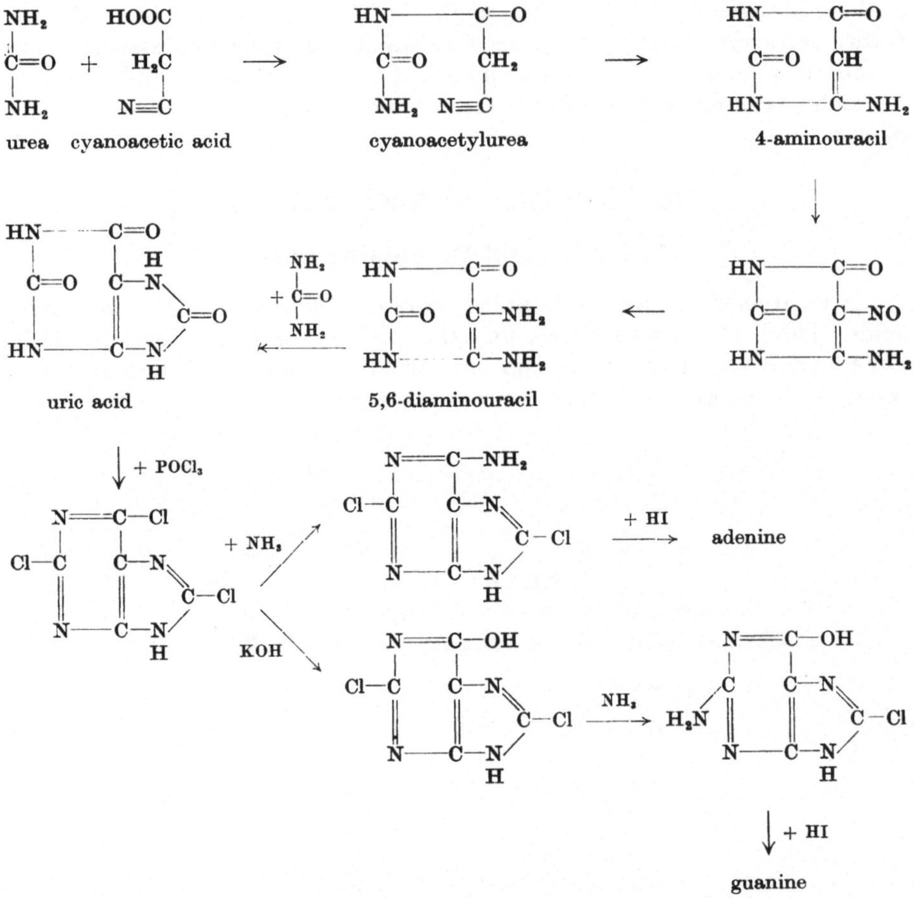

Fig. 2. Synthesis of the purines.

A few formula schemes are described here (Fig. 2 and 3) which lead to the chemical synthesis of the 5 bases and which can be used for the labelling of these compounds by isotopes. A detailed literature may be found in the review by Bendich (1955). If in Fig. 2 guanidine is used instead of urea for the first condensation and formic acid instead of urea for the second ring condensation, one obtaines guanine directly. Further, if one takes in Fig. 3 2-formylethylacetate, instead of 2-formylethylpropionate, than uracil results instead of thymine. By including other starting compounds into these reaction schemes it is easy to obtain bases which do not occur in nature and which might be used as antimetabolites, like 5-fluorouracil, 8-azaguanine, thiouracil, etc. Other bases do occur in nature and are important precursors, but are not incorporated into nucleic acids, like orotic acid (6-carboxyuracil), a precursor of the uracil moiety.

The amino groups of the bases are converted by nitrous acid into hydroxyl groups. In this way cytosine is converted into uracil, another

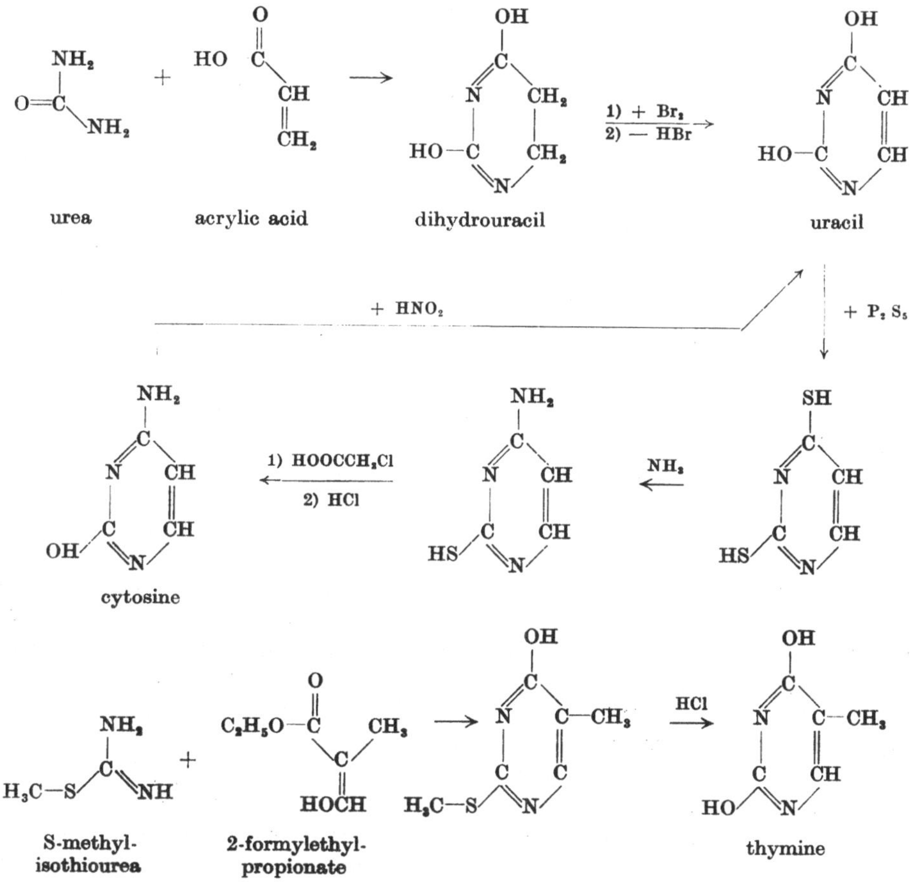

Fig. 3. Synthesis of the pyrimidines.

base occuring naturally in RNA. Adenine is converted into hypoxanthine, and guanine into xanthine, both of which are not found naturally in nucleic acids in appreciable amounts. Those reactions are possible also with highly polymerized nucleic acids and can lead to a mutation or the killing of a virus (SCHUSTER and SCHRAMM 1958, GIERER and MUNDRY 1958). The reaction of the pyrimidine bases in nucleic acids with hydroxylamine may be important in this connection; mutations were also found (SCHUSTER 1961, FREESE et al. 1961, VERWOERD et al. 1961). The formula scheme is shown in Fig. 4. At high pH uridine is destroyed, at low pH cytidine can be converted to uridine.

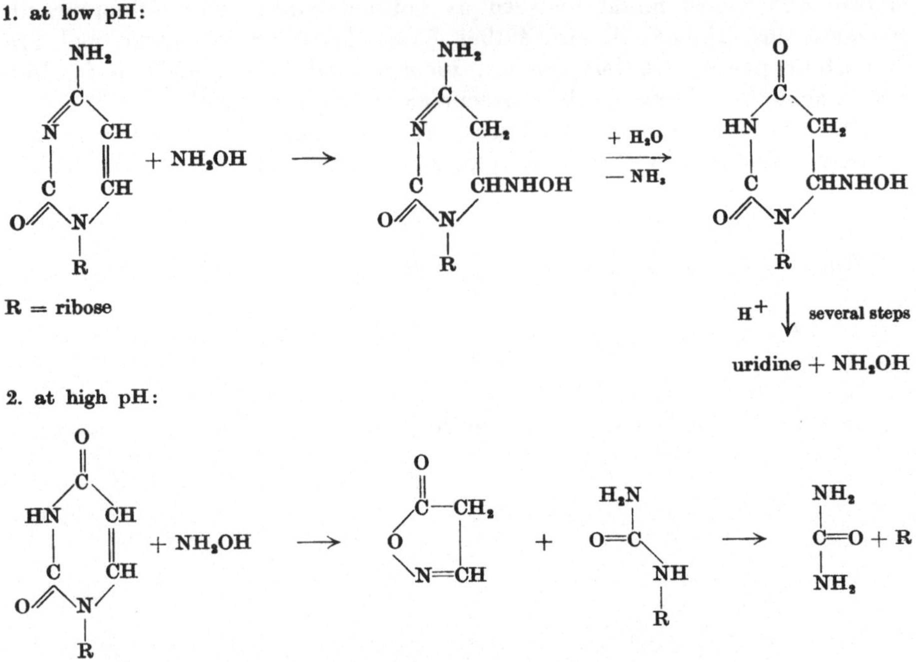

1. at low pH:

R = ribose

uridine + NH$_2$OH

2. at high pH:

Fig. 4. Action of hydroxylamine on pyrimidine nucleosides.

2. Nucleosides

After mild acid hydrolysis of RNA a pentose was obtained which was identified as D-ribose. In order to isolate ribose from pyrimidine nucleosides, it is necessary to reduce them to the 5,6-dihydropyrimidine ribosides, from which the ribose can be split off easily by acid. Much more labile to acid is the sugar component of DNA, which was isolated from the corresponding nucleosides much later under very mild acid conditions. It was identified as D-2-deoxyribose.

The hydrolytic properties of the nucleosides suggested an N-glycosidic linkage of the sugars. Since adenosine and guanosine can be converted by nitrous acid to the hypoxanthine and xanthine ribosides, respectively, the primary amino nitrogens can be excluded. Xanthosine can be converted to theophylline riboside (methyl groups on N-1 and N-3). Spectrophotometric studies on 9-methyl- and 7-methyl-derivatives of adenine finally established the structure of the purine ribonucleosides as 9-glycosides. Similar studies were carried through with the deoxyribonucleosides with corresponding results.

Uridine can be converted to N-methyl-uridine, which yields 3-methyl-uracil after hydrolysis. The positions 5 and 6 can be substituted without cleaving the glycosidic linkage. Cytidine is convertable by nitrous acid to uridine. Similar experiments with corresponding results were performed with thymidine. Therefore the pyrimidine nucleosides are 1-N-(deoxy)-ribosides.

The ring size of the sugar components of the nucleosides was determined by methylation and oxidation. They are furanosides. Further evidence for the furanose structure is the action of periodate on the nucleosides. For the oxidation of ribonucleosides one mole of periodate is consumed leading to a dialdehyde but no formic acid. The cis-glycol group is split under these conditions. The deoxyribosides do not consume periodate.

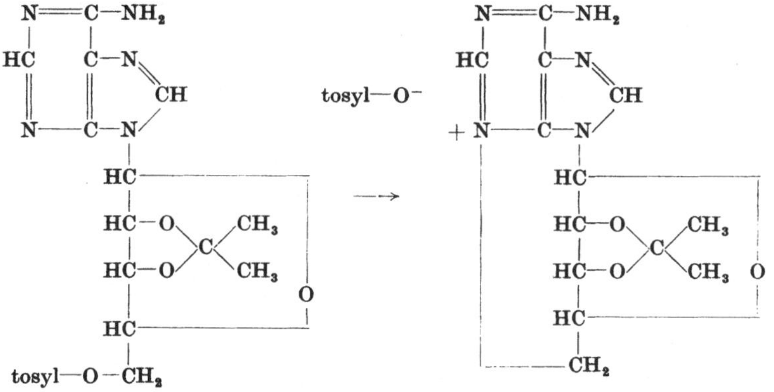

Fig. 5. Isomerization of 2',3'-isopropylidene-5'-p-toluenesulfonyladenosine.

The configuration of the glycosidic linkage of the pyrimidine and purine nucleosides was studied on synthetic samples of 1- or 9-β-D-glucopyranosyl derivatives, respectively, which yielded after periodate oxidation the same dialdehyde as the natural nucleosides. Other evidence for the β-configuration is the isomerization of the 5'-tosyl-derivative of adenosine and cytidine. For steric reasons only the β-glycosides react as shown in Fig. 5.

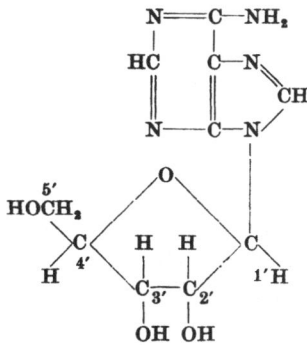

Fig. 6. Adenosine.

X-ray studies have also confirmed the β-configuration of the nucleosides, one of which is presented with the numbering of the sugar C-atoms in Fig. 6.

Three main routes are available for the chemical synthesis of nucleosides. The first one starts with the silver or mercuri salts of the 2,6,8-trichloropurine and acetohalo-sugars. The second one, which leads again to the purine nucleosides, starts with a 4-glycosyl aminopyrimidine. The amino group in position 5 is introduced by coupling with a diazonium salt and reduction. This amino-group is further converted to the 5-thioformamido-pyrimidine and after cyclization the corresponding purine nucleoside results. The most convenient nucleoside synthesis is the third one, which has been discovered recently by Schramm et al. (1962). They condensed the bases with different sugars in the presence of polyphosphoric acid ester and obtained in good yields in one step the corresponding nucleosides.

By the chemical synthesis of nucleosides it is possible to obtain also nucleosides which do not occur in nature and which might be useful in studies with antimetabolites.

A more thorough description of the structure and synthesis of nucleosides may be found in the chapter of "The Nucleic Acids" by Baddiley (1955). More recently Michelson (1961, 1963) reviewed the synthesis of nucleosides again. In these articles he described the chemical conversion of ribosides from the 2'-iodo derivatives into the 2'-deoxyribosides and the synthesis of pyrimidine and purine nucleosides starting from N-glycyl ribofuranosylamine.

3. Nucleotides

Nucleotides—the phosphate esters of the nucleosides— occur free within the cell (acid soluble) or in a polymeric state as nucleic acids (acid insoluble). Mononucleotides are known with one or more phosphate groups which may carry additional substituents (for example coenzymes). Only such mononucleotides which are involved in nucleic acid synthesis or which are necessary for the understanding of the nucleic acid structure will be considered here.

Concerning the ribonucleotides, only 3 hydroxyl groups are available for the esterification with phosphate, those on positions 2', 3', and 5'. All three monophosphates are known. In addition the 2', 3'-cyclic phosphates have been isolated. Furthermore we know of 3', 5'-diphosphates of ribo- and deoxyribonucleosides and their 5'-di and triphosphates (in pyrophosphate linkage), which are the precursors for the biochemical nucleic acid synthesis.

The very efficient analytical methods, like paper chromatography (review by Wyatt 1955), ion exchange chromatography (review by Cohn 1955), and paper electrophoresis (review by Smith 1955), which led to the separation and isolation of the different nucleotides, will not be described in this article, although it is not possible to imagine the marvelous progress in the field of nucleic acids without the application of these tools.

The acid-soluble ribonucleotides which can be extracted from cells have their phosphate in ester linkage at the 5'-position, as established by the following facts:

a) Only the 5'-nucleotides are split by the action of the 5'-phosphatase (see below).

b) They consume 1 mole of periodate; therefore the hydroxyl groups in the 2'- and 3'-positions must be free.

c) Acid hydrolysis of muscle inosinic acid leads to D-ribose-5-phosphate, which can be oxidized to D-ribonic acid-5-phosphate, the structure of which was proven by synthesis.

The ribonucleoside di- and tri-phosphates of cell extracts can be hydrolyzed to 5'-monophosphates and phosphate or pyrophosphate, respectively. It was shown by alkali titration that they contain one secondary and two or three primary phosphoric acid groups, respectively, indicating a linear di- or tri-phosphate in the 5'-position. The positions 2' and 3' must be unsubstituted, since they consume 1 mole of periodate and increase the conductivity of boric acid.

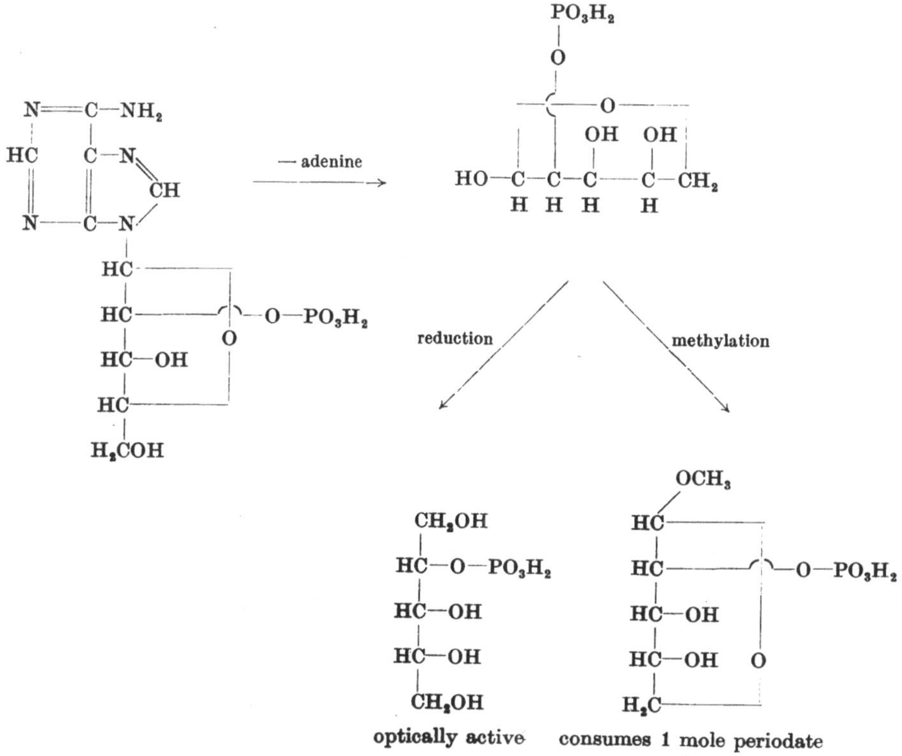

Fig. 7. Degradation of adenosine-2'-phosphate.

The 2'- and 3'-ribonucleotides are obtained by alkali digestion of ribonucleic acids (RNA). The structures of the adenylic acids were established in the following way: The two isomers can be hydrolyzed by an acidic ion exchange resin to the free bases and the corresponding ribose phosphates without isomerization. The methyl glycoside of the ribose-2-phosphate consumes 1 mole of periodate, while the 3-isomere is stable against this oxidation (see Fig. 7, shown for the 2'-isomer). Furthermore the ribose phosphates can be reduced to the ribitol phosphates. The ribitol-3-phosphate is optically inactive, in contrast to the 2-isomer which is optically active (see Fig. 7). By comparison of the physical properties and analysis the structure of the other 2'- and 3'- nucleotides was established.

The pyrimidine-2′, 3′-cyclic phosphates occur among the products of pancreatic ribonuclease digestion of RNA. They are also formed, together with the purine-2′, 3′-cyclic phosphates, by mild alkali digestion of RNA. Their structure is established by synthesis. Starting from the 2′- o r 3′-nucleotide the reaction with trifluoroacetic anhydride results in the formation of the same 2′, 3′-cyclic phosphate.

From enzymic hydrolysis of deoxyribonucleic acid (DNA) the 5′-deoxyribonucleotides of guanine, adenine, cytosine, and thymine were obtained. Their structures were established by the action of 5′-nucleotidase and by synthesis. Acid hydrolysis of DNA also yields in addition to pyrimidine mononucleotides varying amounts of pyrimidine diphosphates, the structures of which were established by synthesis as 3′, 5′-diphosphates.

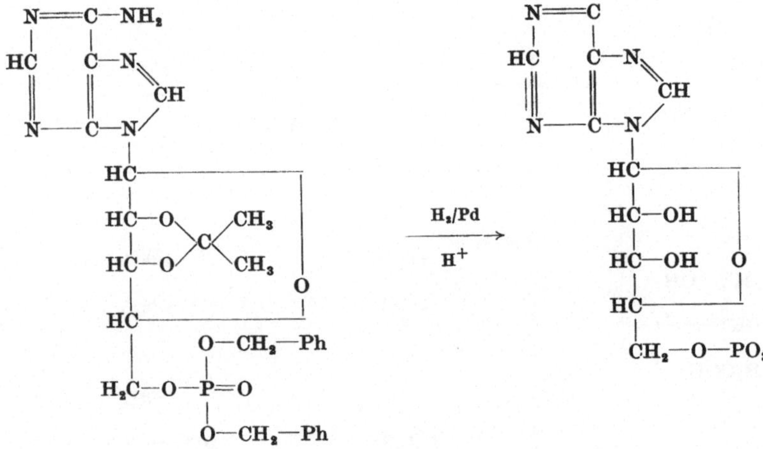

Fig. 8. Step in the synthesis of adenosine-5′-monophosphate.

The 5′-deoxynucleoside di- and triphosphates (pyrophosphate linkage) were also discovered in cell extracts.

The 5′-ribonucleotides can be synthesized by converting the nucleosides into the 2′, 3′-isopropylidene derivatives, which react with dibenzyl phosphorochloridate to give the 5′-dibenzyl phosphates. By catalytic hydrogenation and acid hydrolysis the benzyl-groups and the isopropylidene residue are removed (see Fig. 8). The synthesis of thymidine 5′-monophosphate is accomplished by acetylation of 5′-trityl thymidine in the 3′-position, mild acid hydrolysis, by which the trityl-group is removed, and phosphorylation by dibenzyl phosphorochloridate in the 5′-position followed by removal of the protecting groups. If the 5′-trityl thymidine is phosphorylated directly, the unnatural 3′-thymidylic acid may be obtained.

The 5′-adenosine diphosphate (ADP) is synthesized starting from adenosine- 2′, 3′-isopropylidene-5′-dibenzyl phosphate (Fig. 8), from which the isopropylidene and one benzyl residue are removed by mild acid hydrolysis. To the silver salt, dibenzyl phosphorochloridate is added to give adenosine-5′-tribenzyl pyrophosphate. Catalytic hydrogenation gives ADP. 5′-adenosine triphosphate (ATP) may be synthesized from the disilver salt of 5'-adenosine monophosphate (AMP) and two moles of dibenzyl phosphorochloridate. Passing through an unstable cyclic intermediate and after removal of the benzyl residues, ATP is formed.

Further discussion and detailed literature on the nucleotide synthesis may be found in the review of BADDILEY (1955). In the reviews of MICHELSON (1961, 1963) and KHORANA (1961) other powerful phosphorylating agents are proposed and discussed for the synthesis of nucleotides and their anhydrides.

4. Nucleic Acids

Mild alkali hydrolysis of ribonucleic acid (RNA) leads to the cyclic 2', 3'-phosphates and thereafter to the mixture of the 2'- and 3'-nucleotides of the four bases adenine, guanine, cytosine, and uracil. Theoretically the

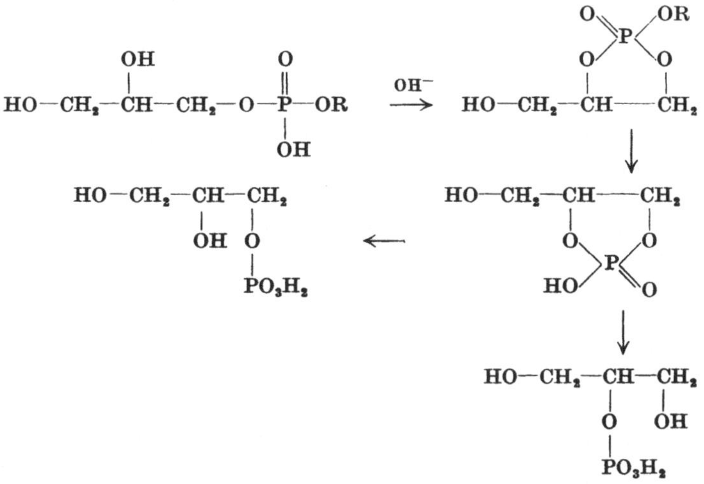

Fig. 9. Conversion of glycerol-α-methylphosphate by alkali. R = methyl.

macromolecules can be formed by phosphate diester bonds between the 2'-, 3'- or 5'-positions of the carbohydrate component. Comparative studies with glycerol-α-methylphosphate have shown that by treatment with alkali the cyclic phosphate triester of Fig. 9 is formed, from which the methyl group is released very easily. The cyclic ester breaks down to a mixture of glycerol-α- and β-phosphate. The action of snake venom phosphodiesterase on RNA leads to the 5'-mononucleotides. From these data the conclusion was drawn that the 2'- or 3'-position of one nucleoside is connected through a phosphate diester linkage to the 5'-position of the next nucleoside. During alkali hydrolysis a cyclic phosphate triester is formed, from which the primary alcohol group (5'-position) is split off. Since the deoxyribonucleic acid (DNA) does not contain a 2'-OH-group, it is stable to alkali and in this way DNA can be separated from RNA by destruction of the latter.

Digestion of RNA by pancreatic RNase (see below) gives oligonucleotides and 3'-pyrimidine mononucleotides. Some other enzymes (phosphodiesterases) split RNA in such a way as to release the 3'-nucleotides of all four bases. The benzyl esters of the 2'-pyrimidine mononucleotides are not attacked by these enzymes, but the 3'-isomers are; this is evidence that in RNA the phosphate linkages are between the position 3' and 5'.

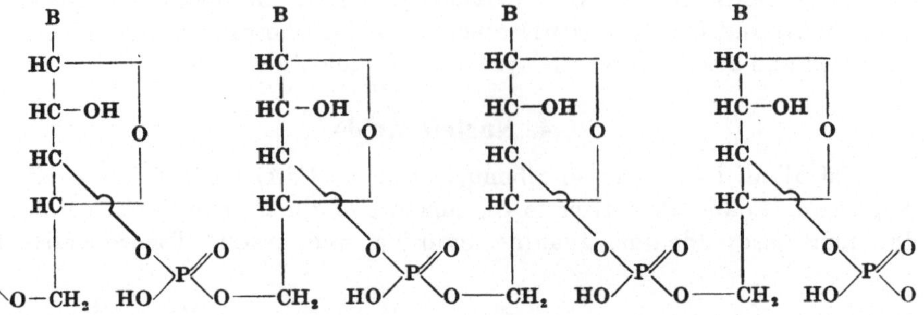

Fig. 10. Scheme of ribonucleic acid. B = adenine, cytosine, guanine or uracil.

Fig. 10 gives a simplified scheme of the structure of RNA. About the sequence of the bases almost nothing is known.

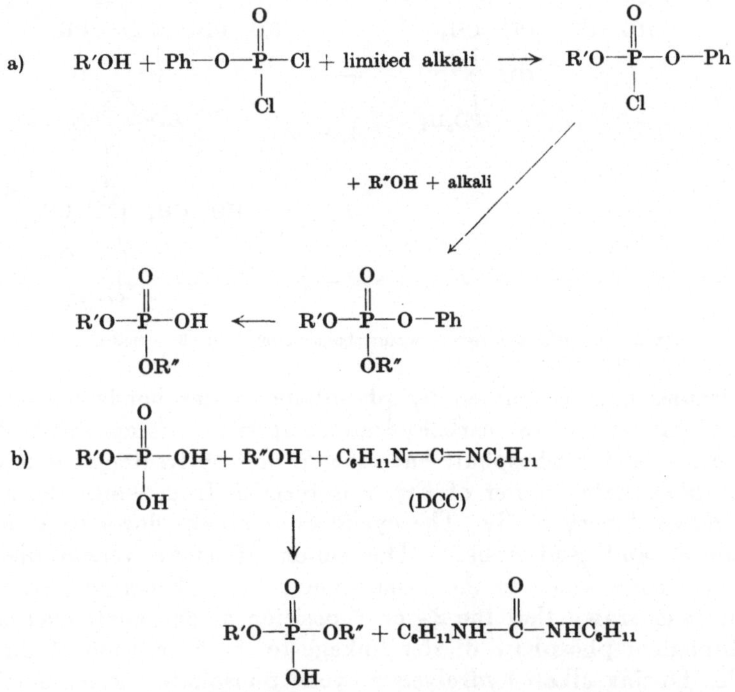

a) $R'OH + Ph-O-\overset{O}{\underset{Cl}{\overset{\|}{P}}}-Cl$ + limited alkali $\longrightarrow$ $R'O-\overset{O}{\underset{Cl}{\overset{\|}{P}}}-O-Ph$

$+ R'OH + alkali$

$R'O-\overset{O}{\underset{OR''}{\overset{\|}{P}}}-OH$ $\longleftarrow$ $R'O-\overset{O}{\underset{OR''}{\overset{\|}{P}}}-O-Ph$

b) $R'O-\overset{O}{\underset{OH}{\overset{\|}{P}}}-OH + R''OH + C_6H_{11}N=C=NC_6H_{11}$

(DCC)

$R'O-\overset{O}{\underset{OH}{\overset{\|}{P}}}-OR'' + C_6H_{11}NH-\overset{O}{\overset{\|}{C}}-NHC_6H_{11}$

Fig. 11. Synthesis of dinucleotides. R = appropriately-protected nucleosides.

Digestion of DNA by pancreatic DNase and snake venom phosphodiesterase yields the 5'-deoxyribonucleotides of the four bases adenine, guanine, cytosine, and thymine. Acid hydrolysis of DNA gives apurinic acid, which is further degraded resulting in the 3', 5'-diphosphates of the pyrimidine nucleosides and pyrimidine oligonucleotides. From these data the structure of DNA can be written in the same way as that of RNA (Fig. 10) except for the lack of an OH-group in the 2'-position and thymine instead of uracil.

The partial degradation of the DNA by acid has been used by Shapiro and Chargaff (1957 b) and by Burton and Petersen (1960) for the characterization of DNA's of different sources. The pyrimidines can be taken off

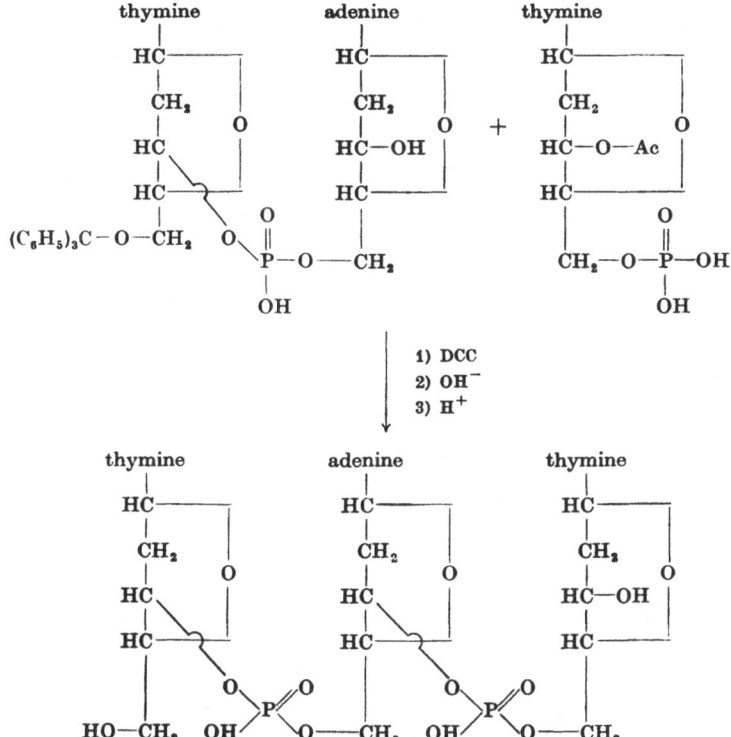

Fig. 12. Synthesis of a trinucleotide.

from nucleic acids by hydrazine (Takemura 1958). In this case apyrimidinic acid results. This method in connection with other specific splitting reactions may be useful in sequence analysis studies (Verwoerd and Zillig 1963).

Concerning the chemical structure of the nucleic acids a better and more extensive survey is given in the first volume of "The Nucleic Acids" by Brown and Todd (1955). Studies on the chemical synthesis of oligonucleotides are summerized by Khorana (1960, 1961) and by Michelson (1961, 1963). Some of these experiments will be described below.

Two different principles of oligonucleotide synthesis will be treated here briefly. The first one uses bifunctional phosphorylating agents like monophenylphosphorodichloridate (Fig. 11 a); for the second one, one nucleotide is activated by dicyclohexylcarbodiimide (DCC) or reactive anhydrides. This activated nucleotide then reacts with another nucleotide (Fig. 11 b).

The most probable mechanism of the latter reaction is discussed by Weimann and Khorana (1962). In order to get the right 3′, 5′-linkages it is necessary to protect properly the other hydroxyl groups of the carbohydrate moiety.

In this way deoxyribooligonucleotides can also be obtained in a higher degree of polymerization (e.g. Fig. 12). The method is applicable also to ribonucleotides, if ribonucleosides or ribonucleotides protected correspondingly are used. The yield of such reactions depends very much on the protecting groups. Thus in most cases it is better to activate the phosphate group in the 3′-position and to combine it with a nucleoside or nucleotide with a free 5′-OH-group than the reverse. If the phosphate group in the 3′-position is combined with a benzhydrol residue, the 2′-OH group in a ribonucleotide is also protected by this voluminous molecule (Cramer and Scheit 1962).

Michelson (see review of 1961) was able to synthesize ribooligonucleotides with a maximum of 12 nucleotides using tetraphenylpyrophosphate or diphenylphosphorochloridate plus tri-n-butylamine. Khorana et al. (see review 1960) obtained a similar degree of polymerization in the deoxy series using DCC. Polymerization of protected dinucleotides resulted in oligonucleotides with an alternating sequence (Schaller et al. 1963). Cramer (1961) described a method for condensing adenosine diethylpyrophosphate ester to polyadenylic acid with a maximum of 5 nucleotides per molecule. Schramm et al. (1962) used polyphosphoric acid ester as the condensing agent for the synthesis of polynucleotides of the deoxyribose and ribose types. Starting with ribomononucleotides, products of a relatively high molecular weight were synthesized, but it is not yet clear whether they contain all phosphate linkages in the 3′, 5′-positions.

III. Physical Properties of Nucleic Acids

1. Bases, Nucleosides, and Nucleotides

X-ray diffraction studies and dipole measurements on pyrimidine derivatives have shown that the pyrimidine ring is planar. The C–N and C–C bond distances in the ring correspond to about 50 % double bond character, as has been found for the benzene ring. In the case of uracil and cytosine three tautomeric forms are possible (Fig. 13). X-ray studies of the crystalline compounds have revealed the amino character in cytosine derivatives. In contrast the C–O link has considerable double bond character, although the hydrogen atom is still covalently bound to the oxygen. As shown by UV-measurements in aqueous solutions at neutral pH, the keto form prevails, and therefore in solution uracil should be written as 2,4-pyrimidinedione. At higher pH-values ionization takes place.

Similar structures have been found for the purines. The C_6–O bond in guanine has double bond character. The short length of the C_6–N_{10} bond of adenine may be due to the contribution of resonance forms in which the N_{10} has a positive charge. Therefore the C_6–N_{10} bond may have some double bond character. In guanine, hydrogen bonds exist between N_3 and N_{10} and between O and N_7.

X-ray studies on nucleosides have confirmed the furanose structure of the ribose and the attachment of the sugar at N_1 for pyrimidines and at N_9 for purines in β-glycosidic configuration. The atoms of the ribose ring lie almost in one plane which is nearly perpendicular to the plane of the bases.

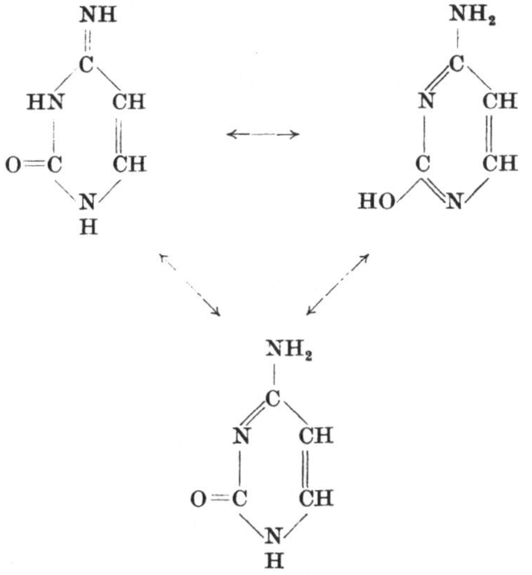

Fig. 13. Tautomeric forms of cytosine.

The acid-base properties of the pyrimidines and the purines are expressed by their pK'_a-values, some of which are listed in Table 1. The pK'_a-value of 9.45 in uracil belongs to the $C=O$ group in position 4 $\left(\text{>COH} \longleftrightarrow \text{>CO}^- + H^+\right)$, which is not markedly changed by the introduction of a methyl group in position 5 (thymine). The dissociation constant

Tab. 1. *pK'a values of bases, nucleosides, and nucleotides.* [Taken from JORDAN (1955)].

	$pK'\alpha_1$	$pK'\alpha_2$	$pK'\alpha_3$	$pK'\alpha_4$
uracil	9.45			
thymine	9.82			
cytosine	4.60	12.16		
adenine	4.15	9.80		
guanine	3.3	9.20	12.3	
uridine	9.2	12.5		
cytidine	4.2	12.3		
adenosine	3.5	12.5		
guanosine	1.6	9.2	12.3	
uridylic acid	1.0	5.9	9.4	
cytidylic acid	0.8	4.2	6.0	
adenylic acid	0.9	3.7	6.0	
guanylic acid	0.7	2.4	6.0	9.3

of the $C = O$ group in position 2 is too weak to be measurable in uracil. In cytosine it has the value of 12.16. The acidic dissotiation of $pK'_a = 12.3$ in guanine must represent the dissociation of the -NH-group to $-\overline{N}-$ in the imidazol ring. In adenine the corresponding value is surprisingly low (9.80), probably because of a considerable contribution of this group to the resonance of the total structure. In the nucleosides the pK'_a-values are somewhat lowered and additional values appear corresponding to those of the sugar components (around 12.5). The nucleotides show again additional pK'_a-values for the first (about 0.9) and second (about 6.0) dissociations of the phosphates.

2. Nucleic Acids

Several important contributions to the structure of DNA were made before Watson and Crick (1953) proposed the helical model for DNA. Astbury and Bell (1938) showed that the X-ray fiber photograph gave reflections corresponding to a spacing of 3.4 Å along the fiber axis. By ultraviolet dichroism studies it was found that the ring planes of the bases were oriented perpendicular to the long axis of the macro-molecule (Caspersson 1940). The results of Wilkins et al. (1953) suggested a helical structure for nucleic acids. Chargaff and his group found that the ratios of adenine to thymine and of guanine to cytosine in DNA-samples examined by them was always close to unity. Taking together all available experimental data, Watson and Crick formulated their model (Fig. 14) in the following way: The bases are located at the inside of two helical chains each coiled around the same axis, but the sequences of the ... sugar-3'-phosphate-5'-sugar linkages in the two chains run in opposite directions. The ring planes are perpendicular to the axis with a repeat distance of 3.4 Å. There are 10 nucleotides (34 Å) to one turn of the helix, so that the angle between adjacent nucleotides in the same chain is 36°. The phosphate groups are located at the outside of the molecule. The two chains are held together by hydrogen bonds between the bases. Only certain base pairs are acceptable. One member of the pair must be a purine, the other one a pyrimidine. If one considers all tauto-meric forms, the following base pairs are possible:

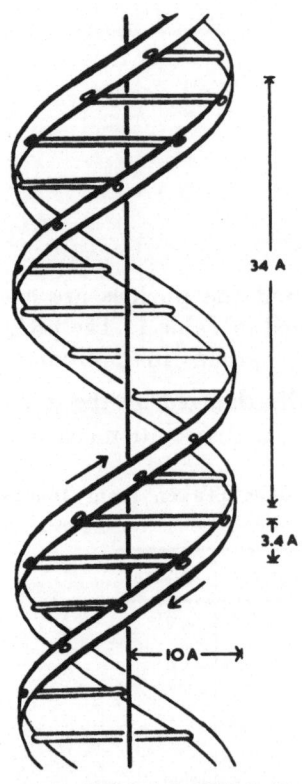

Fig. 14. Model of the DNA double helix proposed by Watson and Crick (1953).

A–T   T–A

G–C   C–G

Fig. 15 shows the permitted hydrogen bonds of the Watson-Crick structure.

Evidence that the amino group of the guanine might not take part in hydrogen bonding has been obtained by titration experiments (see review of Jordan 1955) and more recently by the reaction of native DNA with nitrous acid (Schuster 1960). The reaction with formaldehyde has also been taken as a measure of hydrogen bonding in native and denatured DNA (Haselkorn and Doty 1961).

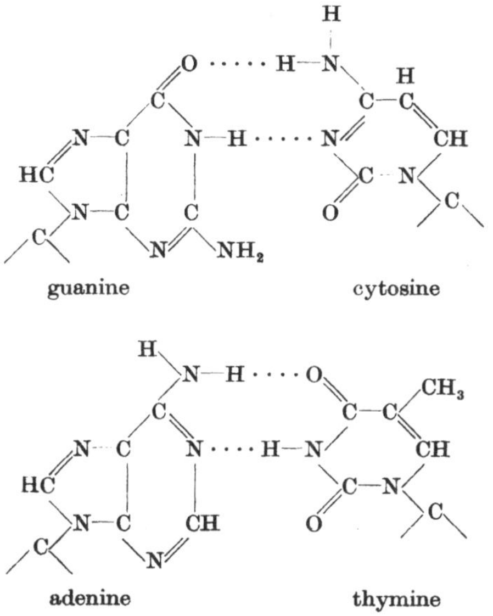

Fig. 15. Permitted hydrogen bonds in the Watson-Crick model.

Double-stranded DNA forms long and relatively rigid molecules which show a high viscosity in aqueous solution, while single-stranded nucleic acids form random coils. In nature single-stranded DNA has been found only in some phages ($\phi\times174$) (Sinsheimer 1959), while RNA appears in almost all cases to be single-stranded. An exception is e.g. a reovirus, which contains a double-stranded RNA (Gomatos and Tamm 1963).

The two strands of DNA can be separated by heating in solution, which gives rise to an increase in optical density (see below). This heat denaturation is reversible, if a relatively homogeneous DNA preparation is used (Doty et al. 1960). It has been found that the melting temperature rises with a higher G-C-content of the DNA. From the melting temperature of a double-stranded DNA the base composition can be determined (Marmur and Doty 1959). A linear relationship also exists between the density of DNA and its guanine-cytosine content (Sueoka et al. 1959).

Double- or triple-stranded RNA can be prepared from synthetic polymers, e.g. from poly-U and poly-A. Whether a double- or a triple-stranded RNA occurs depends on the ratio of the amounts of the two polymers and the conditions used (Rich and Davis 1956, Felsenfeld and Rich 1957). Recently a doublestranded RNA of a complex base composition has been isolated from an in vitro synthesized phage-specific RNA (Geiduschek et al. 1962). The melting point of this RNA is considerably lower (63⁰ C) than that of a corresponding DNA. A double-stranded RNA has been obtained very recently from reovirus, the melting point of which is 99⁰ C (Gomatos and Tamm 1963).

Hybrid complexes between RNA and DNA can be formed from synthetic or natural polymers (Schildkraut et al. 1961, Hall and Spiegelman 1961). A special feature of such hybrids and of double-stranded RNA is that they are stable against pancreatic RNase. This property can be used to decide whether one is dealing with a single-stranded RNA or a hybrid or double-stranded RNA. Another possibility to differentiate between two complementary or non-complementary strands is the determination of the base composition as mentioned above. In a complementary double-stranded nucleic acid the ratios G/C and A/U (or A/T) are unity; in non complementary nucleic acid strands they should be in most cases different from unity. In this way it is easy also to find out whether a single-stranded nucleic acid functions as a template for the synthesis of a newly synthesized nucleic acid. The newly formed nucleic acid should have in this case the reciprocal G/C-, A/U-, or A/T-values of the template.

Fig. 16. Duplication scheme for DNA.

Methods which are concerned with the determination of the molecular weight and the shape of nucleic acid molecules will not be treated in this article. Those methods and a more precise discussion of the problems mentioned above may be looked for in the review articles by Jordan (1955) and Sandron (1960).

For DNA, molecular weights up to about 10^9 have been found. In order to prepare a non-degraded DNA it is necessary to avoid any shearing forces during its isolation. Recently, radioautography has been used for such molecular weight determinations (Cairns 1962, 1963). Concerning RNA, different classes of molecular weights have been found:

1. Soluble RNA with a molecular weight of about 30,000.
2. Ribosomal RNA of 0.55×10^6 and 1.1×10^6.
3. Messenger RNA of different molecular weights. Since messenger RNA (see below) is metabolically very unstable, the method of its preparation might be very important.
4. Viral RNA of molecular weights in most cases around 2,000,000.

The Watson-Crick model gives us an understanding of a self-reproducing unit, a property which we have to postulate for the genetic material of the cell. If the double-helix is opened at one end, as shown schematically in Fig. 16, and on each strand a new complementary molecule is synthesized, one obtains after the total separation and synthesis two identical double-strands. Each strand of the original helix functions as a template for the newly synthesized molecule. JOSSE et al. (1961) have presented evidence that the new strands grow in the direction opposite to the original strands. Good evidence for the semiconservative duplication of DNA as shown schematically in Fig. 16 has been presented by MESELSON and STAHL (1958). Recently it has been shown that the synthesis of the DNA in bacteria starts on one end and runs through the total length of the molecule (CAIRNS 1963, NAGATA 1963, YOSHIKAWA and SUEOKA 1963 a, b). Ring structures of bacterial DNA also have been seen by the autoradiographic technique.

3. Optical Properties of the Nucleic Acids and their Constituents

The purine and pyrimidine bases and their derivatives have a characteristic UV-absorption band between 240 and 290 mμ. Since the bases possess ionizing groups, the UV-spectrum in this region changes markedly with the pH at which the measurements were performed. Keto-enol tautomerism does not play an important role in these changes. Because of the relationship between the changes in the UV-spectrum at a particular pH and the ionization of suitable groups, the pK-values for such an ionization can be determined optically (see Tab. 1). It is preferable to present an UV-spectrum of a base or its derivative at a pH which lies at least 1.5 or 2 units from the nearest pK-value.

There are considerable spectral differences between the bases and their nucleosides, while the spectra of the corresponding ribo- and deoxyribo-nucleosides or nucleotides are very similar. In Table 2 some data are presented.

Very characteristic for each compound are the ratios at selected wave length as shown in Table 2. The millimolar absorbencies (ε) also are listed.

Little is known yet about the infrared spectra of the bases and their derivatives in solution, mainly due to technical difficulties because of their insolubility in organic solvents. More recently D_2O has been used as a solvent, since H_2O strongly absorbs in the infrared. Infrared spectra in films or in KBr might not be the same as in solution.

Concerning the nucleic acids it has been found that the absorption at 260 mμ rises considerably after degradation of the macromolecules. This has been called the hyperchromic effect. This means that the absorption of a nucleic acid is much lower than that calculated from the sum of the absorptions of its constituent nucleotides. This is due to some modification of the chromophoric groups when incorporated into the macromolecules. Since the isolated nucleic acids are more or less degraded, CHARGAFF and ZAMENHOF (1948) introduced the term ε(P), the molar absorbency for nucleic acids, based on one gram-atom of phosphorus per liter. The lower the value is the less is the degradation of the nucleic acid. For the best preparations of DNA this value is about 7,000. Salts and pH have an influence on this value. The increase in the absorbancy after degradation has been used for the determination of the enzyme activity of nucleases.

Tab. 2. *Optical properties of nucleic acid constituents.* [Taken from Beaven et al. (1955)].

			ε_{260}	$\dfrac{A_{250}}{A_{260}}$	$\dfrac{A_{280}}{A_{260}}$	$\dfrac{A_{290}}{A_{260}}$
Adenine	pKα	4.1	13.0	0.76	0.375	0.035
		9.8	13.3	0.76	0.125	0.005
			10.45	0.57	0.60	0.025
Guanine	pKα	3.2	8.0	1.37	0.84	0.495
		9.6	7.2	1.42	1.04	0.54
		12.5	6.4	0.985	1.135	0.585
			7.3	0.805	1.24	0.605
Cytosine	pKα	4.5	6.0	0.48	1.53	0.78
		12.2	5.55	0.78	0.58	0.08
			2.35	0.595	3.28	2.6
Uracil	pKα	0.5	7.8	0.795	0.30	0.05
		9.5	8.2	0.84	0.175	0.01
			4.1	0.71	1.40	1.27
Thymine	pKα	9.9	7.4	0.67	0.53	0.09
			3.7	0.65	1.31	1.41
Adenosine	pKα	3.4	14.3	0.84	0.215	0.03
			14.9	0.78	0.144	0.002
Guanosine	pKα	2.2	11.75	0.94	0.695	0.50
		9.5	11.7	1.15	0.67	0.275
			11.3	0.89	0.61	0.13
Cytidine	pKα	4.1	6.4	0.45	2.10	1.55
		13	7.55	0.86	0.93	0.28
			7.0	0.865	1.17	0.56
Cytosine deoxyri-boside	pKα	4.3	6.15	0.42	2.15	1.61
		13	7.35	0.83	0.965	0.305
			7.05	0.81	1.09	0.405
Uridine	pKα	9.25	9.95	0.74	0.35	0.03
			7.35	0.83	0.29	0.02
Uracil deoxri-boside	pKα	9.3	10.1	0.72	0.375	—
			7.55	0.81	0.31	—

			ε_{260}	$\dfrac{A_{250}}{A_{260}}$	$\dfrac{A_{280}}{A_{260}}$	$\dfrac{A_{290}}{A_{260}}$
Thymidine	pKα	9.8	8.75	0.65	0.72	0.235
			6.65	0.75	0.67	0.16
5'-Adenylic acid	pH	2	14.2	0.84	0.22	0.038
		7	15.0	0.79	0.15	0.009
		12	15.0	0.79	0.15	—
5'-Guanylic acid	pH	1	11.8	0.99	0.70	0.51
		7	11.4	1.15	0.68	0.285
		12	11.2	0.89	0.60	0.11
5'-Cytidylic acid	pH	2	6.2	0.46	2.10	1.55
		7	7.4	0.84	0.99	0.33
		12	—	0.84	0.99	0.33
5'-Uridylic acid	pH	2	10.0	0.74	0.38	0.03
		7	10.0	0.73	0.40	0.03
		12	7.4	0.82	0.33	0.02
Thymidylic acid	pH	2	8.4	0.64	0.72	0.23
		7	8.4	0.65	0.73	0.24
		12	6.7	0.74	0.67	0.17

In an undenatured DNA about 40 to 50 % of the hyperchromicity appears when the hydrogen bonds of the base pairs in the Watson-Crick model are broken by heat or other agents. Using synthetic single-stranded polymers a wide variety of values have been obtained. Poly-U, e.g., shows at neutral pH only 6 % hyperchromicity, poly-C 40 %, poly-A 50–55 %, and poly-I 63 % (WARNER 1957). Since thymidine-3',5'-diphosphate has a molar extinction coefficient 8 % lower than TMP (SHAPIRO and CHARGAFF 1957 a), the low hyperchromicity of poly-U may be explained similarly. The hyperchromicity of oligonucleotides increases with the chain length reaching a limiting value at 5 to 6 nucleotides (MICHELSON 1958, 1959). The hyperchromic effect of single-stranded nucleic acids may be explained in part by some hydrogen bonding between amino groups and phosphate oxygen, but the major cause is due to the ordered stacking of the bases parallel to each other. The interaction between the π-electron orbitals of adjacent rings leads to the formation of electron orbitals extending over several rings with a lower total energy and therefore with a smaller "chromophoric area."

For a more detailed discussion of this field see BEAVEN et al. (1955) and SHUGAR (1960).

IV. Biochemistry of Nucleic Acids

1. Biosynthesis of Pentoses

Since the d e n o v o synthesis of the ribonucleotides starts with ribose-5-phosphate, it is important to know by which pathways this component can be obtained by the living cell. Two major routes are known: The first one is the hexose monophosphate shunt starting with glucose-6-phosphate, which is converted by glucose-6-phosphate dehydrogenase using TPN to 6-phosphogluconate. The latter substance is further oxidized by TPN in the presence of 6-phosphogluconate dehydrogenase to ribulose-5-phosphate

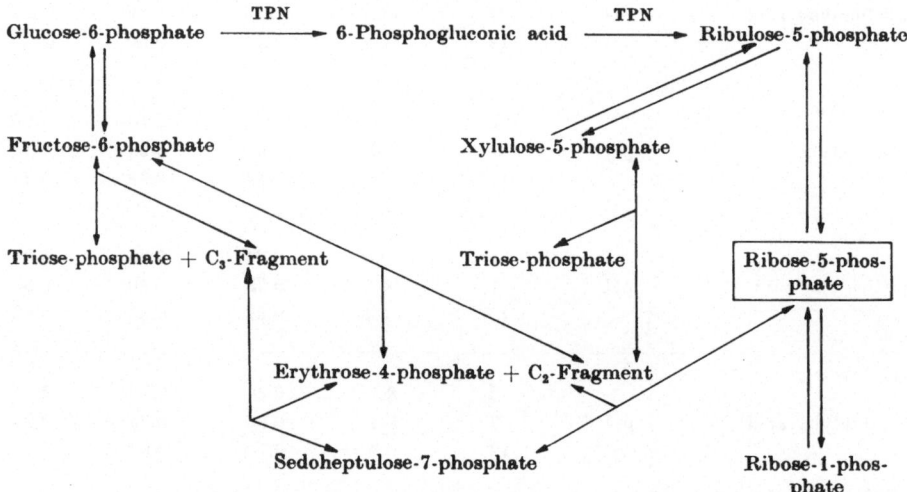

Fig. 17. Alternative pathways leading to ribose-5-phosphate. (Taken from Potter [1960].)

plus CO_2. Ribulose-5-phosphate is in equilibrium with ribose-5-phosphate. This reaction is catalyzed by a pentosephosphate isomerase.

The second route (non-oxidative) involves an activated C_2-fragment, which is derived from a C_5-, C_6-, or a C_7-ketosugar-phosphate by splitting it with transketolase, and glyceraldehyde-3-phosphate. Sedoheptulose-7-phosphate is converted by this enzyme directly to ribose-5-phosphate. After the condensation of the C_2-fragment with the triose-3-phosphate, xylulose-5-phosphate is formed, which is converted to ribulose-5-phosphate. These different pathways leading to ribose-5-phosphate are combined in Fig. 17.

Since ribose-1-phosphate is produced from ribonucleosides by the action of nucleoside phosphorylase and can be converted to the 5-phosphate, it contributes also to the pool of ribose-5-phosphate. Similar enzymes were found for the deoxyribonucleosides, by which deoxyribose-1- and -5-phosphates are formed, respectively:

purine + deoxyribose-1-phosphate $\rightleftarrows$ purine deoxyriboside + inorganic phosphate deoxyribose-1-phosphate $\rightleftarrows$ deoxyribose-5-phosphate

Deoxyribose-5-phosphate can also be synthesized by an enzymic reaction starting from glyceraldehyde-3-phosphate and acetaldehyde.

For a detailed discussion and literature on this field see GLOCK (1955) and POTTER (1960).

2. Biosynthesis of Nucleotides

By injection of simple radioactively-labelled precursors into animals, or addition of such compounds as bicarbonate-C^{14} or glycine-N^{15} to the culture medium of bacteria, and isolation and specific chemical degradation of the bases of the nucleic acids or their derivatives, it is possible to decide

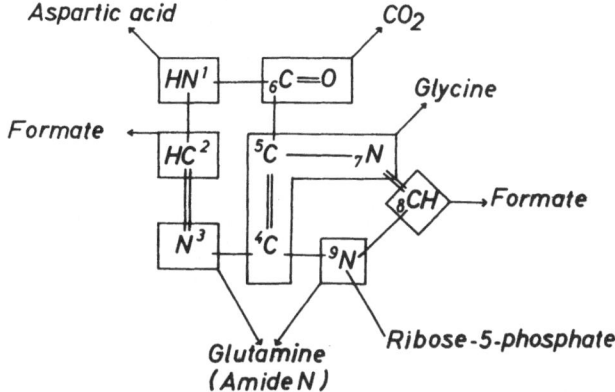

Fig. 18. Precursors of the various atoms in the purine ring of inosinic acid.

from which simple precursor the different atoms of the bases are derived. Using different labelled compounds the precursor scheme for inosinic acid presented in Fig. 18 was established.

A specific chemical degradation for uric acid and uracil is given in Figs. 19 and 20. The precursors for the pyrimidine ring are ammonia (N_1), CO_2 (C_2), and aspartic acid (the rest of the molecule).

Fig. 21 presents a scheme for the d e n o v o synthesis of inosinic acid. The reactions in which tetrahydrofolic acid is involved can be counteracted by folic acid antimetabolites like aminopterin and amethopterin (HANDSCHUMACHER and WELCH 1960). Inosinic acid is converted by aspartate, GTP, and Mg^{++} to adenylic acid. On the other hand inosinic acid can be oxidized to xanthylic acid, which is aminated by glutamine, ATP, and Mg^{++} to guanylic acid. All steps which involve glutamine are inhibited by azaserine or DON (6-diazo-5-oxo-1-norleucine).

The d e n o v o synthesis of uridylic acid is given in Fig. 22. An inhibitor for the decarboxylation of orotidylic acid is 6-azauracil (HANDSCHUMACHER and WELCH 1960). For the enzymatic conversion of UMP to CMP, glutamine, ATP, and GMP are required.

YATES and PARDEE (1956) have found a feed-back mechanism, by which the condensation of aspartic acid with carbamylphosphate is competitively inhibited by cytidine and CMP.

As mentioned above the nucleosides can be synthesized also from pre-formed purines or pyrimidines and the pentose-1-phosphates. Besides the ribonucleoside phosphorylase (Fig. 23, A) there exists a ribonucleotide pyrophosphorylase (C), by which the base plus 5-phosphoribosyl-1-pyrophosphate is converted to the mononucleotide plus pyrophosphate, and a ribonucleoside phosphokinase (B), by which the nucleoside plus ATP results in the mononucleotide plus ADP. Fig. 23 demonstrates such inter-conversions for UMP.

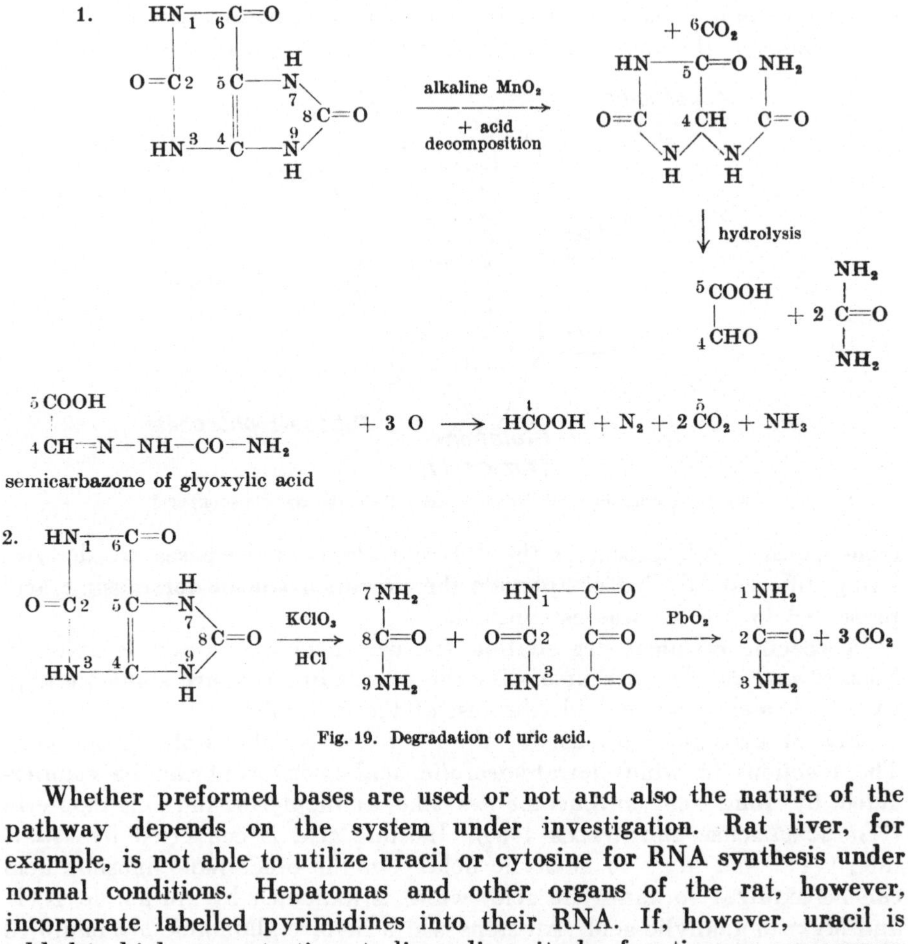

Fig. 19. Degradation of uric acid.

Whether preformed bases are used or not and also the nature of the pathway depends on the system under investigation. Rat liver, for example, is not able to utilize uracil or cytosine for RNA synthesis under normal conditions. Hepatomas and other organs of the rat, however, incorporate labelled pyrimidines into their RNA. If, however, uracil is added in high concentrations to liver slices, it also functions as a precursor for RNA synthesis. In the latter case the catabolic enzymes of pyrimidines are saturated by the high amount of the substrate.

The deoxyribonucleosides or -nucleotides can be synthesized from the preformed bases plus deoxyribose-1-phosphate as mentioned above, or by conversion of the 5'-ribonucleotides by reduction, as has been shown in the case of the synthesis of 5'-deoxycytidine phosphate. The latter conversion may take place at the level of the diphosphate.

As precursors for thymidine within the DNA, uridine and cytidine can function without fission of the N-glycosidic linkage. The first step is the conversion of the ribosides to deoxyuridine-5′-phosphate, which reacts further with N^5, N^{10}-methylene-tetrahydrofolic acid using thymidylate synthetase to form thymidine-5′-phosphate. This reaction can be inhibited by aminopterin, amethopterin, or 5-fluorodeoxyuridine (HANDSCHUMACHER and WELCH 1960).

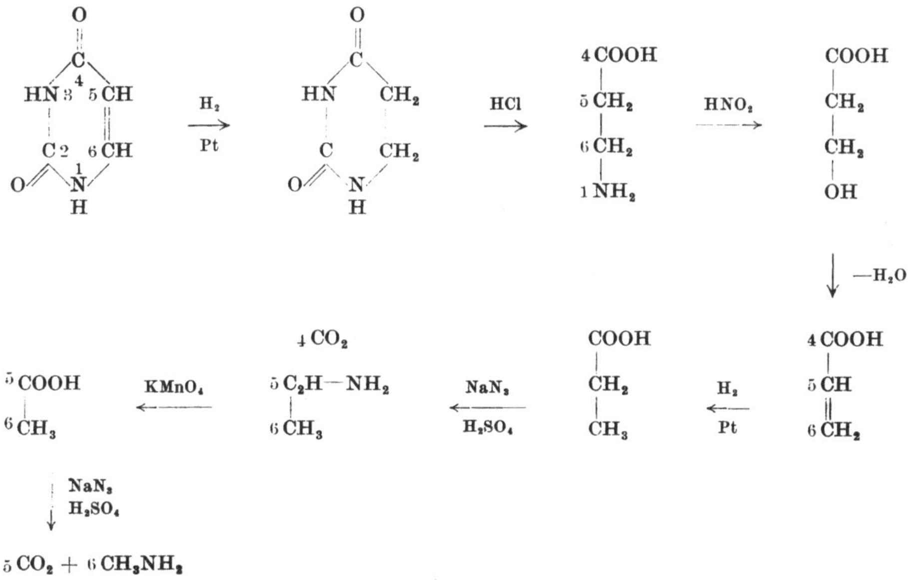

Fig. 20. Degradation of uracil.

Enzymes for the conversion of the nucleoside monophosphates to the corresponding di- and triphosphates are present in all living cells. ATP plays an important role as a phosphate donor. A few examples are given in the following formulas:

$$ATP + UMP \rightleftarrows ADP + UDP$$
$$UTP + AMP \rightleftarrows UDP + ADP$$
$$UTP + UMP \rightleftarrows 2\ UDP$$

For a detailed discussion and literature in the field of the biosynthesis of the nucleotides see REICHARD (1955), SCHLENK (1955), BUCHANAN (1960), and CROSBIE (1960).

It has been shown in this chapter that often there exist different pathways leading to the same compound. In this way regulatory mechanisms are possible which may direct cell duplication and organ growth. They also may play an important role in carcinogenesis. Thus, BOLLUM and POTTER (1959) have found in normal rat liver almost no thymidine kinase, while in regenerating rat liver this enzyme increases markedly at the time of maximal DNA synthesis. After regeneration it drops again to an insignificant level.

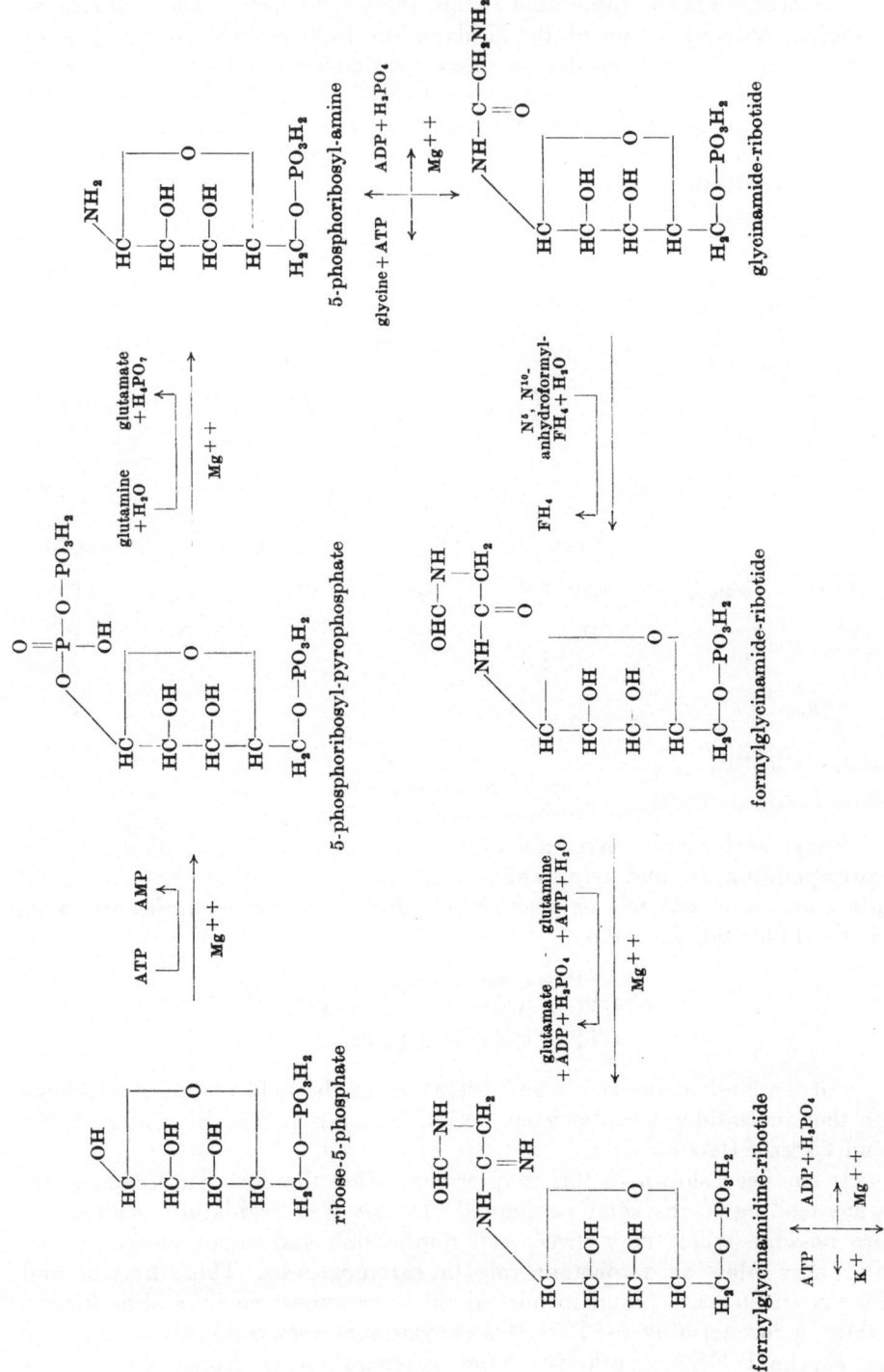

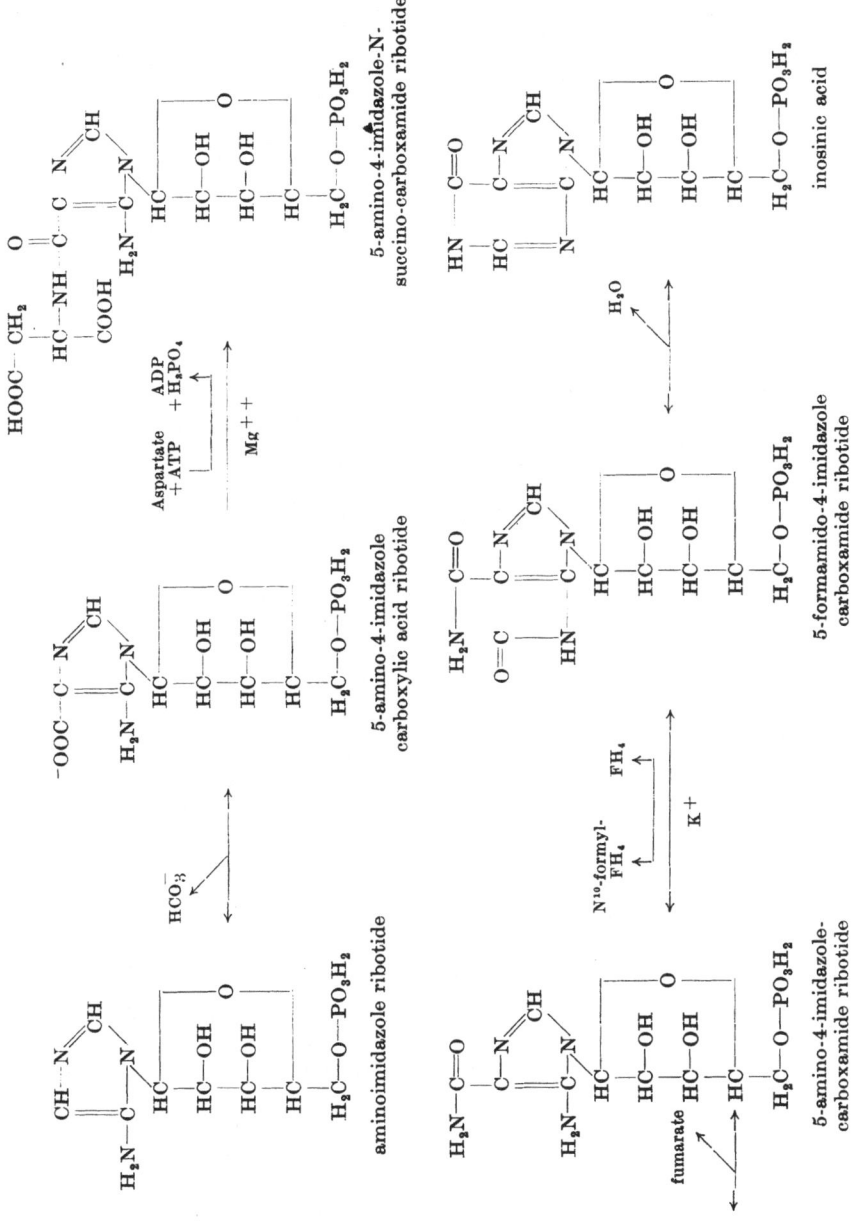

Fig. 21. *De novo* synthesis of inosinic acid. FH_4 = tetrahydrofolic acid. (Taken from BUCHANAN [1960].)

In this connection the importance of the balance between anabolic enzymes and those which catabolize the building blocks for nucleic acid synthesis should be stressed (see above). The catabolic pathways are pre-

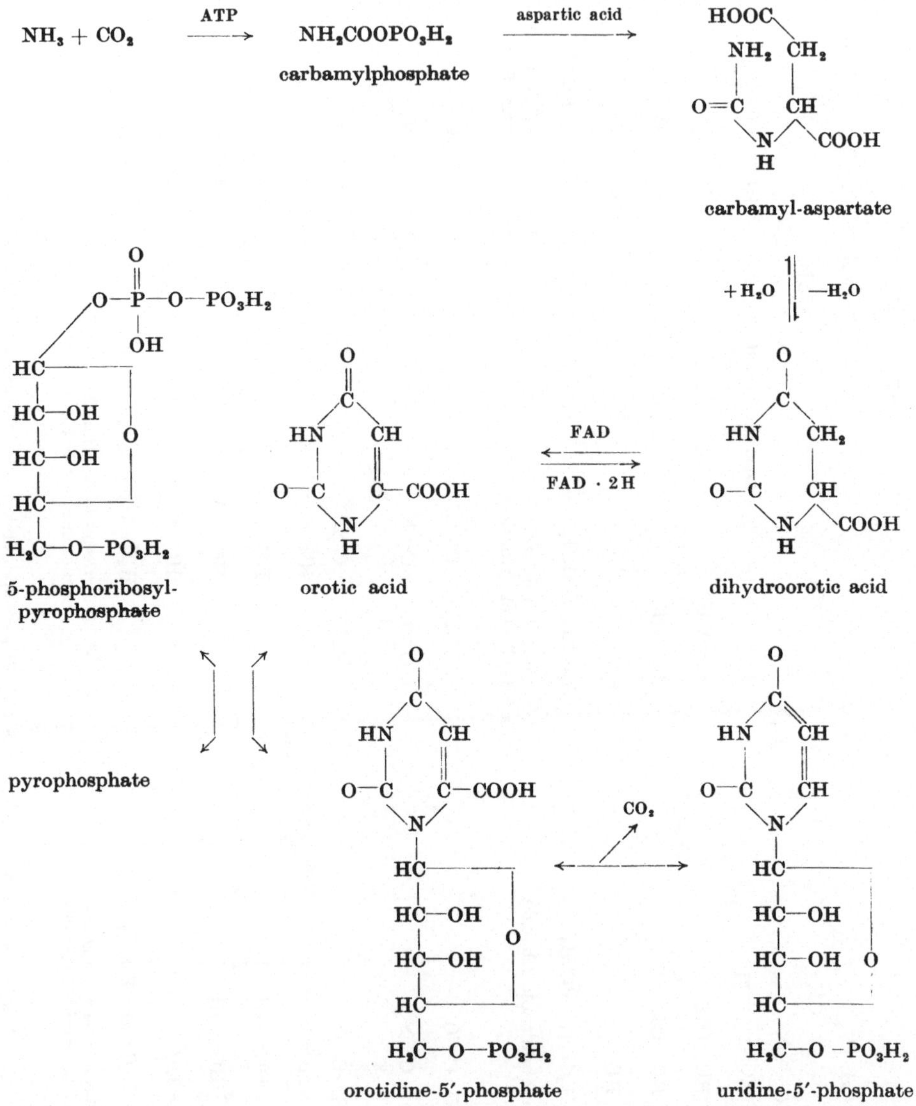

Fig. 22. *De novo* synthesis of uridine-5'-phosphate. FAD = Flavin adenine dinucleotide. (Taken from Crosbie [1960].)

sented in Fig. 24. The nature of the pathway used by the cell depends on the organism under investigation. The oxidation of the pyrimidines to barbituric acid or its derivatives occurs in bacteria, while the reductive pathway is found in animals.

3. Biosynthesis of Nucleic Acids

According to GRUNBERG-MANAGO and OCHOA (1955, 1956) an enzyme was isolated from *Azotobacter vinelandii* which catalyzed in the presence of Mg^{++}-ions the polymerization of the four ribonucleoside diphosphates to RNA. By this reaction orthophosphate was set free. Soon similar enzymes were isolated from other microorganisms, and also such a polynucleotide

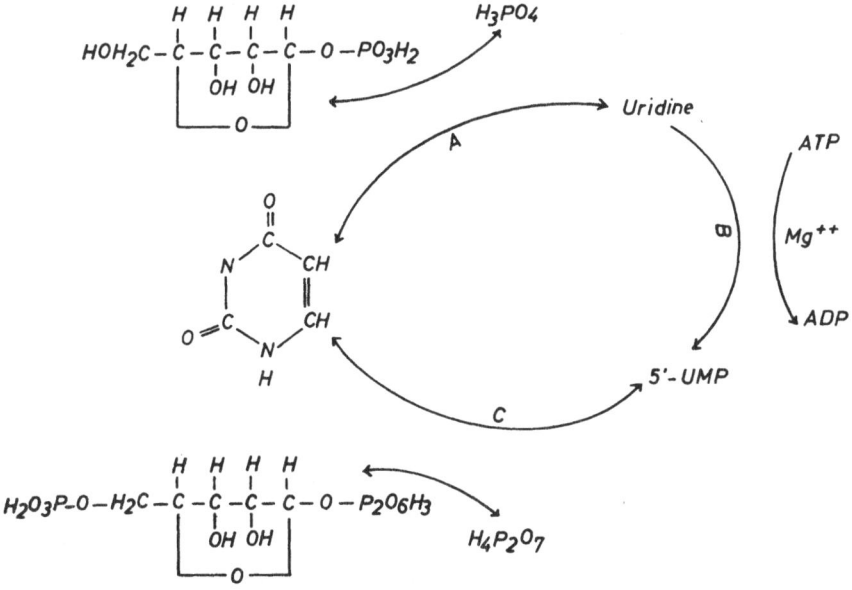

Fig. 23. Formation of uridine and UMP.

phosphorylase has been demonstrated in guinea pig liver nuclei (HILMOE and HEPPEL 1957):

$$\text{n nucleoside diphosphates} \xleftarrow{\quad\overset{\text{Enzyme}}{\text{(primer)} + Mg^{++}}\quad} \text{RNA} + nP$$

The enzyme is unspecific regarding the purine or pyrimidine bases in the ribonucleoside diphosphates. A single diphosphate can be polymerized to the corresponding polynucleotide, e.g. polyadenylic acid. If a highly purified enzyme is investigated, the polymerization starts after a certain lag period. This lag phase can be overcome by addition of RNA of different sources, in which case the RNA of the corresponding organism primes the best. If a single diphosphate is polymerized, the homopolymer is the best primer, while another polymer may inhibit the synthesis. For example, the polymerization of ADP is primed by polyadenylic acid, but is inhibited by polyuridylic acid. SINGER et al. (1960) have studied also the priming capacities of oligonucleotides.

The products obtained by polymerization have molecular weights up to one million. They have a repeating 3'-5'-interribonucleotide linkage like natural RNA and cannot be distinguished from natural RNA preparations.

Although under suitable conditions the polynucleotide phosphorylase favors polynucleotide synthesis, it might have in the living cell rather a

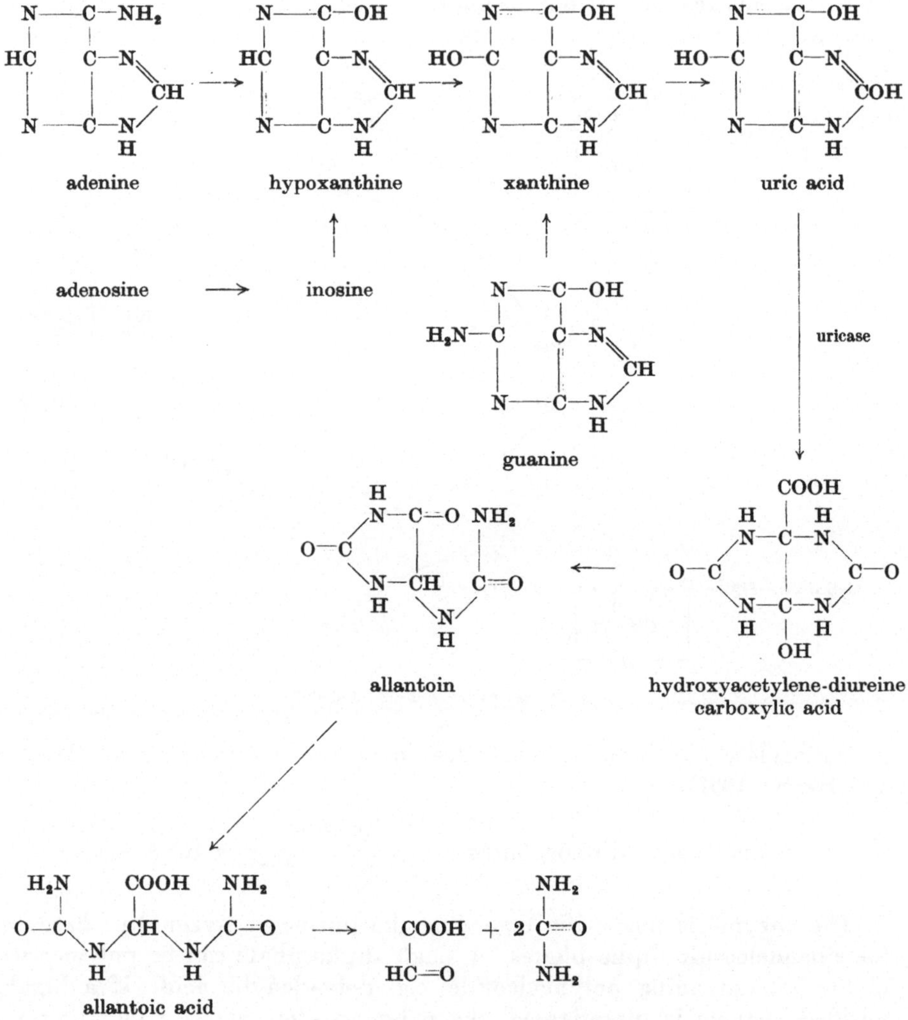

Fig. 24. a) The catabolism of the purines.

catabolic function, since there is too little specificity and no connection to the DNA template. Nevertheless, this enzyme has become highly important for the synthesis of test substances (see chapter V). Concerning phosphorolysis, oligonucleotides with the free 3'-hydroxyl group are degraded easily, while oligonucleotides with the 3'-phosphate group are rather resistant.

Enzymes which incorporate nucleoside-5'-triphosphates into the end region of an RNA of relatively low molecular weight (about 80 nucleotides, e.g. transfer-RNA) are found mainly in the cytoplasm of the cell. Such an RNA with the specific end group -C-C-A has been discovered by HECHT et al. (1958). It functions as an acceptor for amino acids, which are then incorporated into proteins. HERBERT et al. (1962) have described an

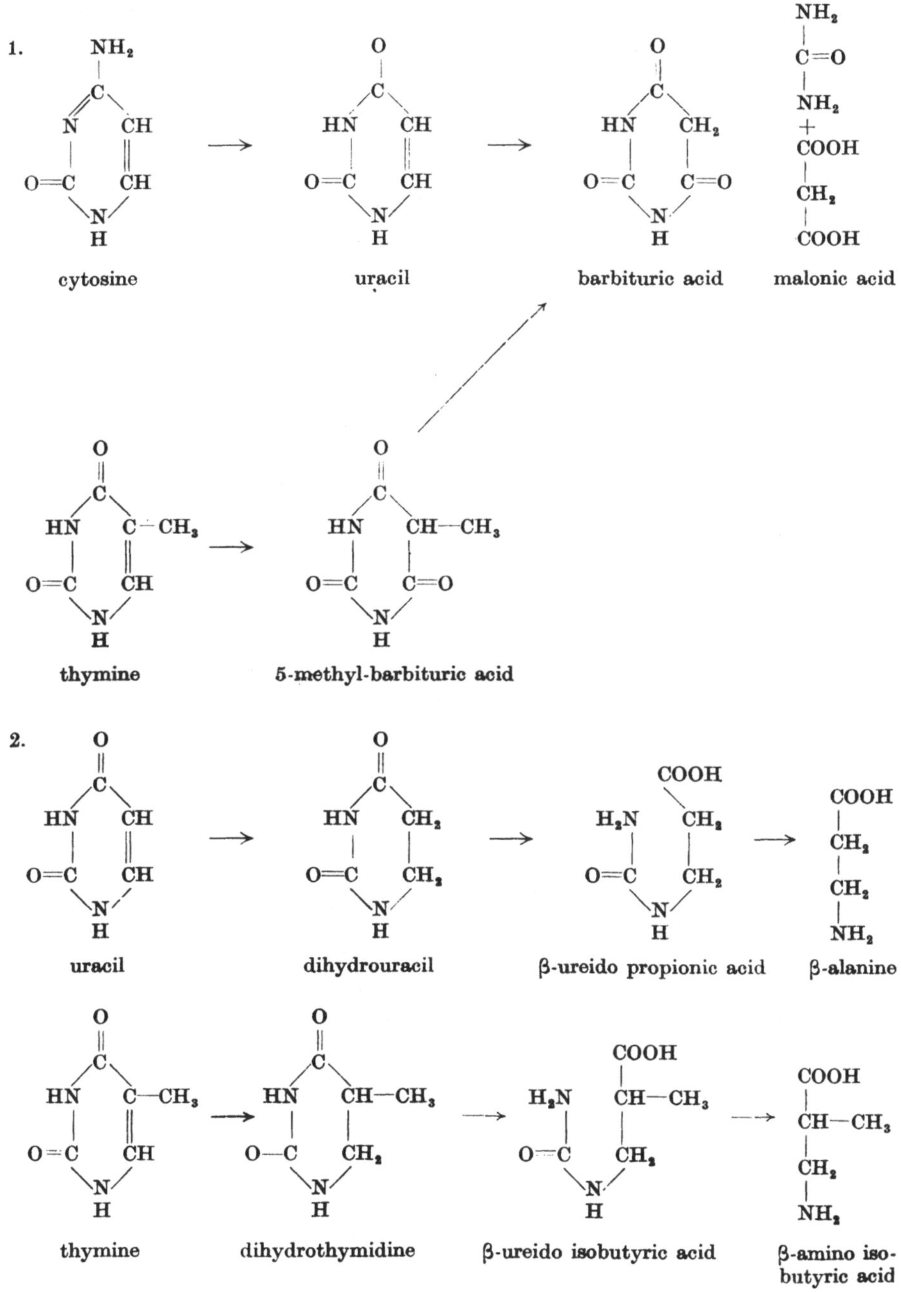

Fig. 24. b) The catabolism of the pyrimidines.

enzyme in rat liver which removes specifically the last three nucleotides and adds them again i n v i t r o under suitable conditions. A high turnover of only the last nucleotide of this transfer-RNA in rat liver i n v i v o has been found by Scholtissek (1959, 1960, 1962). In yeast similar results have been obtained by Rosset and Monier (1963). A review on the function of the transfer-RNA has been given recently by Berg (1961). This RNA also shows a peculiarity, since it has an intramolecular refolded structure (McCully and Cantoni 1962, see there for further literature).

Enzyme systems which form RNA d e n o v o from triphosphates have been described by several authors (see Khorana 1960). Weiss et al. (Weiss 1960, Weiss and Nakamoto 1961 a, 1961 b), Hurwitz et al. (1960, 1962), Stevens (1960, 1961), Ochoa et al. (1961), and Chamberlin and Berg (1962) have shown that this incorporation is DNA-dependent. All four triphosphates and Mg^{++} or Mn^{++} are required:

$$\text{n nucleoside triphosphates} \xrightleftharpoons[\text{DNA} + \text{Mg}^{++}]{\text{Enzyme}} \text{RNA} + \text{nPP}$$

The RNA which is synthesized in this system has complementary base ratios to the primer-DNA. Thus, DNA functions in this case as a template.

This DNA-dependent RNA synthesis can be inhibited by actinomycin (Reich et al. 1961, 1962), which combines with the guanine of the DNA (Goldberg et al. 1962, Hurwitz et al. 1962, Kersten and Kersten 1962). Many RNA-containing viruses, however, can multiply in the presence of actinomycin. Exceptions are the influenza viruses (Barry et al. 1962, Rott and Scholtissek 1964), reovirus (Gomatos and Tamm 1963), and the Rous sarcoma virus (Temin 1963). Therefore, the RNA synthesis of most RNA-containing viruses is not DNA-dependent. For some of these viruses it has been shown that they induce an RNA-dependent RNA synthetase (Baltimore and Franklin 1962, Weissmann et al. 1963).

Kornberg et al. (1956) have purified a DNA-polymerase from *Escherichia coli* which requires for DNA synthesis all four deoxyribonucleoside-triphosphates, Mg^{++}, and primer-DNA:

$$\text{n deoxyribonucleoside triphosphates} \xrightleftharpoons[\text{DNA} + \text{Mg}^{++}]{\text{Enzyme}} \text{DNA} + \text{nPP}$$

Omission of one of the deoxyribonucleotides or digestion of the primer DNA by DNase reduces the incorporation to below 1 % under normal conditions. Net synthesis up to 20 times the amount of the primer has been obtained. The chemical and physical properties of the product are identical to natural occuring DNA (except melting profile). The base composition resembles always that of the primer used, independent of the relative concentration of the triphosphates. Either single-stranded or double-stranded DNA works as a primer under the conditions employed.

If only a single triphosphate is present in the system, it is incorporated on the end of the primer DNA. The triphosphates of deoxyadenosine and thymidine polymerize in the absence of any primer after a lag period of 3 to 6 hours (SCHACHMAN et al. 1960). This polymer can be used again as primer for a copolymer of deoxyadenosine and thymidine even in the presence of all four deoxynucleoside triphosphates. The polymer contains the two deoxynucleosides in alternating sequence. The deoxynucleoside triphosphates of cytosine and guanine also can be polymerized. But in this case a double-stranded product results, which is composed of one strand each of poly-C and poly-G (RADDING et al. 1962).

BOLLUM and POTTER (1957, 1958) demonstrated a similar DNA polymerase in regenerating rat liver (starting 18 hours after partial hepatectomy), in other organs of the rat, and in carcinoma. In their system only single-stranded DNA functions as primer (BOLLUM 1959). For a detailed discussion of the RNA- and DNA-polymerizing enzymes see KHORANA (1960), KORNBERG (1961), and STEVENS (1963).

4. Enzymes which Split the Nucleic Acids and their Components

a) Ribonucleases

Pancreatic ribonuclease was first crystallized by KUNITZ (1940). It specifically splits the phosphoester linkage between a 3'-pyrimidine nucleotide and the 5'-hydroxyl group of the adjacent purine or pyrimidine nucleotide. In this way 3'-pyrimidine nucleotides and oligonucleotides consisting of one or more purine nucleotides and one pyrimidine nucleotide on the end can be obtained. 2',3'-Cyclic phosphates are formed as intermediates.

The acid-soluble oligonucleotides can be separated by a combination of paper electrophoresis and paper chromatography (see SMITH 1955, RUSHIZKY and KNIGHT 1960) or by ion exchange chromatography (VOLKIN and COHN 1953, SCHOLTISSEK 1959 a, STAEHELIN 1961 a). The acid-insoluble core has been separated into a guanine- and uracil-rich and a adenine- and cytosine-rich fraction by paper electrophoresis (SCHOLTISSEK 1957).

If the RNA is treated with hydroxylamine at pH 10, at which uridine is specifically destroyed, RNase splits only after cytidine (VERWOERD et al. 1961, 1963). After treatment of the RNA with hydrazine the enzyme does not split on those points where uracil or cytosine has been removed (TAKEMURA 1958).

The digestion of RNA by pancreatic RNase has been used by several authors as a powerful tool for the characterization of an RNA (SCHOLTISSEK 1959 a, REDDI 1959, SCHOLTISSEK and ROTT 1961, STAEHELIN 1961 b), since it takes not only the base composition of the RNA into account but also the arrangement of the building blocks within the macromolecule. STAEHELIN (1961 b) has shown that the base sequence is not random.

Pancreatic RNase is exceptionally heat-stable and is very difficult to inactivate. ROTH (1958, and earlier publications) has found an inhibitor in rat liver. Bentonite (BROWNHILL et al. 1959) and also polyvinylalcohol (SCHERRER and DARNELL 1962) have been used as inhibitors.

Another ribonuclease, isolated from soy beans (ribonuclease T_1), splits the RNA always behind a guanylate residue leading to guanosine-3'-phosphate and oligonucleotides terminated by guanosine-3'-phosphate (Sato and Egami 1960). This enzyme has been used in studies of sequence analysis (Herbert et al. 1962 b) of transfer RNA.

RNases of other specificities and properties have been isolated from several sources. For reviews see G. Schmidt (1955) and Heppel and Rabinowitz (1958).

b) Deoxyribonucleases

Two different kinds of DNases were found in nature. One (DNase I) has an optimal activity at neutral pH in the presence of divalent cations leading to 5'-mono- and oligonucleotides. The second one (DNase II) has an acid pH-optimum and does not require divalent cations; 3'-mono- and oligonucleotides are formed. Both enzymes do not distinguish in general between pyrimidine and purine bases. In many organs both enzymes were found. Recently Lehman et al. (1962) described a DNase in *E. coli* which splits DNA, leading to oligonucleotides terminated by a 5'-phosphate group. This enzyme can be inhibited by RNA.

c) Phosphodiesterases

Two different kinds of phosphodiesterases are known. Both are mainly exonucleases. They split mononucleotides sequentially from one end of the macromolecule. The first one, isolated mainly from snake venoms, starts from that end having a free 3'-hydroxyl group, producing 5'-mononucleotides (Razzell and Khorana 1958, 1959). Oligonucleotides with a 3'-phosphate group are almost resistant. The second enzyme, isolated from spleen, reacts exactly in the opposite way. Starting from the 5'-hydroxyl end of the RNA, it produces 3'-mononucleotides (Razzell and Khorana 1959). DNA has to be digested first by the corresponding DNases before the diesterases can act. None of these enzymes shows any specificity which respect to the purine or pyrimidine bases. Snake venom diesterase can be used to some extent also in sequence analyses (e.g. Scholtissek 1962 b). Lehman (1960) obtained a very pure exonuclease which attacks only single-stranded DNA.

d) Phosphomonoesterases

In this chapter only the specific phosphomonoesterases will be treated. Two different kinds are in question: The first one splits the 5'-nucleotides. This kind of enzyme has been found in high concentrations in snake venoms, in muscle and nerve tissues, in potatoes, and in bull testicle. The second enzyme splits 3'-mononucleotides. It was discovered mainly in plant material like germinating rye grass or barley. With this enzyme 2'- or 5'-mononucleotides are not touched.

Other enzymes, which have also synthetic activities, are treated already above. Review articles concerning these enzymes have been published by Schmidt (1955) and Heppel and Rabinowitz (1958).

V. Metabolism and Function of Nucleic Acids

1. Localization of Nucleic Acids and their Synthesis within the Cell

By cytochemical means and homogenization techniques it has been shown a long time ago that the DNA of the cell is restricted to the nucleus, while RNA is contained in the nucleus and in the cytoplasm (for a review see DOUNCE 1955, HOGEBOOM and SCHNEIDER 1955). Considerable discussion has been concerned with the site of nucleic acid synthesis. There is no question that DNA duplicates within the nucleus, since DNA never has been found in normal somatic cells of higher animals and plants within the cytoplasm. More recently considerable evidence has been accumulated that all cellular RNA is synthesized within the nucleus.

1. The autoradiographic studies of several authors (GOLDSTEIN et al. 1955, PRESCOTT 1959, McMASTER-KAYE 1960) have shown a rapid labelling of nuclear RNA, which moved, after chasing with non-radioactive precursor, to the cytoplasm. The excellent experiments of ZALOCAR (1959) gave identical results. But these experiments do not say anything about a slowly-labelled RNA, which might possibly be synthesized later in the cytoplasm.

2. The experiments of HARBERS and MÜLLER (1962) demonstrated that the total RNA production can be inhibited by actinomycin D. Actinomycin is known to inhibit only DNA-dependent RNA synthesis (REICH et al. 1961, 1962) by reacting with the DNA-template (GOLDBERG et al. 1962, HUR-WITZ et al. 1962, KERSTEN and KERSTEN 1962).

3. If Earle's L-cells are infected with fowl plague virus, an abortive cycle of virus multiplication is observed. Under these conditions no significant amounts of RNA leave the nucleus. In addition, the nucleolus is kept almost free of label if relatively short pulses of the radioactive precursor are used. Nevertheless, the oligonucleotide pattern of the labelled RNA of the infected cells does not change considerably—except for a small shift of the pattern toward that of viral RNA—compared with non-infected cells, independent of the pulse length of the radioactive precursor. In both cases almost the same amount and the same composition of RNA is synthesized with the only exception, that in the non-infected cells at long pulses most of the labelled RNA was found in the cytoplasm, while in the infected cells it was restricted to the nucleus (SCHOLTISSEK et al. 1962). Therefore one has to assume that the total cellular RNA is synthesized within the nucleus, but does not leave it in the infected L-cells.

4. YANKOFSKY and SPIEGELMAN (1962 a, 1962 b, 1963) reported on a hybrid formation i n v i t r o between ribosomal RNA and the homologous DNA. GOODMAN and RICH (1962) found the same for the soluble RNA of the cytoplasm. This is again evidence that the ribosomal and soluble RNA, normally constituents of the cytoplasm, have templates on the cellular DNA.

From these data it can be concluded that in a normal cell the DNA as well as the total RNA is synthesized in the cell nucleus using the DNA as template. The end turnover of the soluble RNA in the cytoplasm i n v i v o (SCHOLTISSEK 1962 b) cannot be regarded as a real RNA synthesis, but might obscure results obtained only by autoradiography.

2. Metabolism

a) DNA-metabolism

Since the DNA is regarded as the genetic material of the cell, it is clear that it must be duplicated before a cell divides (except before the second meiotic division). This duplication may occur immediately at the onset of mitosis, immediately after mitosis is over, or sometime during interphase. Practically all possibilities exist in nature depending on the cell type under investigation. In most cases the phase of DNA synthesis takes only a certain time during interphase, which shows at this time another sensitivity to antibiotics or irradiation. An excellent tool for such studies is the synchronization of the cell cycle by cold shock, antibiotics, or partial hepatectomy. A few examples are listed by Lajtha (1960).

Another feature of the DNA is that it is completely stable during the life span of the cell. Therefore there is practically no incorporation of radioactive precursors into the DNA of resting cells in contrast to fast growing cell populations. For a review on this matter see Smellie (1955).

b) RNA-metabolism

About 15 years ago the first publications appeared dealing with metabolic inhomogeneities within different cell fractions. It was found that the nuclear RNA was labelled by radioactive precursors much faster than cytoplasmic fractions. In rat liver, e.g., a maximal labelling of nuclear RNA was found about 3 hours after addition of the precursor. Later, when the specific radioactivity of the nuclear RNA had already dropped, the radioactivity in the cytoplasmic RNA was still rising. The conclusions and discussions on this subject might be read in the review of Smellie (1955). Some of them have only historical value.

Some more precise conclusions can be drawn from the short-time autoradiographic studies of Goldstein et al. (1955), McMaster-Kaye (1960), and others, concerning possible precursor relationships of nucleolar and chromosomal RNA and the localization of their synthesis. Although the nucleolar RNA is synthesized sometimes faster, sometimes slower than the chromosomal RNA, it seems that the nucleolar RNA is made first around the organizer and spreads later over the whole nucleolus (Pelling 1959). The results obtained with fowl plague virus-infected L-cells (see above) favour the idea that the nucleolus rather stores some RNA, which is made at the DNA of the organizer, than that it is an RNA-synthesizing center of its own. In the infected cells an accumulation of RNA within the nucleolus does not occur, in contrast to non-infected cells, although this RNA is synthesized normally. The oligonucleotide pattern of the short-pulse RNA is not changed significantly by infection (Scholtissek et al. 1962). Since the base composition of the nucleolar RNA in Chironomus salivary gland cells resembles that of the cytoplasm rather than that of the different chromosomes (Edström and Beermann 1962), it may be concluded that the nucleolar RNA is the precursor of the main cytoplasmic RNA.

3. Nucleic Acids as Genetic Material

For a long time cytologists have assumed from their observations of mitosis, meiosis, and fertilization that the genes—the units transmitting heritable properties from cell to cell—may reside in the chromosomes. Since the chromosomes were the only units in the cell, which contained DNA, this compound soon came into discussion as the genetic material. Then more evidence accumulated on chemical grounds:

1. Mutations are caused by UV-light at a wavelength at which nucleic acids absorb and also by substances which are known to react with nucleic acids, like mustard gas and others (see HOTCHKISS 1955).

2. The amount of DNA per cell nucleus in a diploid animal is always the same, except for sperm and egg cells which contain half this amount. Polyploid cells have the corresponding multiple of DNA (see VENDRELY 1955). But there are great variations in DNA per nucleus from one species to the other.

3. The first direct evidence for the genetic properties of DNA was the transformation experiment in bacteria by AVERY et al. (1944). These authors were able to transform an unencapsulated strain (rough colony type) of pneumococcus into an encapsulated (smooth) one by DNA extracted from an encapsulated strain. This property was then inheritable.

4. Many viruses contain only nucleic acid and protein. It has been possible to extract an infectious nucleic acid from several viruses (SCHUSTER 1960 b). The first one was isolated from tobacco mosaic virus by GIERER and SCHRAMM (1956). In this case the infectious principle was RNA. But now also infectious DNA has been isolated from corresponding viruses (for example: DIMAYORCA et al. 1959, GUTHRIE and SINSHEIMER 1960, HOFSCHNEIDER 1960, ITO 1960). Most of these infectious samples contained protein to an insignificant amount and were sensitive to the respective nucleases.

5. According to SCHUSTER and SCHRAMM (1958) the bases carrying an amino group were deaminated by nitrous acid without destruction of the macromolecular structure. Therefore mutations of infectious nucleic acids in the test tube were expected. GIERER and MUNDRY (1958) were the first who succeeded in demonstrating such mutations of an infectious RNA in vitro. Nitrous acid has been used since this time to induce mutations not only in vitro, but also in vivo (see FREESE 1963). Other possibilities for the induction of mutations are the incorporation of base analoges (e.g. halo-genated bases) into the nucleic acids, the insertion of acridine dyes, and the lowering of the pH (formation of apurinic acid). Freese has summarized the possible mutations and back-mutations using the Watson-Crick-model of DNA (FREESE 1963).

According to all evidence available—the helical structure of DNA, its metabolic behavior, its constancy per nucleus in an organism, and the other properties just mentioned—DNA is the most favored as being the genetic material of the cell of higher organisms.

4. The Messenger Theory

The genetic material of the cell, the DNA, resides in the nucleus, while the main protein synthesis proceeds in the cytoplasm on the RNP particles. The question now is how the genetic information is carried from the DNA to the ribosomes.

The possibility has been raised for a long time that RNA might take over this messenger function, for RNA is synthesized in the nucleus rapidly and might be transported to the cytoplasm.

. The most reasonable theory interpreting all data available was proposed by Jacob and Monod (1961) working in the field of bacterial genetics. A messenger for the genetic information should fulfill the following conditions:

1. The site of its synthesis should be the DNA of the cell.

2. The site of its function should be the ribosomes in the cytoplasm. the "factories" of cell proteins.

3. It should be metabolically very active in order to be regulated genetically.

4. In a protein synthesizing system which is depleted of this messenger (and DNA) one should be able to restore protein synthesis by addition of this messenger. The new proteins formed should be specific for the cell from which the messenger is derived.

5. If this messenger is RNA, its base sequence should be complementary to that part of DNA on which it is synthesized.

Brenner et al. (1961) have contributed direct evidence to the messenger concept using the *E. coli*-T$_4$-phage system. In T$_4$-infected *E. coli*, DNA synthesis stops immediately after infection and resumes 7 min. later. Protein is made at a constant rate, while the amount of RNA does not increase, although it shows a high rate of turnover. The rapidly-turning-over RNA has a base composition corresponding to phage DNA (Astrachan and Volkin 1958). Brenner et al. (1961) labelled the *E. coli* ribosomes with heavy isotopes (C^{13} and N^{15}) and after infection with T$_4$-phage they transfered the bacteria to normal (light, radioactive) medium. They found that the newly made (light, labelled by r a d i o a c t i v e isotope) RNA stayed with the heavy particles and could be chased from there by non-radioactive precursors. The phage protein was synthesized at the same site. The interpretation of these results was that the ribosomes are non-specialized structures which receive the genetic information from the gene in the form of an unstable "messenger"-RNA.

An RNA fraction with similar physical and metabolic properties was found by Gros et al. (1961) in non-infected *E. coli* using very short pulses of C^{14}-uracil. First it appeared free in the cell sap, later it was bound to the 70 S ribosomes into which incorporation of labelled amino acids occurs (Tissières et al. 1960).

Spiegelman and coworkers (Hall and Spiegelman 1961, Hayashi and Spiegelman 1961) have shown a direct correlation of this fast-labelled RNA fraction to the DNA by forming or isolating the corresponding DNA-RNA hybrids (see chapter III).

Further evidence for the messenger character of an RNA during the transmission of the genetic information from the DNA to the proteins has been contributed by the work of ZILLIG and coworkers (DOERFLER et al. 1962). They found that in an in vitro protein synthesizing system, double-stranded DNA or messenger RNA has to be present. Single-stranded DNA forms a hybrid with newly synthesized RNA, which is not removed from the template and is therefore inactive.

Taking together all available results, there exists strong evidence that the genetic information is carried from the DNA to the ribosomes by an RNA fraction called messenger RNA.

Concerning animal cells, a metabolic inhomogeneity of the nuclear RNA has been observed by several authors (e. g. ALFREY and MIRSKY 1957). SCHOLTISSEK et al. (1958, SCHOLTISSEK and POTTER 1960, SCHOLTISSEK 1962 a) have shown that during incubation of isolated rat liver nuclei labelled i n v i v o in their RNA lose under certain conditions only a very small but heavily-labelled fraction of their RNA to the surrounding medium. This nuclear RNA-fraction has been found to have some properties of a messenger RNA:

1. It is different in its oligonucleotide pattern from all other RNA fractions of the cell.

2. In the living animal one finds this fraction first in the nuclei (maximal labelling about two hours after administration of the radioactive precursor), later in a relatively low amount in the post-microsomal fraction of the cytoplasm, and after about 24 hours it cannot be recognized anymore within the cell.

3. This RNA fraction is extremely sensitive to incubation of a homogenate i n v i t r o. During 4 hours incubation about 75% of the RNA labelled i n v i v o for 2 hours was degraded, but less than 5% of the total RNA disappeared.

These findings have been interpreted in this way: that a small RNA fraction can take up the genetic information from the DNA and carry it to the protein producing "factories" of the cell. Its lability makes it favorable for regulations during transmission of this information (SCHOLTISSEK 1959 b, 1960 a, 1962 a).

The existence of a rapidly-labelled RNA fraction in rat liver nuclei has been confirmed by HIATT (1962). In an in vitro system of E. coli, developed by MATTHAEI and NIRENBERG (1961), the ability of an RNA to enhance protein synthesis can be tested. It has been shown by BARONDES et al. (1962) that the nuclear RNA of rat liver is especially active in this system. But it has not yet been demonstrated whether the protein formed is rat liver protein.

Fast-labelling RNA fractions have been found now in several animal cells (e.g. ref. SIBATANI et al. 1962, MARKS et al. 1962, SCHERRER and DARNELL 1962, TAMAOKI and MUELLER 1962, SCHERRER et al. 1963). LEVINTHAL et al. (1962) have calculated how many times on the average a messenger RNA may function as a template for protein synthesis in *B. subtilis*. They obtained a figure between 10 and 20. But, in contrast, in reticulocytes the RNA is completely stable and protein synthesis cannot by influenced by actinomycin as it can in *B. subtilis* (von Ehrenstein, cited by Levinthal et al. 1962). Therefore it can be assumed that there exist messenger RNA's of different stabilities.

This idea is somewhat supported by experiments of Scholtissek (1960 b. 1962 a), who incubated rat liver nuclei labelled i n v i v o for 2 hours in non-radioactive cytoplasm of rat liver, rat kidney, or in a salt-sucrose medium. In all cases about 75 % of the labelled RNA was broken down. But the remaining, almost stable RNA's differed from each other in their oligonucleotide pattern depending on the incubation medium used. This result was interpreted as a selection and stabilization of some messenger RNA molecules by the cytoplasm under investigation.

Wettstein et al. (1963), Warner et al. (1963), Gierer (1963), and Spyride; and Lipmann (1962) have found complexes between messenger RNA's and several ribosomes in animal and bacterial cells (polysomes or ergosomes), which were the only structures for the incorporation of amino acids in an i n v i t r o system. Similar complexes were found with poliovirus RNA (Penman et al. 1963).

Although in thymus nuclei an RNA fraction has been found with a base composition corresponding to the cell DNA (Sibatani et al. 1962), from the theoretical standpoint one should not expect such a correspondence in differentiated cells with different functions of an organism. Scholtissek (1960 a, 1962 a) found significant differences in the oligonucleotide pattern of the fast-labelled RNA's of rat liver, rat kidney, and rat spleen. According to his data each organ seems to synthesize its own messenger RNA pattern. More precise studies were performed by Edström and Beermann (1962), who determined the base ratios of the RNA of different puffs of the giant chromosomes of the salivary glands from Chironomus tentans. The RNA's of each puff were different. This may be interpreted that each active gene makes its own messenger RNA. Since the A/U- and G/C-ratios are not equal to 1, the RNA found in these puffs cannot be the product of both DNA-strands of the gene.

5. The Coding Problem

In the last chapter considerable evidence has been presented that RNA is an intermediate in the transfer of the genetic information from the DNA in the nucleus to the ribosomal particles in the cytoplasm, the main site of protein synthesis. Since to each base in one strand of nucleic acid there exists a complementary base in the other strand, it is easy to understand how the DNA functions as a template for the messenger RNA. But it is difficult to imagine how the RNA codes for the sequence of the amino acids in the proteins. Four bases are available for about 20 amino acids. From all the discussed coding ratios and codes only that one which fits the available data best will be treated here.

For a long time it was suggested that at least 3 bases may be necessary to define one amino acid, because with combinations of two bases only 16 amino acids can be coded. BRENNER (1957) pointed out that in nature no overlapping triplet code (three bases for one amino acid) may exist, since such a code demands restrictions in amino acid sequences in proteins which were not found. Direct evidence was contributed by WITTMANN (1960) by his experiments on the mutations of tobacco mosaic virus with nitrous

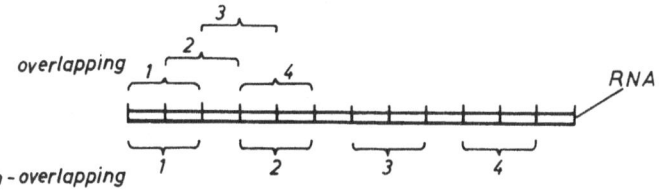

Fig. 25. Scheme of an overlapping and non-overlapping code of RNA. Three bases determine one amino acid (numbers).

acid. If the code is of the overlapping type (see Fig. 25), one alteration of a base should correspond to at least 2 alterations in adjacent amino acids. In a non-overlapping code each change in a single base leads to a change in only one amino acid. The latter has been found to be true.

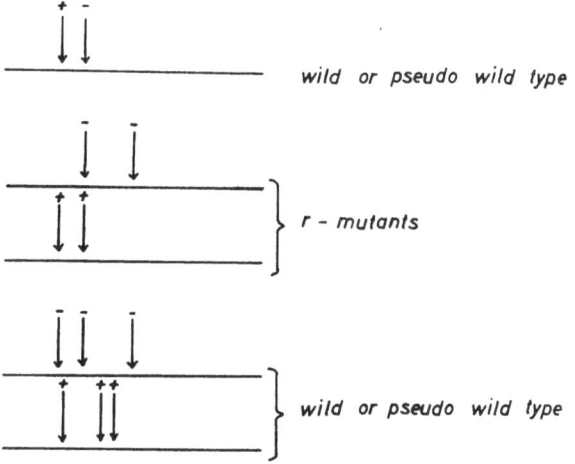

Fig. 26. Scheme of the action of acridines on the DNA during its multiplication (CRICK et al. 1961). The arrows mean an addition (+) or deletion (—) of a base.

The first direct evidence for a triplet code (or, less likely, a multiple of three) comes from the genetic studies of CRICK et al. (1961) on the B cistron of the r_{II} region of the bacteriophage T_4. The wild type grows on both E. coli B and E. coli K; mutants in the r_{II} region grow only on E. coli B, where they produce r plaques. Acridines are thought to add or delete a base during the multiplication of the DNA (see review of FREESE 1963). A mutant which has, e.g., added a base, can be reverted to wild type or pseudo wild type by deleting another base close to it. If a pseudo wild type is produced, only a small (and generally not very important) part of the DNA between these two mutation points (see Fig. 26) may be read wrong. If the second mutational event is not a deletion but another addition of a base, then the phages will not grow on *E. coli* K. But, if a third base is added, one obtains again a pseudo wild type phage progeny. This is explained in more detail in Fig. 26.

Since three additions or three deletions produce pseudo wild type, the coding ratio must be three (or a multiple of three).

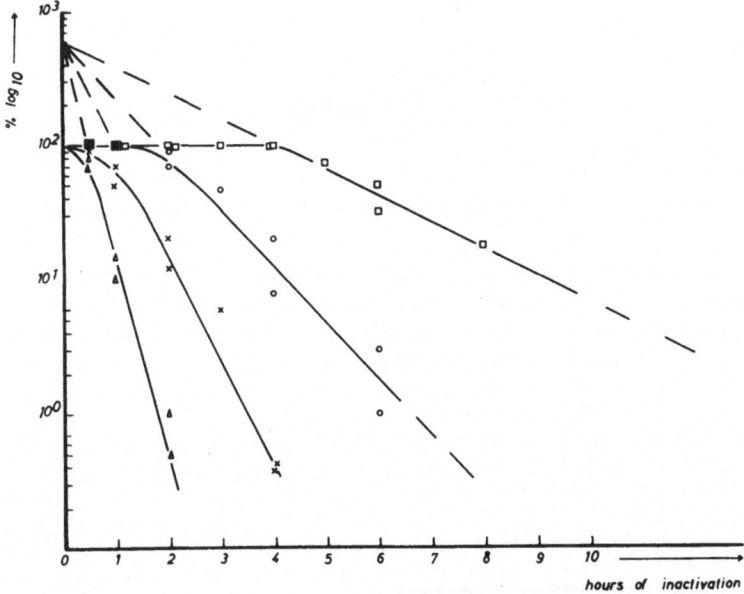

Fig. 27. Stepwise inactivation of fowl plague virus by Bayer A 139 (SCHOLTISSEK and ROTT 1964). The ordinate shows the percentage of the yield of non-treated virus. △ = infectivity; x = hemagglutinating activity; ◯ = neuraminidase activity; □ = s-antigen. The curve of viral RNA synthesis follows that of the s-antigen.

Furthermore these authors (CRICK et al. 1961) found, that the base sequence is read from a fixed starting point. There are no special commas to show how to select the right triplets. These authors suggest that the triplet code is degenerate; this means that more than one triplet exists which codes for one amino acid.

Another possible approach to the determination of the coding size comes from the inactivation studies of fowl plague virus (SCHOLTISSEK and ROTT 1964). Bayer A 139 destroyes specifically nucleic acids leaving proteins intact. By this compound fowl plague virus can be inactivated stepwise as shown in Fig. 27. A short treatment renders the virus non-infectious, but hemagglutinin—this is the outer shell of the virus—and the other viral components are still being synthesized. Further inactivation leads to the loss of the ability to synthesize hemagglutinin. Thereafter the capacity for the synthesis of neuraminidase (the only enzyme of the viral coat), then that of the viral inner component—the s-antigen—and viral RNA is lost. Viral RNA synthesis depends on the presence of an "early protein", which is determined in this way. From the slopes of the curves in Fig. 27

the target size for the different viral components can be calculated. Evidence has been presented that the targets are arranged in an overlapping fashion as shown in Fig. 28. Thus it is possible to calculate the size of the different cistrons from this scheme. The molecular weight of the viral RNA is about 2,000,000 corresponding to 6,600 nucleotides. Only 25 % of the viral genome codes for the viral component responsible for

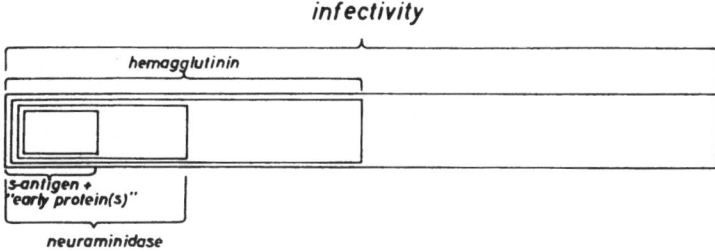

infectivity

Fig. 28. Scheme for the overlapping arrangement of the targets for the different viral components along the viral genome. (Scholtissek and Rott 1964).

the hemagglutinating activity of the virus, corresponding to 1,650 nucleotides. The average molecular weight of an amino acid in a normal protein is about 110. Laver (cited by Fazekas de St. Groth during the Ciba Symposium on myxoviruses, London, February 1964) determined the molecular weight of this Viral component to be 60,000. Thus one has to assume a

Tab. 3. *Tentative summary of code words* [Jones and Nirenberg (1962)].

Amino acid	Corresponding triplets			
Alanine	CCG			
Arginine	CGC			
Aspartic acid	ACA			
Asparagine	UAC	UAA		
Cysteine	UUG	UGG		
Glutamic acid	ACA	AGA	AGU	
Glutamine	UAC*	AAC*	AGG*	
Glycine	UGG			
Histidine	ACC			
Isoleucine	UUA	UAA*	UAC*	
Leucine	GUU	CUU	AUU	UCC*
Lysine	AAA	AAC	AAG	AAU
Methionine	UGA			
Phenylalanine	UUU	UUC*		
Proline	CCC	CCU	CCA	CCG
Serine	UCG	UCU	ACG*	UCC*
Threonine	CAC	CAA	UAC*	AAC*
Tryptophan	UGG			
Tyrosine	UAU	UAC*		
Valine	UGU			

* see Wahba et al. (1963).

t r i p l e t   c o d e—this means that three bases code for one amino acid—in order to calculate from the figures mentioned above a molecular weight of about 60,000.

Direct evidence for a degenerate code and the relation of several triplets to defined amino acids was provided by the Nirenberg group (see survey of Jones and Nirenberg 1962), Ochoa and his group (see Wahba et al. 1962, 1963), Wittmann (1961), and Bretscher and Grunberg-Manago (1962). Nirenberg and Matthaei (1961) have found, that in an i n   v i t r o system, protein synthesis could be stimulated by some kind of messenger RNA. If they added instead of this RNA enzymatically synthesized polyuridylic acid (poly U), they obtained a completely insoluble polyphenylalanine.

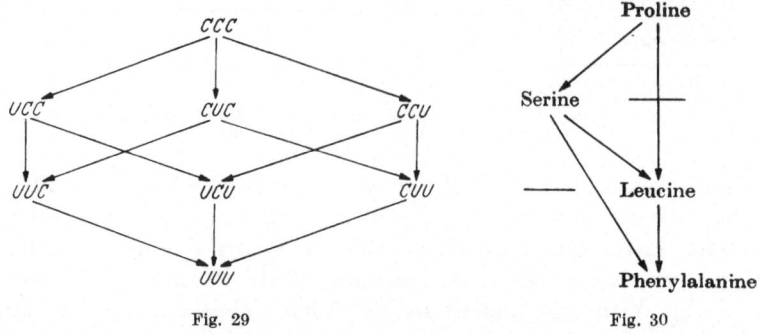

Fig. 29   Fig. 30

Fig. 29. Possible changes of the triplet CCC with nitrous acid (Wittmann 1961).
Fig. 30. Exchange of amino acids in the TMV-protein after treatment of TMV-RNA with nitrous acid (Wittmann, 1961).

With polycytidylic acid (poly C) they got polyproline. These experiments were continued using polymers of mixtures of several nucleotides in different proportions. The results obtained are summarized in Table 3.

Wittmann (1961) found another approach to this problem. He treated infectious tobacco mosaic virus (TMV) RNA with nitrous acid, which converts cytidine to uridine, adenosine to inosine, and guanosine to xanthosine. During the replication of the RNA inosine leads to guanosine and xanthosine also to guanosine. This means that only the conversion of cytidine and adenosine leads to mutations, which were demonstrated by Gierer and Mundry (1958). If in the triplet CCC one base is altered, the three triplets shown in Fig. 29 can be expected. Each of these triplets can be changed by nitrous acid again to two other triplets. If these triplets are attacked again by nitrous acid, only one triplet is possible, namely UUU. 7 other octets of this kind are possible. Wittmann found mutants of TMV after treatment with nitrous acid, which had one proline exchanged by one serine or by one leucine. The serine mutant again could be converted to a leucine mutant or to a phenylalanine mutant, which is demonstrated in Fig. 30. Since phenylalanine is coded by poly U, the other amino acids should be coded by UCU (leucine), UCC (serine), and CCC (proline), which is in accordance with the findings presented in Table 3. Other examples are described in the publication of Wittmann (1961).

But, as has been thought first, it is not necessary that each triplet contains U. Now more and more evidence accumulates that the code for the amino acids is highly degenerate. The sequence of the nucleotides within a triplet is—with a few exceptions—not yet known.

Acknowledgement

The author wishes to thank Dr. D. A. MARVIN, Dr. H. SCHALLER, and Dr. E. ECKERT for helpful suggestions and for correcting the manuscript.

References

ALLFREY, V. G., and A. E. MIRSKY, 1957: Some aspects of ribonucleic acid synthesis in isolated cell nuclei. Proc. Nat. Acad. Sci., U. S. **43**, 821—826.

ASTBURY. W. T., and F. O. BELL. 1938: X-ray study of thymonucleic acid. Nature **141**, 747—748.

ASTRACHAN, L., and E. VOLKIN, 1958: Properties of ribonucleic acid turnover in T₂-infected *Escherichia coli*. Biochim. Biophys. Acta **29**, 536—544.

AVERY, O. T., C. M. MACLEOD, and M. McCARTY, 1944: Studies on the chemical nature of the substance inducing plaque formation of pneumococcal types. Induction of transformation by a desoxyribonucleic acid fraction isolated from pneumococcus type III. J. Exper. Med. **79**, 137—158.

BADDILEY, J., 1955: Chemistry of nucleosides and nucleotides. In: CHARGAFF and DAVIDSON, The Nucleic Acids I, Academic Press: New York, pp. 137—190.

BALTIMORE, D., and R. M. FRANKLIN, 1962: Preliminary data on a virus-specific enzyme system responsible for the synthesis of viral RNA. Biochem. Biophys. Res. Comm. **9**, 388—392.

BARONDES, S. H., C. W. DINGMAN, and M. B. SPORN. 1962: In vitro stimulation of amino-acid incorporation into protein by liver nuclear RNA. Nature **196**, 145—147.

BARRY, R. D., D. R. IVES, and J. G. CRUICKSHANK, 1962: Participation of deoxyribonucleic acid in the multiplication of influenza virus. Nature **194**, 1139—1140.

BEAVEN, G. H., E. R. HOLIDAY, and E. A. JOHNSON, 1955: Optical properties of nucleic acids and their components. In: CHARGAFF and DAVIDSON, The Nucleic acids I, Academic Press: New York, pp. 493—553.

BENDICH, A., 1955: Chemistry of purines and pyrimidines. In: CHARGAFF and DAVIDSON, The Nucleic Acids I, Academic Press: New York. pp. 81—136.

BERG, P., 1961: Specificity in protein synthesis. Ann. Rev. Biochem. **30**, 293—324.

BOLLUM, F. J. 1959: Thermal conversion of nonpriming deoxyribonucleic acid to primer. J. Biol. Chem. **234**, 2733—2734.

— and V. R. POTTER, 1957: Thymidine incorporation into deoxyribonucleic acid of rat liver homogenates. J. Amer. Chem. Soc. **79**, 3603—3604.

— — 1958: Incorporation of thymidine into deoxyribonucleic acid by enzymes from rat tissues. J. Biol. Chem. **233**, 478—482.

— — 1959: Nucleic acid metabolism in regenerating rat liver VI. Soluble enzymes which convert thymidine to thymidine phosphates and DNA. Cancer Res. **19**, 561—565.

BRENNER, S., 1957: On the impossibility of all overlapping triplet codes in information transfer from nucleic acid to proteins. Proc. Nat. Acad. Sci., U. S. **43**, 687—694.

— F. JACOB, and M. MESELSON, 1961: An unstable intermediate carrying information from genes to ribosomes for protein synthesis. Nature **190**, 576—581.

BRETSCHER, M. S., and GRUNBERG-MANAGO, 1962: Polyribonucleotide-directed protein synthesis using an *E. coli* cell-free system. Nature **195**, 283—284.

BROWN, D. M., and A. R. TODD, 1952: Nucleotides X. Some observations on the structure and chemical behavior of the nucleic acids. J. Chem. Soc. **1952**, 52—58.

— — 1953: Nucleotides XXI. The action of ribonuclease on simple esters of the monoribonucleotides. J. Chem. Soc. **1953**, 2040—2049.

— — 1955: Evidence on the nature of the chemical bonds in nucleic acids. In: CHARGAFF and DAVIDSON, The Nucleic Acids I, Academic Press: New York, pp. 409—445.

BROWNHILL, T. J., A. S. JONES, and M. STACEY, 1959: The inactivation of ribonuclease during the isolation of ribonucleic acids and ribonucleoproteins from yeast. Biochem. J. **73**, 434—438.

Buchanan, J. M., 1960: Biosynthesis of purine nucleotides. In: Chargaff and David-son, The Nucleic Acids III, Academic Press: New York, pp. 303—322.

Burton, K., and G. B. Petersen, 1960: The frequences of certain sequences of nucleo-tides in deoxyribonucleic acid. Biochem. J. 75, 17—27.

Cairns, J., 1962: The application of autoradiography to the study of DNA viruses. Cold Spring Harbor Symp. Quant. Biol. 27, 311—318.

— 1963: The bacterial chromosome and its manner of replication as seen by auto-radiography. J. Mol. Biol. 6, 208—213.

Caspersson, T., 1940: Nucleinsäureketten und Genvermehrung. Chromosoma 1, 605—619.

Chargaff, E., 1950: Chemical specificity of nucleic acids and mechanism of their enzymatical degradation. Experientia 6, 201—209.

— and S. Zamenhof, 1948: The isolation of highly polymerized desoxyribonucleic acid from yeast cells. J. Biol. Chem. 173, 327—335.

Chamberlin, M., and P. Berg, 1962: Deoxyribonucleic acid-directed synthesis of ribonucleic acid by an enzyme from *Escherichia coli*. Proc. Nat. Acad. Sci., U. S. 48, 81—94.

Cohn, W. E., 1955: The separation of nucleic acid derivates by chromatography on ion-exchange columns. In: Chargaff and Davidson, The Nucleic Acids I, Academic Press: New York, pp. 211—241.

Cramer, F., 1961: Probleme der chemischen Polynucleotid-Synthese. Angew. Chem. 73, 49—56.

— and K. H. Scheit, 1962: Eine neue Methode zur Synthese von Ribo-oligonucleo-tiden: Die Darstellung von Uridyl-(3′ → 5′)-uridin-3′-phosphat. Angew. Chem. 74, 717.

Crick, F. H. C., L. Barnett, S. Brenner, and R. J. Watts-Tobin: General nature of the genetic code for proteins. Nature 192, 1227—1232.

Crosbie, G. W., 1960: Biosynthesis of pyrimidine nucleotides. In: Chargaff and Davidson, The Nucleic Acids III, Academic Press: New York, pp. 323—348.

Davidson, J. N., and E. Chargaff, 1955: Introduction. In: Chargaff and Davidson, The Nucleic Acids I, Academic Press: New York, pp. 1—8.

DiMayorca, G. A., B. E. Eddy, S. E. Stewart, W. S. Hunter, C. Friend, and A. Bendich, 1959: Isolation of infectious deoxyribonucleic acid from SE polyoma-infected tissue cultures. Proc. Nat. Acad. Sci., U. S. 45, 1805—1808.

Doerfler, W., W. Zillig, E. Fuchs, and M. Albers, 1962: Untersuchungen zur Bio-synthese der Proteine V. Die Funktion von Nucleinsäuren beim Einbau von Aminosäuren in Proteine in einem zellfreien System aus *E. coli*. Z. physiol. Chem. 330, 96—123.

Doty, P., J. Marmur, J. Eigner, and C. Schildkraut, 1960: Strand separation and specific recombination in deoxyribonucleic acids: Physical chemical studies. Proc. Nat. Acad. Sci., U. S. 46, 461—476.

Dounce, A. L. 1955: The isolation and composition of cell nuclei and nucleoli. In: Chargaff and Davidson, The Nucleic Acids II, Academic Press: New York, pp. 93—153.

Edström, J. E., and W. Beermann, 1962: The base composition of nucleic acids in chromosomes, puffs, nucleoli, and cytoplasm of Chironomus salivary gland cells. J. Cell Biol. 14, 371—380.

Felsenfeld, G., and A. Rich, 1957: Studies on the formation of two- and three-stranded polyribonucleotides. Biochim. Biophys. Acta 26, 457—468.

Freese, E., 1963: Molecular mechanism of mutations. In: Taylor, Molecular Genetics I, Academic Press: New York and London, pp. 207—269.

— E. Bautz, and E. Bautz-Freese, 1961: The chemical and mutagenic specificity of hydroxylamine. Proc. Nat. Acad. Sci., U. S. 47, 845—855.

Furth, J. J., J. Hurwitz, and M. Anders, 1962: The role of deoxyribonucleic acid in ribonucleic acid synthesis. I. The purification and properties of ribonucleic acid polymerase. J. Biol. Chem. 237, 2611—2619.

Geiduschek, E. P., J. W. Moohr, and S. B. Weiss, 1962: The secondary structure of complementary RNA. Proc. Nat. Acad. Sci., U. S. 48, 1078—1086.

Gierer, A., 1963: Function of aggregated reticulocyte ribosomes in protein synthesis. J. Mol. Biol. 6, 148—157.

— and K. W. Mundry, 1958: Production of mutants of tobacco mosaic virus by chemical alteration of its ribonucleic acid in vitro. Nature 182, 1457—1458.

— and G. Schramm, 1956: Infectivity of ribonucleic acid from tobacco mosaic virus. Nature 177, 702—703.

GLOCK, G. E., 1955: Biosynthesis of pentoses. In: CHARGAFF and DAVIDSON, The Nucleic Acids II, Academic Press: New York, pp. 247—275.

GOLDBERG, I. H., M. RABINOWITZ, and E. REICH, 1962: Basis of actinomycin action I. DNA binding and inhibition of RNA-polymerase synthetic reactions by actinomycin. Proc. Nat. Acad. Sci., U. S. 48, 2091—2101.

GOLDSTEIN, L., and W. PLAUT, 1955: Direct evidence for nuclear synthesis of cytoplasmic ribonucleic acid. Proc. Nat. Acad. Sci., U. S. 41, 874—880.

GOMATOS, P. J., and I. TAMM, 1963: The secondary structure of reovirus RNA. Proc. Nat. Acad. Sci., U. S. 49, 707—714.

GOODMAN, H. M., and A. RICH, 1962: Formation of a DNA-soluble RNA hybride and its relation to the origin, evolution, and degeneracy of soluble RNA. Proc. Nat. Acad. Sci., U. S. 48, 2101—2109.

GROS, F., H. HIATT, W. GILBERT, C. G. KURLAND, R. W. RISEBROUGH, and J. D. WATSON, 1961: Unstable ribonucleic acid revealed by pulse labelling of Escherichia coli. Nature 190, 581—585.

GRUNBERG-MANAGO, M., and S. OCHOA, 1955: Enzymatic synthesis and breakdown of polyribonucleotides; polynucleotide phosphorylase. J. Amer. Chem. Soc. 77, 3165—3166.

GUTHRIE, G. D., and R. L. SINSHEIMER, 1960: Infection of protoplasts of Escherichia coli by subviral particles of bacteriophage Φ X 174. J. Mol. Biol. 2, 297—305.

HALL, B. D., and S. SPIEGELMAN, 1961: Sequence complementarity of T_2-DNA and T_2-specific RNA. Proc. Nat. Acad. Sci., U. S. 47, 137—146.

HANDSCHUMACHER, R. E., and A. D. WELCH, 1960: Agents which influence nucleic acid metabolism. In: CHARGAFF and DAVIDSON, The Nucleic Acids III, Academic Press: New York; pp. 453—526.

HARBERS, E., and W. MÜLLER, 1962: On the inhibition of RNA synthesis by actinomycin. Biochem. Biophys. Res. Comm. 7, 107—110.

HASELKORN, R., and P. DOTY, 1961: The reaction of formaldehyde with polynucleotides. J. Biol. Chem. 236, 2738—2745.

HAYASHI, M., and S. SPIEGELMAN, 1961: The selective synthesis of informational RNA in bacteria. Proc. Nat. Acad. Sci., U. S. 47, 1564—1580.

HECHT, L. I., P. C. ZAMECNIK, M. L. STEPHENSON, and J. F. SCOTT, 1958: Nucleoside triphosphates as precursors of ribonucleic acid end groups in a mammalian system. J. Biol. Chem. 233, 954—963.

HEPPEL, L. A., and J. C. RABINOWITZ, 1958: Enzymology of nucleic acids, purines, and pyrimidines. Ann. Rev. Biochem. 27, 613—642.

HERBERT, E., and C. W. WILSON, 1962 a: Determination of nucleotide sequences in soluble ribonucleic acid I. Pyrophosphorolysis and reconstitution of a fraction of soluble ribonucleic acid. Biochim. Biophys. Acta 61, 750—761.

— — 1962 b: Determination of nucleotide sequences in soluble ribonucleic acid II. Determination of nucleotide sequences in oligonucleotides derived from the acceptor end of pyrophosphorolyced soluble ribonucleic acid. Biochim. Biophys. Acta 61, 762—774.

HERSHEY, A. D., and M. CHASE, 1953: Independent functions of viral protein and nucleic acid in growth of bacteriophages. J. Gen. Physiol. 36, 39—56.

HIATT, H. H., 1962: A rapidly labeled RNA in rat liver nuclei. J. Mol. Biol. 5, 217—229.

HILMOE, R. J., and L. A. HEPPEL, 1957: Polynucleotide phosphorylase in liver nuclei. J. Amer. Chem. Soc. 79, 4810—4811.

HOFSCHNEIDER, P. H., 1960: Über ein infektiöses Desoxyribonucleinsäure-Agens aus dem Phagen Φ X 174. Z. Naturforsch. 15 b, 441—444.

HOGEBOOM, G. H., and W. C. SCHNEIDER, 1955: The cytoplasm. In: CHARGAFF and DAVIDSON, The Nucleic Acids II, Academic Press: New York, pp. 199—246.

HOTCHKISS, R. D., 1955: The biological role of the deoxypentose nucleic acids. In: CHARGAFF and DAVIDSON, The Nucleic Acids II, Academic Press: New York, pp. 435—473.

HURWITZ, J., A. BRESLER, and R. DIRINGER, 1960: The enzymatic incorporation of ribonucleotides into polyribonucleotides and the effect of DNA. Biochim. Biophys. Res. Comm. 3, 15—19.

— J. J. FURTH, M. MALAMY, and M. ALEXANDER, 1962: The role of deoxyribonucleic acid in ribonucleic acid synthesis III. The inhibition of the enzymatic synthesis of ribonucleic acid and deoxyribonucleic acid by actinomycin D and proflavin. Proc. Nat. Acad. Sci., U. S. 48, 1222—1230.

ITO, Y., 1960: A tumor-producing factor extracted by phenol from papillomatous tissues (Shope) of cotton tail rabbits. Virology 12, 596—601.

Jacob, F., and J. Monod, 1961: Genetic regulatory mechanisms in the synthesis of proteins. J. Mol. Biol. 3, 318—356.

Jones jr., O. W., and M. W. Nirenberg, 1962: Qualitative survey of RNA code words. Proc. Nat. Acad. Sci., U. S. 48, 2115—2123.

Jordan, D. O., 1955: The physical properties of nucleic acids. In: Chargaff and Davidson, The Nucleic Acids I, Academic Press: New York, pp. 447—492.

Josse, J., H. D. Kaiser, and A. Kornberg, 1961: Enzymatic synthesis of deoxyribonucleic acid VIII. Frequences of nearest neighbor base sequences in deoxyribonucleic acid. J. Biol. Chem. 236, 864—875.

Kersten, W., and H. Kersten, 1962: Zur Wirkungsweise von Actinomycinen III. Bindung von Actinomycin C an Nucleinsäuren und Nucleotide. Z. physiol. Chem. 330, 21—30.

Khorana, H. G., 1960: Chemical and enzymatic synthesis of polynucleotides. In: Chargaff und Davidson, The Nucleic Acids III, Academic Press: New York, pp. 105—146.

— 1961: Some Recent Developments in the Chemistry of Phosphate Esters of Biological Interest. J. Wiley & Sons: New York.

Kornberg, A., 1961: Enzymatic Synthesis of DNA. Ciba Lectures in Microbial Biochemistry J. Wiley & Sons: New York and London.

— I. R. Lehmann, M. J. Bessman, and E. S. Simms, 1956: Enzymic synthesis of deoxyribonucleic acid. Biochim. Biophys. Acta 21, 197—198.

Kunitz, M., 1940: Crystalline ribonuclease. J. Gen. Physiol. 24, 15—32.

Lajtha, L. G., 1960: The effect of radiations on nucleic acid metabolism. In: Chargaff and Davidson, The Nucleic Acids III, Academic Press: New York, pp. 527—546.

Lehman, I. R., 1960: The deoxyribonucleases of *Escherichia coli* I. Purification and properties of a phosphodiesterase. J. Biol. Chem. 235, 1479—1487.

— G. G. Roussos, and E. A. Pratt, 1962: The deoxyribonucleases of *Escherichia coli* II. Purification and properties of a ribonucleic acid-inhibitable endonuclease. J. Biol. Chem. 237, 819—828.

Lengyel, P., J. F. Speyer, and S. Ochoa, 1961: Synthetic polynucleotides and the amino acid code. Proc. Nat. Acad. Sci., U. S. 47, 1936—1942.

Levinthal, C., A. Keynan, and A. Higa, 1962: Messenger RNA turnover and protein synthesis in B. subtilis inhibited by actinomycin D. Proc. Nat. Acad. Sci., U. S. 48, 1631—1638.

Marks, P. A., C. Willson, J. Kruh, and F. Gros, 1962: Unstable ribonucleic acid in mammalian blood cells. Biochem. Biophys. Res. Comm. 7, 9—14.

Marmur, J., and P. Doty, 1959: Heterogeneity in deoxyribonucleic acids I. Dependence on composition of the configurational stability of deoxyribonucleic acids. Nature 183, 1427—1429.

Matthaei, J. H., and M. W. Nirenberg, 1961: Characteristics and stabilization of DNase-sensitive protein synthesis in E. coli extracts. Proc. Nat. Acad. Sci., U. S. 47, 1580—1588.

— O. W. Jones, R. G. Martin, and M. W. Nirenberg, 1962: Characteristics and composition of RNA coding units. Proc. Nat. Acad. Sci., U. S. 48, 666—677.

McCully, K. S., and G. L. Cantoni, 1962: Studies on soluble ribonucleic acid (s-RNA) of rabbit liver VII. A base sequence model of s-RNA. J. Mol. Biol. 5, 497—505.

McMaster-Kaye, R., 1960: The metabolic characteristics of nucleolar, chromosomal, and cytoplasmic ribonucleic acid of Drosophila salivary glands. J. Biophys. Biochem. Cytol. 8, 365—378.

Meselson, M., and W. Stahl, 1958: The replication of DNA in *Escherichia coli*. Proc. Nat. Acad. Sci., U. S. 44, 671—682.

Michelson, A. M., 1958: Hyperchromicity and nucleic acids. Nature 182, 1502—1503.

— 1959: Polynucleotides I. Synthesis and properties of some polyribonucleotides. J. Chem. Soc. 1959, 1371—1394.

— 1961: Chemistry of the nucleotides. Ann. Rev. Biochem. 30, 133—164.

— 1963: The Chemistry of Nucleosides and Nucleotides. Academic Press: London and New York.

Nagata, T., 1963: The molecular synchrony and sequential replication of DNA in *Escherichia coli*. Proc. Nat. Acad. Sci., U. S. 49, 551—559.

Nirenberg, M. W., and J. H. Matthaei, 1961: The dependence of cell-free protein synthesis in E. coli upon naturally occuring or synthetic polyribonucleotides. Proc. Nat. Acad. Sci., U. S. 47, 1588—1602.

Ochoa, S., D. P. Burma, H. Kröger, and J. D. Weill, 1961: Deoxyribonucleic acid-dependent incorporation of nucleotides from nucleoside triphosphates into ribonucleic acid. Proc. Nat. Acad. Sci., U. S. 47, 670—679.

PELLING, C., 1959: Chromosomal synthesis of ribonucleic acid as shown by the incorporation of uridine labelled with tritium. Nature 184, 655—656.

PENMAN, S., K. SCHERRER, Y. BECKER, and J. E. DARNELL, 1963: Polyribosomes in normal and poliovirus-infected HeLa cells and their relationship to messenger-RNA. Proc. Nat. Acad. Sci., U. S. 49, 654—662.

POTTER, V. R., 1960: Nucleic Acid Outlines, Burgess Publishing Company: Minneapolis, Minn.

PRESCOTT, D. M., 1959: Nuclear synthesis of cytoplasmic ribonucleic acid in Amoeba proteus. J. Biochem. Biophys. Cytol. 6, 203—206.

RADDING, C. M., J. JOSSE, and A. KORNBERG, 1962: Enzymatic synthesis of deoxyribonucleic acid XII. A polymer of deoxyguanylate and deoxycytidylate. J. Biol. Chem. 237, 2869—2876.

RAZZELL, W. E., and H. G. KHORANA, 1958: The stepwise degradation of thymidine oligonucleotides by snake venom and spleen phosphodiesterases. J. Amer. Chem. Soc. 80, 1770—1771.

— — 1959: Studies on polynucleotides IV. Enzymatic degradation. The stepwise action of venom phosphodiesterase on deoxyribo-oligonucleotides. J. Biol. Chem. 234, 2114—2117.

REDDI, K. K., 1959: Structural differences in the nucleic acids of some tobacco mosaic virus strains II. Di- and trinucleotides in ribonuclease digests. Biochim. Biophys. Acta 32, 386—392.

REICH, E., R. M. FRANKLIN, A. J. SHATKIN, and E. L. TATUM, 1961: Effect of actinomycin D on cellular nucleic acid synthesis and virus production. Science 134, 556—557.

— — — 1962: Action of actinomycin on animal cells and viruses. Proc. Nat. Acad. Sci., U. S. 48, 1238—1245.

REICHARD, P., 1955: Biosynthesis of purines and pyrimidines. In: CHARGAFF and DAVIDSON, The Nucleic Acids II, Academic Press: New York, pp. 277—308.

RICH, A., and D. R. DAVIS, 1956: A new two stranded helical structure: Polyadenylic acid and polyuridylic acid. J. Amer. Chem. Soc. 78, 3548—3549.

ROSSET, R., and R. MONIER, 1963: On the instability of transfer-RNA terminal nucleotide sequence in yeast. Biochem. Biophys. Res. Comm. 10, 195—199.

ROTH, J. S., 1958: Ribonuclease VIII. Studies on the inactive ribonuclease in the supernatant fraction of rat liver. J. Biol. Chem. 231, 1097—1105.

ROTT, R., and C. SCHOLTISSEK, 1964: Einfluß von Actinomycin auf die Vermehrung von Myxoviren. Z. Naturforsch. 19 b, 316—323.

RUSHIZKY, G. W., and C. A. KNIGHT, 1960: An oligonucleotide mapping procedure and its use in the study of tobacco mosaic virus nucleic acid. Virology 11, 236—249.

SANDRON, C. L., 1960: Deoxyribonucleic acids as macromolecules. In: CHARGAFF and DAVIDSON, The Nucleic Acids III, Academic Press: New York, pp. 1—37.

SATO, K., and F. EGAMI, 1960: Ribonuclease in takadiastase. Nature 185, 462—463.

SCHACHMAN, H. K., J. ADLER, C. M. RADDING, I. R. LEHMAN, and A. KORNBERG, 1960: Enzymatic synthesis of deoxyribonucleic acid VII. Synthesis of a polymer of deoxyadenylate and deoxythymidylate. J. Biol. Chem. 235, 3242—3249.

SCHALLER, H., G. WEIMANN, and H. G. KHORANA, 1963: The synthesis of deoxyribopolynucleotides containing specific nucleotide sequences. J. Amer. Chem. Soc. 85, 355—356.

SCHERRER, K., and J. E. DARNELL, 1962: Sedimentation characteristics of rapidly labelled RNA from HeLa cells. Biochem. Biophys. Res. Comm. 7, 486—490.

— H. LATHAM, and J. E. DARNELL, 1963: Demonstration of an unstable RNA and of a precursor to ribosomal RNA in HeLa cells. Proc. Nat. Acad. Sci., U. S. 49, 240—248.

SCHILDKRAUT, C. L., J. MARMUR, J. R. FRESCO, and P. DOTY, 1961: Formation and properties of polyribonucleotide-polydeoxyribonucleotide helical complexes. J. Biol. Chem. 236, PC 2-PC 4.

SCHLENK, F., 1955: Biosynthesis of nucleosides and nucleotides. In: CHARGAFF and DAVIDSON, The Nucleic Acids II, Academic Press: New York, pp. 309—339.

SCHMIDT, G., 1955: Nucleases and enzymes attacking nucleic acid components. In: CHARGAFF and DAVIDSON, The Nucleic Acids I, Academic Press: New York, pp. 555—626.

SCHOLTISSEK, C., 1957: Auftrennung des bei der Ribonucleaseverdauung anfallenden, höhermolekularen Anteils (Core) der Ribonucleinsäure durch Papierelektrophorese. Z. physiol. Chem. 309, 129—135.

— 1959 a: Charakterisierung von Ribonucleinsäuren verschiedener Herkunft, die gleiche Basenverhältnisse besitzen, durch enzymatischen Abbau. Biochem. Z. 331, 138—143.

Scholtissek, C., 1959 b: Beziehungen zwischen der Ribonucleinsäuresynthese im Zellkern und in verschiedenen Cytoplasmafraktionen. Biochem. Z. 331, 365—374.
— 1960 a: Verwandtschaftsbeziehungen zwischen den Ribonucleinsäuren des Zellkerns, der Mitochondrien, Mikrosomen und des Zellsaftes bei Rattenleber, Milz und Nieren. Biochem. Z. 332, 458—466.
— 1960 b: Veränderungen von in vivo mit P³² markierter Ribonucleinsäure aus Rattenleberkernen während der Inkubation der Kerne in vitro. Biochem. Z. 332, 467—476.
— 1962 a: An unstable ribonucleic acid in rat liver nuclei. Nature 194, 353—355.
— 1962 b: End-turnover of rat liver soluble RNA in vivo. Biochim. Biophys. Acta 61, 499—505.
— and V. R. Potter, 1960: Austritt von Ribonucleinsäure aus isolierten Rattenleber-Zellkernen während der Inkubation in vitro. Z. Naturforsch. 15 b, 453—460.
— and R. Rott, 1961: Untersuchungen über die Vermehrung des Virus der Klassischen Geflügelpest. Die Synthese der virusspezifischen Ribonucleinsäure (RNS) in infizierten Gewebekulturen embryonaler Hühnerzellen. Z. Naturforsch. 16 b, 109—115.
— — 1964: Behavior of virus-spezific activities in tissue cultures infected with myxoviruses after chemical changes of the viral ribonucleic acid. Virology 22, 169—176.
— — P. Hausen, H. Hausen, and W. Schäfer, 1962: Comparative studies of RNA and protein synthesis with a myxovirus and a small polyhedral virus. Cold Spring Harbor Symp. Quant. Biol. 27, 245—257.
— J. H. Schneider, and V. R. Potter, 1958: Transfer of ribonucleic acid from nuclei to cytoplasm in vitro. Fed. Proc. 17, 306.
Schramm, G., H. Grötsch, and W. Pollmann, 1962: Nicht-enzymatische Synthese von Polysacchariden, Nucleosiden und Nucleinsäuren und die Entstehung selbstvermehrungsfähiger Systeme. Angew. Chem. 74, 53—59.
Schuster, H., 1960 a: Die Reaktionsweise der Desoxyribonucleinsäure mit salpetriger Säure. Z. Naturforsch. 15 b, 298—304.
— 1960 b: The ribonucleic acids of viruses. In: Chargaff and Davidson, The Nucleic Acids III, Academic Press: New York, pp. 245—301.
— 1961: The reaction of tobacco mosaic virus ribonucleic acid with hydroxylamine. J. Mol. Biol. 3, 447—457.
— and G. Schramm, 1958: Bestimmung der biologisch wirksamen Einheit in der Ribonucleinsäure des Tabakmosaikvirus auf chemischem Wege. Z. Naturforsch. 13 b, 697—704.
Shapiro, H. S., and E. Chargaff, 1957 a: Studies on the nucleotide arrangement in deoxyribonucleic acids I. The relationship between the production of pyrimidine nucleoside 3',5'-diphosphates and specific features of nucleotide sequences. Biochim. Biophys. Acta 26, 596—608.
— — 1957 b: Studies on the nucleotide arrangement in deoxyribonucleic acids II. Differential analysis of pyrimidine nucleotide distribution as a method of characterization. Biochim. Biophys. Acta 26, 608—623.
Shugar, D., 1960: Photochemistry of nucleic acids and their constituents. In: Chargaff and Davidson, The Nucleic Acids III, Academic Press: New York, pp. 39—104.
Sibatani, A., S. R. deKloet, V. G. Allfrey, and A. E. Mirsky, 1962: Isolation of a nuclear RNA fraction resembling DNA in its base composition. Proc. Nat. Acad. Sci., U. S. 48, 471—477.
Singer, M. F., L. A. Heppel, and R. J. Hilmoe, 1960: Oligonucleotides as primers for polynucleotide phosphorylase. J. Biol. Chem. 235, 738—750.
Sinsheimer, R. L., 1959: A single-stranded deoxyribonucleic acid from bacteriophage Φ X 174. J. Mol. Biol. 1, 43—53.
Smellie, R. M. S., 1955: The metabolism of the nucleic acids. In: Chargaff and Davidson, The Nucleic Acids II, Academic Press: New York, pp. 393—434.
Smith, J. D., 1955: The electrophoretic separation of nucleic acid components. In: Chargaff and Davidson, The Nucleic Acids I, Academic Press: New York, pp. 267—284.
Spyrides, G. J., and F. Lipmann, 1962: Polypeptide synthesis with sucrose gradient fractions of E. coli ribosomes. Proc. Nat. Acad. Sci., U. S. 48, 1977—1983.
Staehelin, M., 1961 a: Studies on nucleotide sequences in ribonucleic acids I. Separation of oligonucleotides on DEAE-cellulose. Biochim. Biophys. Acta 49, 11—19.

STAEHELIN, M., 1961 b: Studies on nucleotide sequences in ribonucleic acids III. Amounts of oligonucleotides in pancreatic ribonuclease digests. Biochim. Biophys. Acta **49**, 27—35.

STEVENS, A., 1960: Incorporation of the adenine ribonucleotide into RNA by cell fractions from *E. coli* B. Biochem. Biophys. Res. Comm. **3**, 92—96.

— 1961: Netformation of polyribonucleotides with base compositions analogous to deoxyribonucleic acid. J. Biol. Chem. **236**, PC 43-PC 45.

— 1963: Ribonucleic acids. — Biosynthesis and degradation. Ann. Rev. Biochem. **32**, 15—42.

SUEOKA, N., J. MARMUR, and P. DOTY, 1959: Heterogeneity in deoxyribonucleic acids II. Dependence of the density of deoxyribonucleic acids on guanine-cytosine content. Nature **183**, 1429—1431.

TAKEMURA, S., 1958: Hydracinolysis of herring-sperm deoxyribonucleic acid. Biochim. Biophys. Acta **29**, 447—448.

TAMAOKI, D. T., and G. C. MUELLER, 1962: Synthesis of nuclear and cytoplasmic RNA of HeLa cells and the effect of actinomycin D. Biochem. Biophys. Res. Comm. **9**, 451—454.

TEMIN, H. M., 1963: The effect of actinomycin D on growth of Rous sarcoma virus in vitro. Virology **20**, 577—582.

TENER, G. M., H. G. KHORANA, R. MARKHAM, and E. H. POL, 1958: Studies on polynucleotides II. The Synthesis and characterization of linear and cyclic thymidine oligonucleotides. J. Amer. Chem. Soc. **80**, 6223—6230.

TISSIÈRES, A., D. SCHLESSINGER, and F. GROS, 1960: Amino acid incorporation into proteins by *Escherichia coli* ribosomes. Proc. Nat. Acad. Sci., U. S. **46**, 1450—1463.

VENDRELY, R., 1955: The deoxyribonucleic acid content of the nucleus. In: CHARGAFF and DAVIDSON, The Nucleic Acids II, Academic Press: New York, pp. 155—180.

VERWOERD, D. W., and W. ZILLIG, 1963: A specific partial hydrolysis procedure for soluble RNA. Biochim. Biophys. Acta **68**, 484—486.

— H. KOHLHAGE, and W. ZILLIG, 1961: Specific partial hydrolysis of nucleic acids in nucleotide sequence studies. Nature **192**, 1038—1040.

— W. ZILLIG, and H. KOHLHAGE, 1963: Die Reaktion von Nucleinsäuren mit Hydroxylamin als Hilfsmittel für ihre definierte Partialhydrolyse. Z. physiol. Chem. **332**, 184—203.

VOLKIN, E., and L. ASTRACHAN, 1956: Phosphorus incorporation in *Escherichia coli* ribonucleic acid after infection with bacteriophage T₂. Virology **2**, 149—161.

— and W. E. COHN, 1953: On the structure of ribonucleic acids II. the products of ribonuclease action. J. Biol. Chem. **205**, 767—782.

WAHBA, A. J., C. BASILIO, J. F. SPEYER, P. LENGYEL, R. S. MILLER, and S. OCHOA, 1962: Synthetic polynucleotides and the amino acid code VI. Proc. Nat. Acad. Sci., U. S. **48**, 1683—1686.

— R. S. MILLER, C. BASILIO, R. S. GARDNER, P. LENGYEL, and J. F. SPEYER, 1963: Synthetic polynucleotides and the amino acid code IX. Proc. Nat. Acad. Sci., U. S. **49**, 880—885.

WARNER, R. C., 1957: Studies on polynucleotides synthesized by polynucleotide phosphorylase III. Interaction and ultraviolet absorption. J. Biol. Chem. **229**, 711—724.

WARNER, J. R., P. M. KNOPF, and A. RICH, 1963: A multiple ribosomal structure in protein synthesis. Proc. Nat. Acad. Sci., U. S. **49**, 122—129.

WATSON, J. D., and F. H. C. CRICK, 1953: Molecular structure of nucleic acids. A structure for deoxyribose nucleic acid. Nature **171**, 737—738.

WEIMANN, G., and H. G. KHORANA, 1962: Studies on polynucleotides XVII. On the mechanism of internucleotide bound synthesis by the carbodiimide method. J. Am. Chem. Soc. **84**, 4329—4341.

WEISS, S. B., 1960: Enzymatic incorporation of ribonucleoside-triphosphates into the interpolynucleotide linkages of ribonucleic acid. Proc. Nat. Acad. Sci., U. S. **46**, 1020—1030.

— and L. GLADSTONE, 1959: A mammalian system for the incorporation of cytidine triphosphate into ribonucleic acid. J. Amer. Chem. Soc. **81**, 4118—4119.

— and T. NAKAMOTO, 1961 a: Net synthesis of ribonucleic acid with a microbial enzyme requiring deoxyribonucleic acid and four ribonucleoside triphosphates. J. Biol. Chem. **236**, PC 18-PC 20.

— — 1961 b: The enzymatic synthesis of RNA: nearest-neighbor base frequences. Proc. Nat. Acad. Sci., U. S. **47**, 1400—1405.

Weissmann, C., L. Simon, and S. Ochoa, 1963: Induction by an RNA phage of an enzyme catalyzing incorporation of ribonucleotides into ribonucleic acid. Proc. Nat. Acad. Sci., U. S. 49, 407—414.

Wettstein, F. O., T. Staehelin, and H. Noll, 1963: Ribosomal aggregate engaged in protein synthesis: Characterization of the ergosome. Nature 197, 430—435.

Wilkins, M. F. H., A. R. Stokes, and H. R. Wilson, 1953: Molecular structure of deoxypentose nucleic acid. Nature 171, 738—740.

Wittmann, H. G., 1960: Comparison of the tryptic peptides of chemically induced and spontaneous mutants of tobacco mosaic virus. Virology 12, 609—612.

— 1961: Die Entschlüsselung des genetischen Codes. Naturwiss. 48, 729—734.

Wyatt, G. R., 1955: Separation of nucleic acid components by chromatography on filter paper. In: Chargaff and Davidson, The Nucleic Acids I, Academic Press: New York, pp. 243—265.

Yankofsky, S. A., and S. Spiegelman, 1962 a: The identification of the ribosomal RNA cistron by sequence complementarity I. Specificity of complex formation. Proc. Nat. Acad. Sci., U. S. 48, 1069—1078.

— — 1962 b: The identification of the ribosomal RNA cistron by sequence complementarity II. Saturation of and competitive interaction at the RNA cistron. Proc. Nat. Acad. Sci., U. S. 48, 1466—1472.

— — 1963: Distinct cistrons for the two ribosomal RNA components. Proc. Nat. Acad. Sci., U. S. 49, 538—544.

Yates, R. A., and A. B. Pardee, 1956: Control of pyrimidine biosynthesis in *Escherichia coli* by a feed-back-mechanism. J. Biol. Chem. 221, 757—770.

Yoshikawa, H., and N. Sueoka, 1963 a: Sequential replication of bacillus subtilis chromosome I. Comparison of marker frequences in exponential and stationary growth phase. Proc. Nat. Acad. Sci., U. S. 49, 559—566.

— — 1963 b: Sequential replication of bacillus subtilis chromosome II. Isotopic transfer experiments. Proc. Nat. Acad. Sci., U. S. 49, 806—813.

Zalokar, M., 1959: Nuclear origin of ribonucleic acid. Nature 183, 1330.

Protoplasmatologia
  V. Karyoplasma (Nucleus)
    3. Chemistry and Cytochemistry of Nucleic Acids and Nuclear Proteins
    b) Cytochemistry of the Nucleic Acids

Cytochemistry of the Nucleic Acids

By

BRIAN M. RICHARDS

Department of Biophysics, University of London King's College, London, W. C. 2, England

With 2 Figures

Contents

I. Introduction

Cytochemical methods with quantitative aims are now standard procedures in many laboratories. Following a period of technical progress, quantitative cytochemistry has entered a phase in which the investigation of a cell biological problem, sometimes with direct medical import, is the main goal. A wider use of the methods is welcome, but the need for careful attention to sources of error and limits of accuracy, cannot be too firmly stressed. By far the most encouraging sign is the small but increasing use in combination of two or more methods which give similar information. In particular, a combination of microspectrophotometry, or biochemical methods, with autoradiographic detection of isotopically labelled precursors, has the advantage of reciprocal checking in addition to providing complementary data. Fortunately, the undisputed biological importance of desoxyribose nuclei acid (DNA) is matched to some extent by the number and efficiency of the methods available for its assay. If it had turned out that lipids, for example, were the basis of the genetic material, the lack of quantitative methods would have hampered progress. While it may be argued that methods would be developed, it is significant that we are still without a satisfactory quantitative cytochemical method for the estimation of RNA. Quantitative methods for DNA were available before its biological significance was generally accepted.

Quantitative cytochemical methods for nucleic acids are of two main types: photometric methods and autoradiographic procedures. Photometric or light absorption methods have been developed directly or indirectly from those originated by Caspersson in Stockholm, beginning in the early thirties and culminating today in the present laboratories at the Institute for Cell Research and Genetics, which are the most impressive and comprehensively equipped in the world for cytochemical work. Autoradiography, which became a feasible technique following the demonstration of the stripping film procedure by Pelc (1947) and Doniach and Pelc (1950), is now universally used, partly because of the availability of radioactively labelled precursors, and partly because the apparatus requirements contrast sharply in their simplicity with those of photometric methods. The number of reports of experiments using autoradiography increased markedly when tritiated precursors became available, since these had considerably better resolution than previously available labels. The DNA precursor thymidine was tritiated by HUGHES in 1955, but the high resolution was indicated by FITZGERALD and collaborators as early as 1951.

As in all active fields, quantitative cytochemical methods have frequently been subjected to review. The present author intends to limit the scope of this article, firstly, to brief accounts of practical details with comment on recent technical improvement, and secondly to consideration of examples of the results which have been obtained in the most actively studied problems in cell biology.

a) Methods applicable to single cells

The description "cytochemical" is here used to refer to methods which provide information on the location, and/or the quantity of substances in single cells. It is not intended to include biochemical procedures, the results of which are expressed on a "per cell" basis. All cytochemical methods require the use of the optical microscope, or, occasionally, the electron microscope. It cannot be repeated too often that cytochemical methods have the particular advantage that they permit correlation of amounts of substance, or rate of change in amount, with phase of cell activity; much of their interest and value derives from this feature.

An important development in cytochemistry is that of dry mass measurement (DAVIES 1958). This technique provides a base line against which all other cytochemical measurements are to be assessed. If the dry mass of a cell nucleus is found to be 100×10^{-12} g. then the significance of its DNA content measured as 12×10^{-12} g. is clearer. Theoretically, mass measurement can be used to obtain quantities of nucleic acids by difference measurement, before and after selective extraction, but it is likely to be of low accuracy owing to uncertainties in the extraction procedure, especially as regards their influence on the optical properties of the substances remaining after extraction.

b) Microscopy

A high standard of optical performance and practice is essential for cytochemical methods. Procedures for periodical testing of photometric devices are simple to develop, and should not be neglected. In autoradiography much depends on the ability of the observer to identify accurately labelled cells, or make accurate grain counts. Convenience and comfort, such as the ease of changing from bright field to phase contrast, are not the least important considerations in grain counting, and good optical performance is essential to eliminate false positives or negatives in labelling. Automatic or semi-automatic grain counting devices are regularly described, but little use has been made of them.

c) Information obtainable

The difference in the kind of information obtained with photometric methods and autoradiography, make their combined use on the same cells valuable. Measurement of natural or induced light absorption yields the amounts of nucleic acid in cell structures, or regions, at the time of measurement. Only certain procedures, namely, measurement of ultra-violet light absorption for nucleic acids or haemoglobin absorption, can be applied to living cells, but there is risk of change or damage from exposure to light, especially ultra-violet light. Most photometric methods are used to indicate the distribution, concentration and total amounts of nucleic acids in cells after fixation. Several assumptions are necessary to calculate the changes in amounts with time. WALKER (1954) describes a method for the calculation of "average" synthesis curves from frequency histograms of amounts. In contrast, autoradiographic experiments can be designed to determine relative rates of synthesis, but are unsuitable to indicate absolute or relative amounts of substance. Availability of the labelled precursor, and other factors, are very relevant to the interpretation of results, and have sometimes led to controversy. Photometric methods generally give the static picture, therefore, while autoradiographic ones give the dynamic picture. Neither group of methods is free from error, and increased confidence in their results is obtained if they are applied sequentially to the same cells. Photometry of Feulgen stain together with autoradiography of H³-thymidine gives additional information to that obtained by each technique separately as indicated by, for example, the studies of GALL and JOHNSON (1960) on the mouse seminal vesicle.

II. Photometry of Naturally Absorbed Light

Both DNA and RNA absorb strongly in the ultra-violet region with a λ maximum at 2600 Å. Apart from the aminoacids tryptophan, tyrosine and phenylalanine, which are present in most proteins, if only in small amounts, no other cellular substances which absorb in this region occur in significant quantities, and such proteins have an absorption spectrum with a λ maximum at 2800 Å which is distinct from that of nucleic acids. CASPERSSON (1936) was the first to attempt to measure ultra-violet absorption in cells, and, with reference to the specific absorption of nucleic acids in solution, to determine the concentration of nucleic acids in cell structures. In reviews WALKER (1956 and 1958) has dealt in detail with the principles of measurement, and the design of the often complicated apparatus used. An excellent concise account is given by RUDKIN (1960). A commercial instrument (Fig. 1) is now in production by Carl Zeiss (Oberkochen) which will record absorption spectra from 2400 Å in the ultra-violet to 7000 Å in the visible region in an area as small as ½ μ diameter. Furthermore, this instrument can be used to determine the amount of absorbing substance in a whole cell, or its regions, by scanning the specimen and integrating the absorption in the scanned area.

a) Basic procedure

The ultra-violet absorption spectra of nucleic acids, although dependant on such factors as base composition and pH, is for practical purposes identical for DNA and RNA, and cannot be used to distinguish between them. Measurements of amounts of nucleic acid in cells assumes that the absorptivity, or extinction coefficient, k (at λ maximum), which is measurable only in solutions, is the same for the nucleic acid in the cell. The equation for the calculation is:

Absorbance $A = k\,c\,d = \log_{10} I_o/I_T$
where c = concentration,
d = path length,
I_o = incident light intensity,
I_T = transmitted light intensity.

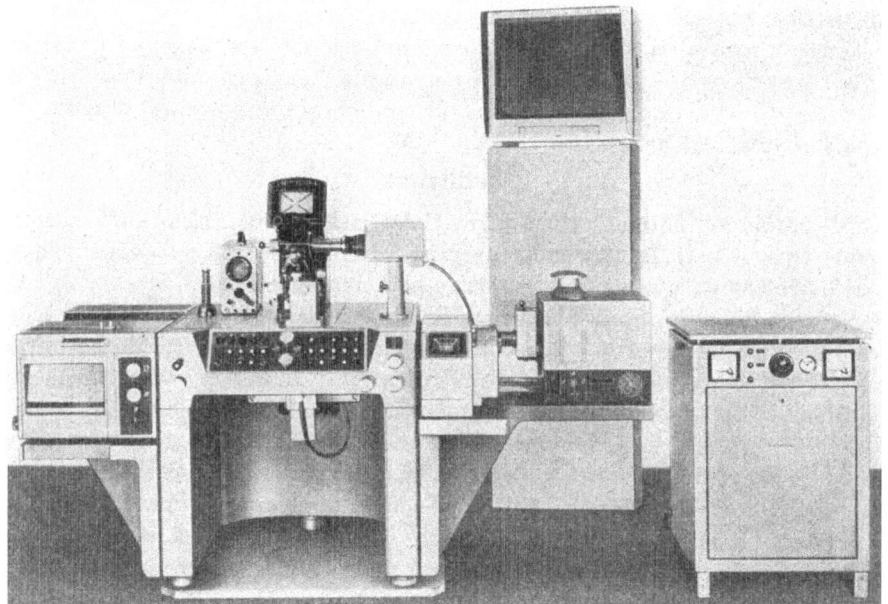

Fig. 1. The universal microspectrophotometer manufactured by Zeiss (Oberkochen) which will record absorption spectra from 7000 Å to 2400 Å with apertures as small as ½ μ diameter. The instrument is also capable of scanning and integrating for total absorbances at single wavelengths over areas of 100 μ² and smaller. Reproduced by permission of the manufacturers.

The result can be obtained directly in absolute amounts, unlike those from measurements of induced absorption which depend on the ability to determine the stoichiometric relationship of the amount of colour to the given mass of substance, as described later. The mass m of absorbing material in a given area a is given by:

$$m = A \times a/k.$$

Absorbance is recorded photographically or photoelectrically. The former provides a permanent record, and has been used recently by JANSEN (1958), ANTENEN (1958), LAURILA et al. (1959) and RUCH (1960). Photographic recording requires simpler apparatus, but is slower than photoelectric recording, since for subsequent integration it requires ancillary measurements of photographic density on the film or plate with a recording densitometer, such as that developed by WALKER (1955), and presently manufactured by Joyce Instruments, Newcastle-upon-Tyne.

b) Effects of exposure on unfixed cells

Radiation effects on living cells have frequently been studied. More recently, FREED et al. (1959) used a flying spot microscope to follow swelling changes in living cells under radiation with ultra-violet light. Using the time before the onset of swelling as a measure of the severity of damage, they found little damage if wavelengths below 2550 Å are excluded, even when very large doses are used. Swelling of the whole cell may not, therefore, be the best indication of cellular damage, since most previous workers have found that exposure to monochromatic ultra-violet light at wavelengths above 2550 Å results in increases in nuclear absorbance.

Exposure to ultra-violet light is sometimes used as a means of physiological dissection of living cells. For example, GAULDEN and PERRY (1958) used an ultra-violet light microbeam to inactivate nucleoli, and observe the resulting effect on mitosis.

c) Errors

Microspectrophotometry in ultra-violet and visible light suffers from errors caused both by the measuring instrument and the specimen. Those which are common to measurements in both regions of the spectrum will be dealt with in section III. Non-specific light scatter is certainly the most serious source of error in ultra-violet microspectrophotometry, because it is highly wavelength-dependant. Furthermore, it increases with increasing gradient or sharpness of refractive index changes in the specimen. For this reason fixed cells show far more light scatter than living ones. Attempts to minimise error from this source are of two types: first, determination of the amount of scattered light and its wavelength dependance (e.g. CASPERSSON 1950, RUDKIN and CORLETTE 1957), and second, search for a specimen mounting medium with a refractive index close to that of the specimen, but which does not absorb.

JANSEN (1958) measured the refractive index of tissue sections from 2400 Å to 6000 Å. He showed that the section had refractive indices greater that the mounting medium of highest refractive index thus far obtained— this is a mixture of 95% glycerol and 5% lanthanum acetate with varying amounts of zinc chloride (RUDKIN and CORLETTE 1957). JANSEN finds that a concentrated solution of chloral hydrate in glycerol has a similar refractive index to this mixture.

Light scatter remains a serious problem in quantitative studies in the ultra-violet region but, fortunately, technical developments have not been entirely discouraged. Some reports of investigation on biological problems using the technique are discussed in section V.

d) Variations in Absorptivity

The most interesting future development of ultra-violet microspectrophotometry may perhaps lie in studying the hyperchromicity effect in living cells. Briefly, this effect is one in which the molar absorptivity (i.e. the absorbance per gram atom of phosphorus (CHARGAFF and ZAMENHOFF 1948) increases in response to changes of temperature, pH, ionic environment etc.

Measurement of temperature dependant hyperchromicity (melting profile) is now an established method for the study of nucleic acids in terms of variations due to base composition or strength of hydrogen bonding (MARMUR, ROWND and SCHILDEKRAUT 1963). WALKER (1957, 1958) and CHAMBERLAIN and WALKER (1965) have investigated both pH-dependant and temperature dependant hyperchromicity in sperm heads before and after acid hydrolysis. Even for sperm heads, which are simple systems in the sense that they contain only DNA and not RNA, the results have proved difficult to interpret. One feature, which is not without interest, is the possibility that free "unreppressed" DNA may give an initial step in the melting profile, which could be an indication of the amount of DNA available for template activity within the cell.

III. Induced Light Absorption

Many chemical components of cells are detectable by means of induced colour—see review WALKER and RICHARDS (1959). Few are sufficiently reliable, qualitatively and quantitatively, to have been used extensively. Colourinduction is possible by two types of procedure: firstly, by dye-binding in which the attachment between molecules of chromophore and molecules of substrate is electrostatic, e.g. gallocyanin chrome alum stain for DNA, and secondly, by chemical reaction which gives a coloured product, e.g. the Feulgen reaction for DNA. Accurate photometry of induced colour requires care in each of the following three stages of technique: (a) Preparation of material and fixation; (b) Dye-binding or colour reacting; (c) Measurement of amount of colour.

a) Preparation of material and fixation

Both tissue sections and tissue squashes or smears are used in microspectrophotometry. Sections are essential if the tissue architecture is important to the observations, but in very many instances it is quite possible to distinguish between cell types in smear or squash preparation by means of detailed differences in the appearance of their stained nuclei. A serious disadvantage of sections is the presence of cut cells and nuclei, and, for certain methods of photometry, which require measurements in clear field outside the nucleus (see section III c), the close packing of nuclei is prohibitive even with careful choice of section thickness. In only few instances are smear preparations impractical.

Chemical fixation techniques are often unsuitable for photometry because they cause heterogeneity in the absorbing material, and the newer physicochemical methods, freeze drying (BELL 1956) and freezing substitution (PATTEN and BROWN 1958, FEDER and SIDMAN 1958), are the methods of choice. A good account of fixation methods in cytophotometry is given by POLLISTER and ORNSTEIN (1955). In the authors laboratory freezing substitution is done routinely as follows: Solid tissues are homogenised with a flat ended rod on a microscope slide, with the addition of a minimal quantity of isotonic saline containing 1% blood plasma (to increase adhesion of cells). If

required they are mixed fluid tissues (e.g. ascites tumours). Drops of
the tissue or tissue mixture are transferred to coverslips and quickly
smeared immediately before fixation. If necessary cell flattening is increased
by shearing the drop with the edge of a coverslip (see Killander et al.
1962 a). Coverslips bearing the smears are held with forceps, which have
had their tips ground smooth to avoid fracture of the glass. Each is plunged
smoothly and rapidly into the quenching bath of liquid propane or iso-
pentane (often mixed 3 : 1), kept near their freezing point in a Duer flask
full of liquid nitrogen. After several seconds, the coverslip is transferred
rapidly to a tube containing Analar quality ethanol or methanol, kept at
— 76⁰ C. by a mixture of solid carbon dioxide and methylated spirits, in
a Duer flask. Subsequently, the tube containing the coverslips is removed
from the cooling mixture and allowed to warm up in air. Above 0⁰ C.
warming can be assisted. Provided that the original smears are sufficiently
thin, this procedure gives remarkably life-like preservation. It is marred
only by occasional shrinkage cracks across cells, and areas of ice-crystal
artefact which have a reticular appearance. The latter is avoided by using
thin smears with rapid quenching and transfer.

b) Colour induction

Unfortunately, the great majority of our knowledge of nucleic acids
obtained by visible light cytophotometry relates to DNA. Reliable methods
for RNA remain to be established. The extensive use of the Feulgen reaction for DNA is no real testimonial for its reliability, since it remains somewhat capricious even in experienced hands.

i) The Feulgen reaction for DNA

The biological importance of DNA, the extensive use of the Feulgen
reaction and the frequent reports of difficulties in establishing its quantitative basis, have resulted in the many reviews on its nature and usage. They
include Stowell (1945), Swift (1953, 1955), Lessler (1953), Kurnick (1955),
Leuchtenberger (1958), Walker and Richards (1959) and Kasten (1960).

The laboratory procedure has two stages: first, hydrolysis of the fixed
cells in N · HCl at 60⁰ C. for between 6 and 20 miuntes, depending on the
type of specimen and fixative used, and second, reaction with Schiff's reagent (containing ½–1 % pararosaniline) for periods of between 1 to 2 hours
followed by removal of excess Schiff with HCl-bisulphite washes. Specimens
can be stored, after alcohol dehydration and passage through xylene, in
non-drying immersion oil. Photometry may be done in this mounting
medium or, if cell crushing is necessary (see later), in glycerol. Many
variations in composition of the Schiff's reagent, or length and type of acid
hydrolysis have been recorded by previous reviewers, but most variations
have been little used.

The reaction mechanism is still incompletely understood. Wieland and
Scheuing (1921) suggested that the leuco-compound of basic fuchsin and
sulphurous acid becomes recoloured when it reacts with groups exposed

after acid hydrolysis of DNA. The coloured product is unlikely to be a single compound, however, since paper electrophoresis of the reaction mixture with the simplest aldehyde (formaldehyde) separates as many as six (BARKA and ORNSTEIN 1959) or ten (HIRAOKA 1960) coloured components. The latter author found four of his ten final products to exist in the original basic fuchsin. Clearly this type of experiment should be done with purified dyes.

Less interest has been shown in the chemistry of the Feulgen reaction than in establishing its quantitative basis. Unlike experiments to demonstrate the specificity of the reaction for DNA, attempts to prove its stoichiometry have presented very great difficulties.

Specificity and Stoichiometry

DNase abolishes the reaction, RNase does not. Unsaturated fats which contain aldehydes give colour with the Feulgen reagent but are almost always eliminated by alcohol as a fixative or in alcohol dehydration. Persistance of other substances which interfere with the reaction, such as polysaccharides, is a rare occurence but not unknown e.g. BELL (1960) finds strong colour in the neck cells of the archegonia of ferns. In the case of such doubtful positives, resistance of the substrate to DNase is a crucial test.

Attempts to establish the stoichiometry of the reaction have generally been of two types:

(1) Test tube experiments in which hydrolysed DNA or aldehyde reagents are quantitatively reacted with Schiff's reagent, with or without the presence of possible interfering substances and the amount of colour per unit mass of reactants measured.

(2) Direct comparison of amounts of colour with DNA content in cells, or, occasionally, in model solids.

Test tube experiments require concentrations of reactants which are much smaller than those which exist in cells. In general they present an alarming picture of wide variations in amounts of colour, and extreme sensitivity to interference particularly by proteins. CASPERSSON (1932), SIBATANI (1950), and SIBATANI and ISHIDA (1959) have shown variations in both amount and absorption spectrum of the reaction products depending on the type and amount of protein added to the reaction mixture. All proteins increased the amounts of colour, and histone and protamines were shown to cause shifts in the λ maximum to longer wavelengths, while gelatin, casein or albumin did not. Such observations contrast with findings of stable amounts and spectral characteristics of the Feulgen colour in cell nuclei despite wide differences in protein/DNA ratio (SIBATANI and NAORA 1953, SIBATANI 1953 and 1954 and KASTEN 1956 and 1957).

Direct comparison experiments on either model systems or cells give a much more favourable picture. Most model systems use concentrations of DNA much lower than that in cell nuclei, which is around 10% or higher in chromocenters or sex chromatin. LESSLER (1951) found a linear relation between amounts of colour and DNA concentrations between 0.2 and 1 mg./ml. in 20% gelatin. WALKER and RICHARDS (1957) measured the-absorbance ratio 2650 Å/5460 Å in Feulgen-stained films of pure DNA, which was directly proportional over a threefold variation in film thickness, but were unable to give an absolute figure for the stoichiometry owing to the hyperchromicity effect shown by hydrolysed DNA.

Recently, PERSIJN and VAN DUIJN (1961) developed a method of incorporating DNA into cellulose films, which approximated to the situation in cell nuclei, in that the DNA is concentrated into microscopic dense granules. After staining hydrolysed films with both Schiff's reagent and thionin-SO₂, these workers found a linear proportionality between absorbance at λ maximum and phosphorus content per unit area of film. This work was done primarily to test the model system, and, unfortunately, did not include studies on DNA nucleoproteins.

All the model system described have demonstrated a linear proportionality between amounts of colour and amounts of DNA, but none has given a practical determination of the stoichiometry which would permit amounts of colour to be converted directly to amounts of DNA in photometry on Feulgen-stained cells.

Direct comparisons of colour and DNA content in cells are obviously crucial. The choice of cell is important; homogeneous populations are to be prefered, and ones in which DNA synthesis is occuring are undesirable, since average DNA values determined biochemically must be compared with average colour values. The cell systems used have rarely met with these requirements. Early attempts using biochemical methods (RIS and MIRSKY 1949) and ultra-violet microspectrophotometry (LEUCHTENBERGER, LEUCHTEN-BERGER, VENDRELY and VENDRELY 1952) did not deny the existence of proportionality. DONDERO, ALDER and ZELLE (1954), having failed to get consistent results with the Feulgen reaction on Azotobacter species, devised an extraction procedure for testing the quantitation of the reaction in several bacterial species. They hydrolysed cell suspensions, reacted them with Schiff's reagent and washed them with SO₂ water until washing would no longer give colour with formaldehyde. The bound Feulgen colour was extracted with acid, re-reacted with formaldehyde, and the amount of colour determined. This procedure gave typical time-hydrolysis curves, it maintained a linear correlation between cell number and amount of extractable colour, it was inhibited by DNase treatment on dried cells, and, finally, it gave a high positive correlation with the DNA content of parallel cell samples determined with the diphenylamine reaction.

McLEISH and SUNDERLAND (1961) and SUNDERLAND and McLEISH (1961) have published the results of very careful studies in ten plant species which varied widely in amount of DNA per cell (0.55 to 31.34 g. × 10⁻¹¹) and concentration (0.06 to 0.36 g./ml.). The results showed an excellent proportionality between DNA content and amount of Feulgen colour, but, consistent with very many workers observations, showed different proportionality factors for cells fixed in formaldehyde and alcohol fixatives.

All the studies mentioned agree in demonstrating that cytophotometry of the Feulgen reaction is a reliable means of determining the relative amounts of DNA in cells. It must be mentioned, however, that the well known variations in amount of colour with type of fixation (e.g. McLeish and Sunderland 1961, Millhouse 1961), time of hydrolysis and pH of staining (e.g. Dutt 1963), and length of exposure to Schiff's reagent and other factors (Srinivasachar and Patau 1959), virtually precludes the comparison of amounts of Feulgen colour in situations other than between cells in the same area of the slide. Fortunately the use of "standard" cells permits comparison between different slides (e.g. Atkin and Richards 1956), but even the DNA constancy relationship is eliminated when staining is done at pH levels above about 1.6 (Walker and Richards 1959), and it is imperative to use fresh Schiff's reagent having a pH between 1.4 and 1.6. Subsequent fading of colour due to unfresh Schiff's reagent (Torre and Salisbury 1963) either after storage of slides in the dark, or after exposure of slides to light in a photometer (Kasten, Kiefer and Sandritter 1962) make it difficult to compare stain values between different laboratories.

ii) Other induced colour methods for DNA

The remaining two colour induction methods for DNA are the only ones, which, apart from the Feulgen reaction, have formed the basis of attempts at quantitative estimation of DNA. They are both dye-binding methods. The dyes used are methyl green and gallocyanin chrome alum, and investigations as to their suitability for quantitative work were reviewed by Walker and Richards (1959). Since that time few reports of their usage have appeared and, consequently, neither can compete with the Feulgen reactions as the method of choice for cytochemical estimation of DNA content.

Like all dye-binding techniques, methyl green and gallocyanin chrome alum, rely for both their specificitiy and stoichiometry on the absence of competing and interfering substances. The former compete with the substrate molecules (i.e. DNA) in that they also bind dye molecules, and the latter occupy sites on the substrate which would otherwise be available to bind dye molecules. Dye coupling at a specific pH and removal of RNA by enzymolysis (for gallocyanin chrome alum) both help to reduce competition and interference.

At one time methyl green staining promised to be a valuable technique in that the amount of colour might be related to the degree of polymerization. Kurnick (1947) suggested that this was the basis of its specificity for DNA and that RNA was not sufficiently polymerized to compete. Later work suggested that one dye molecule became bound to 10 or 13 phosphoric acid residues (Kurnick and Mirsky 1950, Kurnick und Foster 1950). On this basis, RNA may well be expected to bind methyl green since recent work on soluble RNA, which is likely to contain the smallest molecules of RNA in the cell, shows that it has long double helical regions. Chain length is important in determining the amount of dye bound, however. Ebel and Muller (1958) used methyl green binding to study the distribution of inorganic poly-phosphates in cells, and found increased stain intensities with increasing chain length.

KURNICK's interpretation of the mechanism of staining of nucleic acids by methyl green was criticised by ALFERT (1952) on the reasonable grounds that the binding of methyl green by electrostatic forces was likely to be modified by competition from proteins which, in the case of RNA, might completely inhibit the binding. Although KURNICK (1955) claimed to have removed the histones from the DNA before staining, BLOCH and GODMAN (1955) showed that the cold HCl treatment used is not sufficient to remove all protein competition, and that only after acetylation of the amino groups does methyl green staining show similar distribution in the same cells to that of the Feulgen reaction for DNA. ALFERT's (1952) view is thus borne out for DNA, and probably also for RNA, since evidence is accumulating that all "fixable" RNA in the cell exists as ribonucleoproteins.

Interfering substances can also be interesting. DEITCH (1961) showed how amounts of methyl green bound to nuclei of differen iating lymphocytes after acetylation of interfering protein groups, tends to be larger as matu ity progresses. This suggests that proteins are gradually lost as the lymphocytes mature. It is in this kind of experiment that methyl green staining has given valuable information.

The second of the dye-binding methods for DNA, staining with gallocyanin chrome alum after removal of DNA (STENRAM 1953), has not received the attention it deserves. Fortunately, SANDRITTER and his collaborators in Germany have made extensive studies using this technique. After removal of RNA the staining is specific for DNA and unlike methyl green, gives intense colour which is more than adequate for accurate photometry. Interference by proteins may not be serious. SANDRITTER, FISCHER, SUSSENBERGER and SCHIEMER (1959) showed by photometry of fast green, and of the arginine reaction, that added protamines will enter the cell, but this does not affect the nuclear staining by gallocyanin chrom alum. The German group has also used gallocyanin chrome alum in the Papanicoloau technique to indicate general basophilia in studies of the incidence of vaginal tumours (SANDRITTER, CRAMER and MONDORF 1959). TERNER and CLARK (1960), however, have questioned the specificity of gallocyanin for cytoplasmic RNA, after finding that RNase does not remove cytoplasmic basophilia. Contaminating dyes and wide variations between dye batches were among the factors which led these authors to criticize the use of gallocyanin staining for general basophilia.

iii) Methods for RNA

Although the lack of satisfactory methods for RNA has long been deplored, it is clear from the number and complexity of the types of RNA demonstrated by biochemistry, that a simple determination of the total amounts of RNA in a cell will not be of great interest. Indeed, the early work of BRACHET on visual estimation of basic stains and CASPERSSON with ultra-violet absorption, both showing that more protein synthesis meant more RNA, probably established the most valuable result which could be expected from estimation of total cellular RNA. If methods to distinguish between, and possibly measure, the amounts of the individual types of RNA could be developed, this would undoubtedly further our knowledge.

In attempts to estimate RNA photometrically the colour induction technique most used is that of binding of the dye Azure B. FLAX and HIMES (1952) showed that it bound to DNA orthochromatically (blue-green) and to RNA metachromatically (purple). The technique has been used with success by SWIFT, REBHUN, RASCH and WOODWARD (1956) who checked its specificity for RNA with nucleases. HIMES (1961) has recently shown that Azure B can give a metachromatic binding to DNA, resulting in a similar purple colour to that given with RNA, if the cells are subjected to prior acetylation or deamination. This precludes the use of stains with Azure B in the investigation of competition and interference effects.

Minor usages of other dyes for cytophotometry of RNA have been reported; toluidine blue (STENRAM 1953) and cresyl violet RITTER, DI STEFANO and FARAH (1961) are two examples. The latter author showed that cresyl violet has a stable absorption maximum (an unstable secondary peak exists, however) which, after staining at pH 4.2, obeys the absorption law (LAMBERT—BEER LAW—see later) over the range of concentration and thickness found in cells. Furthermore, the dye is specific for RNA, it is resistant to ethanol extraction, and comparison of the natural absorbance at 2570 Å with the induced absorbance at 5850 Å suggests that it is stoichiometric. In practically every respect, according to RITTER et al. (1961) cresyl violet stain is superior to toluidine blue, azure B and gallocyanin as a technique for RNA.

iv) Toluidine blue and metachromacy

The intelligent use of the phenomenon of metachromacy in dye binding to give cytochemical information has often been proposed (for review see BERGERON and SINGER 1958). A dye stains metachromatically if the dye plus substrate has an absorption spectrum different from that of the dye alone. KELLY and BLOOM (1959) made some interesting studies in which they used a microspectrophotometer to analyse the metachromatic reaction of toluidine blue in most cells. They used the ratio of absorbance 5460 Å/6300 Å to compare the absorption spectrum in dye solutions with that in stained sections. For dilute solutions of dye the ratio was between 0.2 and 0.3, compared with a ratio of 2 for heparin and chondroitin stained with toluidine blue. On the other hand, concentrated solutions of the dye and stained cartilage sections had a ratio close to 4. The authors suggest that the potential value of metachromatic dyes lies in such marked deviations from BEER's law. Clearly, however, attempts to exploit metachromatic reactions for cytochemical purposes will depend on rapid and accurate analysis of absorption curves, such as can only be done with a sophisticated microspectrophotometer. The extensive possibilities are indicated by the studies of Love and collaborators, such as LOVE and SUSKIND (1961), on the staining of ribonucleoproteins by toluidine blue—molybdate, where it is claimed that no less than seven types can be identified in mitotically dividing mammalian cells. Such a system would seem to defy analysis by cytophotometry, but if in fact the outcome could be a demonstration of amount and turnover pattern of nucleolar, nuclear, messenger, ribosomal and transfer RNA, the effort would certainly be worthwhile.

c) Measurement of amount of colour by cytophotometry
i) Basic requirements

The amount of colour per cell or per nucleus resulting from a colour induction method is meaningful only if certain conditions are satisfied. For the colour induction these are firstly, that a stoichiometric relationship exists, and, secondly, that the colour obeys the LAMBERT-BEER absorption law. In the measurement of amount of colour, errors can arise from features of the measuring device, or its use with unsuitable objects.

The LAMBERT-BEER absorption law is a combination of two factors:

(1) That absorbance is directly proportional to thickness—LAMBERT's Law,

(2) That absorbance is directly proportional to concentration—BEER's Law.

Attempts to establish adherence to the law of a coloured product by test-tube experiments are impossible at the concentration levels and thicknesses which exist in cells. It is therefore necessary to devise a model system, such as the technique involving thin films of material used by WALKER and RICHARDS (1957), or to incorporate the stained substrate into a non-reacting material such as cellulose (VAN DUIJN and PERSIJN 1960) or gelatin (LESSLER 1953).

In the measuring device the most serious error to arise is likely to be that of stray light, which gives glare in the image, and results in the measured absorbance being lower than the actual one. Stray light arises by unwanted reflections from metal or glass surfaces, and by scattering from dust on the surface of lenses.

NAORA made extensive studies of stray light problems in microspectro-photometry (e.g. NAORA 1957) and warned that the resulting S-V effect (Schwarzchild-Villinger effect) caused serious errors. HOWLING and FITZ-GERALD (1959) have re-examined this problem in a detailed paper. In general they confirm NAORA's findings in that the effect is likely to be present in all optical systems, but may vary in severity. Fortunately, according to HOWLING and FITZGERALD, the error is less than 1% for optical densities less than 1, provided that the area of the specimen illuminated by the condenser is very little larger than the measured area. For practical purposes these conditions can be achieved by choosing specimens of low absorbance, or by using the cell crushing condenser to reduce their absorbance to within the optimum range for measurement i.e. between 0.3—0.6 optical density, and using a field aperture of such size that only the measured area is illuminated.

The cell crushing procedure devised by DAVIES, WILKINS and BODDY (1954) consists in mounting specimens on coverglasses with a flexible cover, such as cellophane, through which individual cells are flattened by the hemispherical top component of the crushing condenser. Flattening occurs within a very restricted area, usually no greater than the field of view of a ×100 oil immersion objective. The mounting medium is usually glycerol;

non-drying immersion oil is totally unsatisfactory in that cells mounted in this will not crush. The procedure gives the following important results: (1) it brings the whole of the object to within the depth of focus of the objective; (2) it reduces error from glare by reducing the absorbance to within the optimum range; and (3) it minimises error from scattered light by squeezing all the absorbing material into a thin sandwich so that scatter occurs only at the boundary of the object with the immersing fluid.

Unsuitable objects are mainly those in which the absorbing material is inhomogeneously distributed. The error which occurs in measurements of inhomogeneous objects is called "distributional error". It is serious only with static aperture measuring devices (see later).

ii) Types of apparatus

Three main groups are recognisable:

(1) S t a t i c a p e r t u r e or "Plug" p h o t o m e t e r s of which two main types exist. The first is that developed by POLLISTER and MOSES (1949) in which the measuring aperture is a large fraction of the projected area of the object. The second is the LISON (1950) photometer in which the measuring aperture is small relative to the projected area of the object. Both types consist of a light source, sometimes a monochromator, a microscope, a photo-cell or photomultiplier preceded by a measuring aperture, and a galvano-meter. Such devices are satisfactory only for measurements on objects in which distributional error is not large. In measurement, the galvanometer deflection with the object in the beam is noted relative to that with clear field. The difference in reading is proportional to transmission, and must be converted to total absorbance thus:

$$A = E/F \times C^2$$

where, $E =$ extinction i.e. $\log_{10}$ Transmission.

$F =$ fraction of total volume included in the "plug", which is calculated from measurements of major and minor axes of the object,

$C =$ radius of "plug".

(2) T w o - w a v e l e n g t h p h o t o m e t e r s are similar to those used with the "plug" method but have a light source giving a continuous emission spectrum and a monochromator. Two-wavelength photometry is based on a method for overcoming distributional error devised by ORNSTEIN (1952), and independently by PATAU (1952). Distributional error increases with optical density increase, and is detectable in terms of distortion of the absorption spectrum. The absorption spectrum is first measured in a region of the specimen of low absorbance where the error is negligible. Two wavelengths are selected for which the absorbance is in the ratio 2 : 1, and, subsequently, the absorbance of all objects is measured at the selected wavelengths. The value obtained will depart from the expected ratio of 2 : 1 by an amount dependant on the extent of the distributional error, and from which the true value for average absorbance can be calculated. MEN-DELSOHN (1958 b) has produced a set of tables to speed the calculations. Excellent agreement between the two wavelength method and a scanning

method described below was obtained by Mendelsohn and Richards (1958), and recently Van Duijn, Tonkelaar and Hardonk (1962) found the two-wavelength method to be satisfactory when tested by the model system developed by Van Duijn and his collaborators, mentioned previously. Using colour transparency photographs, Mendelsohn (1958 a) has extended the method for macrosize objects and natural absorption.

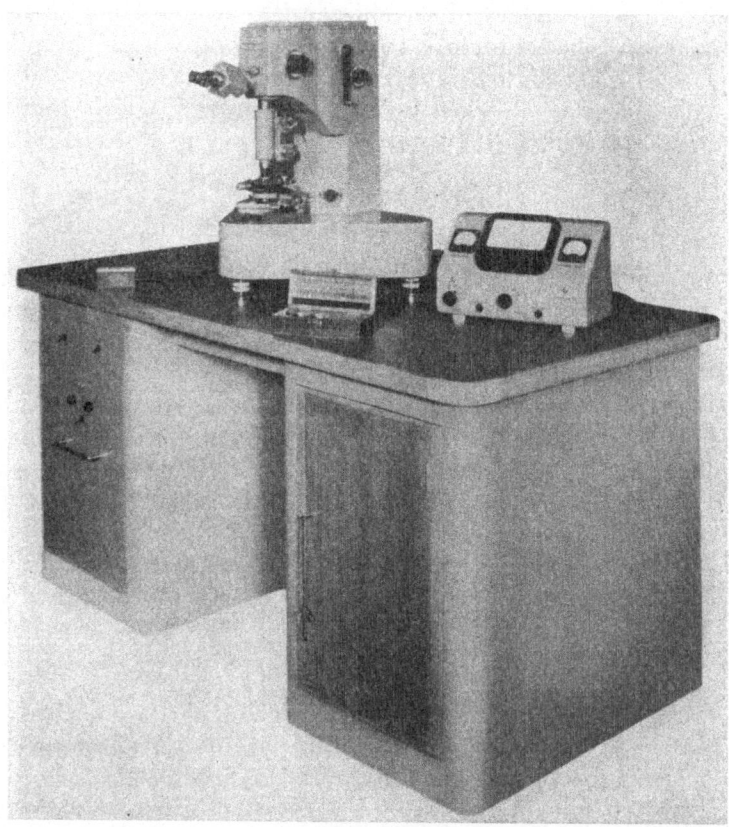

Fig. 2. The photoelectric scanning microdensitometer manufactured by Barr and Stroud (Glasgow). Developed from the design of Deeley (1955) this instrument has a tungsten source replacing the original mercury arc, and is equipped with a continuous interference filter to permit measurements of absorption spectra and total absorbences in the visible range of wavelengths. Reproduced by permission of the manufacturers

(3) Scanning aperture photometers were developed, like the two-wavelength method, to minimise distributional error. They use measuring apertures whose size is small relative to the structural pattern of inhomogeneity. It is necessary to scan the aperture relative to the object to obtain the integrated total absorbance.

In the various microspectrophotometers developed in Caspersson's laboratory (see Caspersson, Lomaka and Caspersson 1960) the object is scanned relative to the measuring beam. This permits the measuring aperture always to be on the well corrected centre line of the optical system. In the instrument designed by Deeley (1955) which is now available commercially from Barr and Stroud, Glasgow (Fig. 2), the measuring aperture scans the image in two directions at right angles, and the anode current of the photo-multiplier is integrated after logarithmic amplification and displayed as a single meter reading. The difference between readings with object and with clear background is directly proportional to total absorbance. This instrument is very rapid, each scan requiring only 3 seconds. Repeat scans, to eliminate minor instrumental instabilities, give a measuring time of approximately 25 seconds per cell. Used in conjunction with the cell crushing condenser, this instrument is the fastest and most accurate yet available.

A very much simpler scanning device has been described by JANSEN (1961). The results with this instrument show an accuracy comparable to that obtained by DEELEY's photometer.

IV. Autoradiography
a) Introduction

Autoradiography, a well established and widely used cytochemical technique, has recently had two significant advances. The first of these concerns the development of methods for the detection of labelled substances in the aqueous components of tissues; it gives the exciting possibility of a completely new approach to old problems such as the maintenance of sodium/potassium balance. It will also permit new studies on the availability of nucleic acid precursors. Extension of autoradiography to the electron microscope is the second of the two recent developments; it aims to take advantage of the increased resolution and magnification of the electron microscope.

This section will briefly discuss the technical aspects of autoradiographic technique, both of the conventional and of the newer developments. Recent reviews include those by PELC (1958) and FICQ (1959).

b) Basic procedure

Autoradiography needs little more apparatus than is normally available in histological laboratories. Metabolic precursors labelled with radioactive isotopes are administered by injection or by feeding to animals or plants, for known intervals of time before fixation. Tissue sections or smears are prepared by standard histological methods, and, if necessary, after selective extraction of substance which might give rise to confusion, the sections or smears are coated with photographic emulsion applied as a fluid or a film in a darkroom. Emulsioned slides are exposed in the dark for periods of time which depend on the half-life of the isotope, and the specific activity and dose level of the precursor. Photographic development then gives silver grains in the emulsion over the region of the specimen where the labelled precursor is incorporated. If the emulsion used is a photographic type then individual grains are produced—the so-called "blackening" method. If a nuclear emulsion is used the path of an emitted particle is visualized by a line of grains—the "track" method. The cells of the tissue are made visible by histological staining, or by phase-contrast microscopy, so that the position of silver grains can be used to indicate the site of incorporation of the radioactive precursor.

The technique is simple, but the difficulties are many. Most difficulties can, however, be avoided. They begin with problems of specificity and availability of the labelled precursors, and possible interference with normal metabolism by auto-irradiation effects. Lack of specificity can be overcome by selective extraction procedures, and tests for auto-irradiation effects can be designed. Levels of labelling in terms of numbers of grains per nucleus or per cell are easily adjusted, but care must be taken to minimise labelling in the background especially from external sources, either during storage of unused emulsion or during exposure of the autoradiographs. Application of the emulsion and the processing procedures, either photographic or histological, can result in poor contact or in relative movement between the emulsion and the specimen. Fogging of the film or artefacts resembling photographic grains occur occasionally. Finally, in experiments where grain counts are necessary, both optical problems and observer errors are encountered. Despite the many possible difficulties, autoradiography has been used with great success, and has yielded much valuable information in cell biology.

c) Administration of labelled precursors

Nucleic acids are made up of nucleotides each of which consists of a phosphate group, sugar ring and nitrogenous base. Radioactive labels are available for each of the components of the nucleotide. The most commonly used for autoradiography are P^{32}, C^{14} and H^3; the physical characteristics of each of these have been discussed (see for example LAJTHA and OLIVER 1959) and are summarized in Table 1 together with a note of their principal advantages and disadvantages.

Despite the fact that in 1951 FITZGERALD, ELDINOFF, KNOLL and SIMMEL had pointed out many of the characteristics which make tritium a highly desirable isotope for autoradiography, the widespread use of this label was delayed until suitable labelled precursors became available. Most of the early studies on nucleic acids by autoradiography employed inorganic phosphate, labelled with P_{32}, as a precursor. The now classic work of HOWARD and PELC (1951), which laid the foundations for all later studies of DNA synthesis by autoradiography, was done with P^{32}-labelled inorganic phosphate. Nowadays this label is rarely used; it suffers from several disadvantages, the most serious of which, is the fact that the high energy β particles it generates have a very long range, and consequently it gives poor resolution. Inorganic phosphate also has the disadvantage that it is incorporated into a wide range of cellular components, such as phospholipids, phosphoproteins and other inorganic phosphates, in addition to DNA, RNA, and nucleotides at all stages of polymerization. Extraction procedures are necessary to remove such competing substances. Acid extraction removes most non-nucleic acid components with the exception of phosphoproteins, and the two nucleic acids can be distinguished by appropriate enzymatic digestion. High background labelling is also a problem with P^{32}.

C^{14} labelled compounds give relatively low energy β-particles and yield autoradiographs having good resolution. The most commonly used precursors, adenine-8-C^{14}, is incorporated into both RNA and DNA, so that selective enzymatic extraction is necessary. Formate C^{14} is often used since it has been shown to be incorporated mainly into the methyl group of

Table 1.

Radio-active label	Usual precursor	Half life	Maximum energy of β particle (MeV)	Range in emulsion (μ)		Advantages	Dis-advantages
				Maximum	Average		
P³²	Inorganic phosphate	14 days	1.7	16×10^6	0.8×10^6	High grain-producing efficiency	Very short half-life. Range very long giving poor resolution. High autoirridiation effect
C¹⁴	Many types e.g. adenine cytidine formate	5600 years	0.16	300	60	Range short giving good resolution	Long half-life-requires long exposures
H³	Usually as thymidine for DNA and cyto-sine for RNA	12.26 years	0.018	8	1.5	Very short range gives excellent resolution	Very short range results in problems of self absorption: Geometry of specimens must be known

thymine (LAJTHA 1954). The pools for both adenine and formate are much smaller than that for phosphorus, so that a higher labelling efficiency can be achieved using these precursors. Both these compounds show a low tendency to non-specific absorption ("stickiness") as compared with inorganic phosphate and consequently give low background labelling. Unfortunately, C^{14} has a very slow rate of disintegration (half-life 5600 years), so that high rates of incorporation are necessary to achieve reasonable exposure periods.

Triated thymidine (H^3-thymidine) has the exclusive advantage of being completely specific for DNA, and no extraction procedures are necessary. Other less specific pyrimidines labelled with H^3 are also available but for almost all recent studies on DNA H^3-thymidine has been used. H^3-labelled compounds give excellent resolution owing to the very short range ($1.5\,\mu$ average) of the β-particles. Anomalous results are obtained, however, if insufficient attention is paid to the fact that only the label in a surface layer of about $1\,\mu$ of the specimen which is closest to the emulsion will be detected. The geometry of the specimen must therefore be known or manipulated (e.g. Prescott and Bender 1962). The uncertainties, which the self-absorption problem causes, make H³-labelled precursors unsuitable for most experiments where grain counting is required (see e.g. Wimber et al. 1960).

An advantage of H³-thymidine, which is now frequently made use of, is its rapid incorporation into DNA. Detectable incorporation has been found, for example, by Painter, Drew and Giadique (1960), as soon as 30 seconds after administration to tissue cultures of Hela cells and rabbit kidney cells. This rapid incorporation, and the very small pools of thymidine in most organisms permits sophisticated experimental design, but care is necessary to check the availability time (Diderholm, Fichtelius and Linder 1962).

The use of H³-thymidine has been criticized on the grounds that the commercially available supplies are usually of such high specific activity that auto-irradiation effects may seriously interfere with the metabolic process being investigated (Lajtha and Oliver 1959, Reid 1960). Very many workers have investigated this hazard in terms of growth and survival of a variety of cell types. Mendelsohn (1960) pointed out that the effect of radiation "from within" by incorporated H³-thymidine was likely to be greater than with other radioactive labels for two reasons: first, the short range of the β-particles means that most of the resulting ionizations occur within the cell, and, second, that the helium atoms which result from the disintegration of tritium leave the residual thymine moiety in an unstable form, which might lead to genetic mutation (see Kaplan and Siskin 1960). Mendelsohn suggests that a wide margin exists between the dose levels which produce observable toxicity, and the amounts needed for autoradiographic studies, but Plaut (1959) calculates that a single H³ β-particle is very likely to produce a chromosome break. Several reports have appeared in which observable effects on growth rates, chromosome aberrations or mutation rates have been produced by dose levels commonly used.

In 1958, Painter, Drew and Hughes, claimed that H³-thymidine caused growth inhibition in Hela cells. As compared with growth rates, induction of chromosome aberrations might be a more sensitive test. Prompted by previous work which showed that even C¹⁴ labels could produce chromosome aberrations, McQuade and Friedkin (1960) compared the effects of C¹⁴ and H³ labels in onion root meristem. They found that at similar concentrations and specific activities the two labels gave similar frequencies of aberrations, and that the effects were markedly dose-dependant. High specific H³-thymidine ($890\,\mu c./\mu\text{mole}$) produced aberrations at a very high frequency (99.45%) when used at a concentration of $0.0225\,\mu\text{mole/ml.}$, but aberrations were absent at concentrations of isotope below $1 \times 10^{-3}\,\mu\text{mole/ml.}$ On the assumption, based on theoretical considerations, that C¹⁴ labels would be less likely than H³ labels to interfere with the normal course of DNA synthesis, Krause and Plaut (1960) in an ingenious experiment used the uptake of C^{14}-thymidine to determine the effect of H^3-thymidine. They found that in plant root meristem the cells exposed to C^{14}-thymidine and H^3-thymidine simultaneously showed an enhanced incorporation of the C^{14} label as compared with cells exposed to C^{14}-thymidine alone. The effect was detectable with specific activities as low as 16 μc./μmole, and led these authors to suggest that the auto-irradiation effect may have been significant in those experiments which suggest the existence of metabolic DNA synthesis in plant root elongating zone (Pelc and La Cour 1959), and polytene chromosome "puffs" (Ficq and Pavan 1957). Separate studies are clearly necessary for each new tissue being studied, and Lajtha and Oliver (1959) have argued that the need for extensive ancillary investigation where H^3-thymidine is employed make the use of C^{14}-thymidine more practicable. Natarajan (1961) suggested that the problem of auto-irradiation can be minimised by the use of low concentrations of isotope and long exposures of autoradiographs. This author claims that the dose levels routinely used produce chromosome breakage and mitotic inhibition. For bean root meristem he found negligible mitotic inhibition at concentrations of 0.375 μc./ml., while a tenfold increase in concentration he found to cause 50% mitotic inhibition 24 hours after administration. It is difficult to compare dose levels used by various workers expressed as per unit body weight, because of the likelihood of differential concentration effects in different tissues. Monesi (1962) quotes dose levels of 0.62–0.71 μc./gm. body weight of mice as producing no observable effect on the relative frequency of normal and necrotic cells in spermatogenesis, and Grisham (1960) showed that a dose of 1 μc./gm. body weight in rats given 20 hours after partial hepatectomy causes transient inhibition in the rate of increase of residual liver mass, a delay of 6 to 12 hours in the appearance of mitosis and a highly significant increase in the frequency of abnormal anaphases. The same dose level of 1 μc./gm. body weight was claimed by Lisco, Baserga and Kiseleski (1961) to increase the incidence of tumors in mice 20–24 months old from 4% in control animals to 16% in experimental animals injected at 3 to 7 days after birth. For such long term experiments Cronkite, Greenhouse, Brecher and Bond (1961) have estimated that significant genetic hazard exists for somatic cells. For each and every experimental system it is clear that the auto-irradiation effect must be investigated in terms of the biological property being studied. In the majority of labelling experiments, it seems likely that dose levels and exposure rates can be adjusted to minimise the effect, while giving satisfactory autoradiographs.

Auto-irradiation as a source of error is entirely absent in the novel use of autoradiography described by Barnard and Marbrook (1961). While other procedures depend on the incorporation of the radioactively labelled substance into the chemical components of the cell, the technique described by these authors employs radioactive stains or reagents. All manipulations are done on fixed cells, and hence labels can be used at the highest levels of concentration and specific activity, with no danger of influence by auto-irradiation. The method has been tested using H^3-labelled acetic-anhydride;

the autoradiographs indicate the number of groups available for acetylation. There is no reason why this technique cannot be used for nucleic acids, for example, with H³-labelled Feulgen reagent.

d) Tissue fixation and detection of water soluble substances

The conventional technique of autoradiography employs standard fixation procedure. Consequently all the problems of retention and stabilization of the component being studied apply equally for autoradiography as for histochemical work. Furthermore, apart from recent developments of methods to retain the water soluble components, all procedures results in the loss of these substances. Attempts to retain the watersoluble intermediates of DNA synthesis led Fitzgerald, Ord and Stocken (1961) to devise a dry-mounting technique for the application of film to specimens. Smears of isolated nuclei were fixed by freeze-drying or freeze substitution with drying down from alcohol at room temperature. Dry film emulsion was then applied directly by thumb pressure, a coverslip being interposed to avoid contamination. The authors claim significant differences in distribution of P³² and H³ labelling in dry-mounted autoradiographs, as compared with conventional technique where no attempts were made to avoid loss of water soluble components. An even simpler procedure was used by Feinendegen and Bond (1962) to demonstrate the presence of a pool of H³-thymidine in the water soluble fraction of cells engaged in DNA synthesis. They simply brushed cells onto slides and covered them with photographic film followed by exposure at 4⁰ C.

More recently Appleton (1964) has rationalized the technique for retention of water soluble substances by making autoradiographs directly on unfixed tissues prepared by sectioning in a cryostat. Similar techniques were used by Reinholz, Belloch-Zimmermann and Wirth (1960) and Cummings and Mitchison (1961). Clearly this technique opens up hitherto unexplored fields to investigation by autoradiography. Pelc and Appleton (1965) have already obtained new information on, for example, the diffusion of Na²² and H³-thymidine through cells and tissues.

e) Detection of radioactivity

All autoradiographic procedures use sensitive emulsions to detect the radioactive material. Although the spacial resolution is mainly a function of the isotope used, the properties of the emulsion also have some influence. The grain size in the autoradiograph varies considerably between different emulsions, and with different developers. Most commonly used photographic emulsions are too thick for use in autoradiography, apart from their unsuitability for reasons of grain size. The application of the emulsion to the specimen has been done in a variety of ways. The coating technique and the stripping film method have been the most widely used; a third method, that of applying the specimen directly onto a photographic film or plate (Evans 1947), is now rarely used in cytochemistry, owing to its relatively poor resolution. Belanger and Leblond (1946) first developed the coating technique which consists simply in covering radioactively labelled tissue sections or smears with fluid emulsion. Like the stripping film technique, the coating method suffers from many minor difficulties, but its principal disadvantage is that of a lack of precise control over the final thickness of the emulsion coat, and the consequent uncertainty for quantitative autoradiography. Recently, however, KOPRIWA and LEBLOND (1962) have undertaken a step-by-step analysis of the coating technique in an attempt to improve its suitability for quantitative autoradiography. The improved method shows good proportionality between exposure time and grain counts with both H^3 and C^{14} for up to 80 days exposure, and satisfactory proportionality for up to 360 days exposure. The convenience of the coating method recommends it for routine work, where quantitation is not required (JOFTES 1959).

The stripping film technique (PELC 1947, DONIACH and PELC 1950, PELC 1958) consists in floating on a water surface emulsions which have been stripped of their original support. When it is thoroughly wetted, the film is attached to the microscope slides or coverslips bearing the specimen by lifting from underneath the floating film. The filmed specimen is then dried before exposure. Since it provides a constant and reproducible emulsion thickness, the stripping film technique is the method of choice for quantitative autoradiography. Latent image fading is a source of error where quantitation is required and HERZ (1959) has shown that the use of drying agents and storage in CO_2 reduces this effect.

Apart from considerations of choice of emulsion, method of application and type of processing, attempts to obtain the highest resolution have occasionally employed such devices as high magnetic fields applied during exposure of the autoradiograph. HARTFORD and HAMLIN (1961) recently used a field of 10 kilogauss to concentrate the β-particles into a small solid angle. Although they could not show much improvement in resolution they claim an increase in grain number. Their explanation of these has been questioned by CARO (1961) who claims that the intensity of the magnetic field required would be about a hundred times greater than that yet acheived. The recent interest in such new approaches has been stimulated by the need for improved resolution for full exploitation of autoradiography at the electron microscope level.

An extension of the conventional technique for the simultaneous and differential detection of H^3 and C^{14} labels was proposed by BASERGA (1961), who employed two superimposed emulsions interposed with a layer of celloidin. BASERGA and NEMEROFF (1962) described the technical aspects in more detail. The first emulsion and the celloidin layer which together are 12 μ thick is applied first. This is exposed and the specimen stained before application of the second emulsion layer and a second exposure. The grains in the emulsion will be from both H^3 and C^{14} β-particles and will have sufficient energy to reach the second emulsion. The first emulsion was shown to reduce the flux of C^{14} β-particles by 38%. The 12 μ gap between the specimen and the second emulsion results in considerably lower resolution, but this disadvantage is minimised by the use of highly specific precursors such as C^{14}-thymidine.

f) Quantitation

Relative quantities of incorporated radioactive label can be determined by counting grains or tracks. Errors arise mainly from variable geometry of the objects or variations in emulsion thickness and its proximity to the specimen. The less the energy of the emitted particles the greater the chances of error. Almost all the published work has been done by straight-forward visual counting of grains, sometimes with the help of photography (Ostrowski and Sawicki 1961). Nevertheless, several specialized devices for grain counting have been described (Dudley and Pelc 1953, Mazia, Plaut and Ellis 1955, Gullberg 1959, Tolles 1959, Rogers 1961).

The determination of absolute quantities of radioactive label, as the number of atoms of isotope present in a given object, is fraught with pit-falls and artefacts. Levi (1954, 1957) and Levi and Nielsen (1959) have made extensive studies to obtain a factor for converting grain counts or track numbers into numbers of disintegrations, and thence numbers of atoms. The number of variables influencing the relationship is so great, however, that no satisfactory conversion factors are as yet available.

g) Extension to the electron microscope

The greatest single advance in autoradiography over the past decade is undoubtedly that of the extension of the technique for use in the electron microscope. The full realization of this development has obviously been dependant on improvements in electron microscopic technique as well as those in emulsion type, in emulsion application and in processing, but the first attempts to demonstrate that the technique was a feasible one were made as early as 1956 by Liquier-Millward. Attention was slight until in 1958 Caro, Van Tubergen and Forro decisively demostrated the locali-zation of DNA in the "nuclear" region of E. coli by electron microscope autoradiography. They used thin sections of a thymine-dependant stain of E. coli grown in a medium containing H^3-thymidine. It was clear from these studies that the limiting factor was the resolution of the autoradiograph, in terms of the characteristics of the isotope, the thickness of the emulsion and the size of the silver halide crystals and not any factor in the electron microscope technique.

The work of Caro et al. (1958) was quickly followed by O'Brien and George (1959) and George and Vogt (1959), who described the application of the technique to thin sections of yeast and radioactive particles trapped on millipore filters, respectively. In several laboratories workers became interested in the technical aspects of electron microscope autoradiography, and attempted to improve the resolution obtainable. Pelc, Coombes and Budd (1961) discussed the limitations as to grain size for different isotopes. They calculated that the samllest grains could be 0.1 μ with S^{35} or C^{14} and between 0.01 μ and 0.05 μ for H^3 as compared with the usual grain size for light microscope autoradiographs of 0.2 μ to 0.3 μ. Van Tubergen (1961) attempted to improve the technique of Caro et al. (1958) in which he had collaborated, by the use of thinner emulsion to reduce the possible distance between grains and their source, from 2.5 μ to within 1 μ in length and less than ½ μ^2 in area. To reduce the effect of the photographic solutions on the electron microscope grids PRZYBYLSKI (1961) used grids made of titanium. He also claimed improved visibility of the specimen if the gelatin of the emulsion was removed by trypsinization. HAY and REVEL (1963) also removed the gelatin emulsion by means of 0.05 N NaOH for 1 hour.

CARO and VANTUBERGEN (1962) discuss the requirements of emulsion suitable for electron microscope autoradiography. For their current work they use Ilford L-4 emulsion which gives grain slightly larger than 0.1 μ. CARO (1962) calculates an expected resolution of around 1000 Å for H³-thymidine labelled bacteriophage and finds close agreement between experimental result and calculation.

The method of application of the emulsion to specimens involves similar difficulties to that experienced in the usual coating method. KOEHLER, MÜHLETHALER and FREY-WYSSLING (1963) have reviewed in detail the various methods used by other workers to obtain thin emulsion layers, and themselves have developed a method of centrifugal spreading which is claimed to produce uniformly thin layers.

The technique of electron microscope autoradiography is now producing valuable information on such aspects of cell biology as non-nuclear localization of DNA and chromosome replication (REVEL and HAY 1961, MEEK, MOSES and EIRING 1961, HAY and REVEL 1963).

V. The Contribution of Quantitative Cytochemistry of the Nucleic Acids to Cell Biological Problems

A vast literature has accumulated over the last fourteen or fifteen years which bears witness to the impact of quantitative cytochemical techniques on progress in cell biology. It cannot be claimed, however, that the contribution of these techniques has always been constructive, since, all too often, results have been published which contradict previous ones. Controversy is always healthy, but when it is not possible to decide on the basis of published reports whether results may be erroneous, due to lack of sufficient attention to or awareness of serious sources of error, the progress of cell biology is impeded. This being so, the concurrent aplication of two or more methods, each being independently based, is the most likely approach to meet with success. The likelihood of achieving significant results is immeasurably increased if the s a m e cells, not merely similar ones, are used at all phases of the investigation. Instances of this type are all too few; those which exist demonstrate that this approach has been applied when the biological problem is uppermost in the mind of the investigator, and not the application of a given technique for its own sake. In the ensuing section examples are given of those biological problems which have been investigated by quantitative cytochemical techniques for nucleic acids; no attempt has been made to provide a comprehensive literature review in this complicated and extensive field.

a) Pioneering contributions

It is not difficult to single out those cytochemical contributions to cell biology which are of the first importance either as technological advances or as conceptual ones. The work of Caspersson and his school is undoubtedly the greatest contribution to cytophotometry made anywhere. The now classical cytophotometric studies on Feulgen stain in animal and plant tissues by Ris and Mirsky (1949), Swift (1950 a, b) and Alfert (1950) confirmed on a single cell basis the DNA constancy hypothesis (predicted by Bovin [1947], demonstrated biochemically by Bovin, Vendrely and Vendrely [1948]). Swift's work established, together with the similarly classic autoradiographic studies on P[32] incorporation in bean roots by Howard and Pelc (1951), that DNA synthesis normally occurs in interphase. DNA constancy and interphase synthesis of DNA helped to establish the view that DNA is the genetic material at a time when such an idea was revolutionary.

The cytophotometric work also demonstrated the existence of multiple series of DNA content (polyploidy) in animal and plant tissues. This provided a cellular basis for the discrepancies from constancy in average DNA values per cell for certain tissues such as liver shown in the biochemical determinations of Bovin and the Vendrelys. Polyploid DNA values may be the result of true chromosome polyploidy, such as in tomato leaf cells (Moses, Agnew and Sparrow 1953) and mammalian liver (Biesele 1944), or may arise from polyteny which is frequently found in insect tissue (e.g. Merriam and Ris 1954). Polyploidy is probably dependent on hormone balance in animals (Leuchtenberger, Helweg-Larsen and Murmanis 1954) and in plants (Deeley, Davies and Chayen 1957).

Synthesis of DNA during interphase in mitotically dividing cells was first suggested by Swift (1950 a) to explain the existence of intermediate values of Feulgen stain lying between the limits of an approximately twofold distribution. Direct confirmation of this result was given by the autoradiographic studies of Howard and Pelc (1951) and the ultraviolet absorption studies by Walker and Yates (1952 a, b). The latter measured the total absorbance at 2650 Å of living cell nuclei, before and after fixation, for cells in which the time since mitosis was known from time-lapse films. Similar results were later obtained for normal mouse cells and mouse ascites tumour cells by Richards, Walker and Deeley (1956).

Confirmation of interphase synthesis can be obtained by demonstrating constancy of DNA content at other stages of the cell cycle, namely at all stages of mitosis. Unfortunately, this was not possible at first since mitotic stages present serious problems of error owing to their extreme inhomogeneity, and techniques to overcome distributional error—scanning and two wavelength cytophotometry—had first to be developed. Thus Patau and Swift (1953) found no change in amount of DNA during mitosis in the root meristem of onion by two wavelength cytophotometry, and Deeley et al. (1954, 1957) and Richards et al. (1956) found similar results for both mouse and chick tissues in tissue culture and bean root meristems and mouse ascites tumours respectively by scanning cytophotometry. In a commendably detailed study PATAU and SRINIVASACHAR (1959) left little room for doubt in the case of mitosis in onion root meristem.

Autoradiography has made and continues to make highly significant contributions to cell biology. If it is permissible to single out any one contribution from the many as a pioneering effort it should probably be the work of TAYLOR, WOODS and HUGHES (1957) which first suggested that chromosomes replicate semi-conservatively. This work was possible only by virtue of the good resolution conferred by tritium. Autoradiography has since contributed to a wide range of problems in cell biology (see LIMA-DE-FARIA 1962) and examples are given in a later section.

b) Controversial topics

Both cytophotometry and autoradiography have had their share of controversial issues. Neither technique is simple to execute without care and attention to detail, and undoubtedly many of the results which initiate controversy originate from faulty practice or interpretation. For obvious reasons it is not possible to apply such a criticism to every published report of an anomalous result, but in many instances later work has failed to produce confirmation.

Owing to its relatively great importance in cell biology, the controversial topic which will be considered here is that of exceptions to DNA constancy obtained by cytophotometry. An analogous problem concerning the auto-radiographic results which demonstrate the existence of "metabolic" DNA, has recently been discussed in the review by LIMA-DE-FARIA (1962).

i) Exceptions to DNA constancy

Reports of exceptions to DNA constancy appear regularly. None of these, with the notable exception of HALE (1964), is sufficiently detailed to satisfy those criteria suggested by RICHARDS et al. (1956) to be necessary before such evidence of exception is acceptable. The criteria are: (1) that the chromosome number and possible aneuploid variations are known, (2) that no increase in DNA content attributable to synthesis has occurred, and (3) that errors in measurement, arising from either faulty handling of the stain reaction or the instrument, are absent. For most instances the first of these will be difficult or impossible to satisfy; this does not preclude the existence of abnormal chromosome numbers, however. The second can be checked by autoradiography of samples exposed to H^3-thymidine, and the third by staining standard "normal" tissues and by employing checking routines for the measuring instrument.

The reported exceptions to DNA constancy fall into two main categories: physiological variations, together with which can be included those due to amitosis or abnormal post-natal growth where chromosome number changes may have occurred, and experimentally induced variations resulting from chemical, physical or surgical interference. Certain investigators have claimed that in some tissues DNA is entirely absent from cell nuclei. Thus MARSHAK and MARSHAK (1953, 1956) claimed that the mature egg nucleus of sea urchins contained much less than the haploid quantity of DNA. CHAYEN and Norris (1953), Chayen (1960) and recently Gahan (1962) claimed that bean root meristem cells contain no DNA in their nuclei. The observations by ultraviolet microscopy of Chayen and Norris have recently failed to be confirmed by Branton and Ruch (1964).

Prominent amongst the reported physiological variations are those in the tissues of female animals which undergo cyclic physiological changes. Thus Vokaer (1951) and Vokaer, Gompel and Ghilain (1953) found an 80% increase in the average DNA content per cell in human endometrium during the proliferative phase with a return to normal values during the secretory phase. Atkin and Richards (1956) failed to confirm this result but found a distribution of DNA values typical of premitotic synthesis. More recently Thiery (1960) found elevated DNA contents in cells of mouse vaginal epithelium during pro-oestrus and oestrus which he claims to be reversed during keratinization. This latter aspect is supported by the results of Pelc (1958) showing reduction in DNA labelling with H³-thy-midine during keratinization. Branez and Roels (1961) found changes in average DNA content in all four zones of the rat adrenal cortex during the oestrual cycle. Secretory activity has been claimed to result in non-multiple increases in DNA content of the salivary glands of the snail (Leuchten-berger and Schrader 1952) but remained unchanged in the silk glands of a spider despite a threefold variation in nuclear volume (Inamdar and Wagh 1959). As Hale (1964) has emphasized, many such observations of major changes can be explained in terms of premitotic synthesis, but this does not explain those instances of possible true physiological variation in "puffs" in dipteran salivary gland chromosomes (e.g. Stich and Naylor 1958) or in early embryonal development where excessive DNA contents are reduced by DNA elimination (Stich 1962). Departures from the chromo-some mechanism as in amitosis (Cleland 1961) or in the behaviour of proto-zoan macronuclei (Walker and Mitchison 1957, Richards 1964, unpublished) automatically result in "abnormal" DNA contents. Correlation of DNA con-tents of mammalian sperm with infertility (Leuchtenberger, Schrader, Weir and Gentile 1953, Leuchtenberger, Murmanis, Murmanis, Ito and Weir 1956) and milk yield in daughter cows (Iversen 1961) has also been attempted.

Surgical intervention which causes hypertrophy has been claimed to induce elevated DNA contents. Partial hepatectomy typically results in a stimulation of premitotic DNA synthesis (Richards 1957, Gerzelli 1957, Laquerriere 1957) but some workers have reported non-synthetic increases in average DNA value in the compensatory hypertrophic reaction in kidney following removal of one of the pair (e.g. Fautrez, Cavalli and Pisi 1955), and now rejected by Becker and Ogawa (1959) and Frank (1960). Uncon-firmed changes in DNA content of cells of adrenal medulla due to the effects of physical stress, splanchnicectomy and injections of insulin have been reported by Leeman (1959 a, b, 1960), while Hutchinson, Burns and Hale (1958) found the adrenal DNA content to remain unchanged following administration of corticotrophin, despite significant increases in nuclear size. Hormonal or steroid treatments are claimed to influence the nuclear content of DNA; thyroxine (Roels 1954 a, b) or cortisones (Roels 1958) affected thyroid cells and liver cells (LOWE, BOX, VENKATARAMAN and SARKARIA 1959), but in the latter DUNN, BASS and McARDLE (1958) demonstrated that the relative frequency of ploidy classes changed. FINALLY, ANTEUNIS and LIU (1960) claims that testicular intestitial cells contained more DNA after steroid stimulation, and less after hormonal inhibition.

Plants grown at lower than normal temperatures were shown by DAR-LINGTON and LA COUR (1941) to exhibit Feulgen-negative segments in the metaphase chromosomes of their root meristems. Cytophotometry of Feulgen stained cells, both mitotic stages and interphases, indicated a real reduction in Feulgen stainable material (LA COUR, DEELEY and CHAYEN 1956, RODKIEWICZ 1960) but this has recently been contradicted (WOODWARD and SWIFT 1961). A 40% decrease in the Feulgen-DNA content in nuclei of the adrenal medulla of rats kept at alternating low and normal temperatures (4^{0} C. for 15 hours per day over 100–300 days), was found by VIOLA (1964).

Most of the reported changes in DNA content are about 10% of the total, and, if substantiated, should certainly be worthy of consideration (see GOVEART 1957). All too often, however, such reports do not satisfy the criteria mentioned earlier, have been unconfirmed, or result from errors in measuring techniques or interpretation. Notable exceptions are those of "puffs" in salivary gland chromosomes (RUDKIN and CORLETTE 1957, FICQ and PAVAN 1957), where, incidentally, amounts of DNA far less than 10% of the total are involved, and the human leucocytes first noted by ATKIN and RICHARDS (1956) in a large number of observations to contain about 10% less Feulgen-stainable DNA that other somatic tissues. Recently, HALE (1964) has confirmed this exceptional DNA content in a careful and systematic study which satisfies all the necessary criteria. He suggests that the most likely explanation is that leucocytes are in the initial stages of a change similar to pyknosis and have lost DNA from their nuclei.

c) Contributions from autoradiography

As discussed earlier, the relatively simple material requirements for autoradiography have permitted its widespread use for a great number of problems. It is impossible here to cover its usage completely, and this account will be confined to examples of investigations in the major problems in cell biology which have been studied. These have included work on synthesis of both DNA and RNA, and, in particular, on "metabolic" synthesis or turnover of DNA and on the translocation of RNA from nucleus to cytoplasm. The relative stability of the H^3-thymidine label when incorporated into DNA has facilitated studies on asynchronous replication of chromosomes, cell turnover and cell migration. Autoradiography, especially at the electron microscope level, has provided some of the evidence for the existence of DNA in non-nuclear sites such as mitochondria (BELL and MÜHLETHALER 1962), chloroplasts (SWIFT, KISLEV and BOGORAD 1964, SAGAN, BEN-SHUAL, SCHIFF and EPSTEIN (1964), basal bodies of cilia (RANDALL et al. 1964) or viral inclusions.

Incorporation of H^3-thymidine into DNA detected by autoradiography has been the method of choice to demonstrate DNA synthesis where it is

6*

of doubtful occurrence (e.g. in muscle by STOCKDALE and HOLTZER 1961) or measure the length of the synthetic period in relation to the remaining parts of the cell cycle (e.g. EDWARDS, KOCH and YOUCIS 1960, FIRKET and VERLY 1958, MENDELSOHN, DOHAN and MOORE 1960, or McDONALD 1962). More recently it has provided a means of identifying DNA-containing structural components in the electron microscope (HAY and REVEL 1963). It is also widely employed to determine the spacial or temporal relationship with other important cellular events such as the duplication of cell centres in sea urchin eggs (BUCHER and MAZIA 1960), or the transition of lysine-rich histone to argenine-rich histone in spermatogenesis of the snail (BLOCH and HEW 1960).

The results of the use of H^3-thymidine are not always clear-cut, however. Detectable incorporation occurs after as short an exposure as 30 seconds (PAINTER, DREW and GIAUQUE 1960) but the application of a simple technique for the detection of label in water soluble components enabled FEINENDEGEN and BOND (1962) to show that a definite pool of water soluble label exists in synthesizing cells, mainly in cell nuclei, while nonsynthesizing cells show no appreciable pool. A variable grain count depending on the time since an exposure of constant duration to H^3-thymidine label was found by HOWARD and DEWEY (1961) who suggest that the rate of DNA synthesis varies throughout the synthetic period in any one nucleus. This may reflect a tendency towards asynchronous replication within or between chromosomes, the extreme example of which is the early or late replication, relative to the remaining chromosomes, of the sex chromosomes shown for example by LIMA-DE-FARIA (1959) in *Melanopus* and *Secale,* TAYLOR (1960) in Chinese hamster cells in culture, WIMBER (1961) in root-tips of *Tradescantia paludosa,* and by several authors in human cells in culture (TAYLOR, MORISHIMA and GRUMBACK 1961, PAINTER, GERMANIS and STORTS 1961, ATKINS, TAFT and DALAL 1962, MOORHEAD and DEFENDI 1963).

The mode of chromosome replication is of fundamental importance in cell biology. The work of TAYLOR, WOODS and HUGHES (1957) showing semi-conservative replication in chromosomes pre-dated the equivalent demonstration at the molecular level for DNA by MESELSOHN and STAHL (1958). The results of TAYLOR et al. (1957) were repeated by several workers and a controversy arose. This has been discussed in detail by LIMA-DE-FARIA (1962), who has pointed out that much of the variance between the results of different investigations may be due to asynchrony of replication in the various chromosome segments, but that there remains little doubt that newly synthesised chromosomal DNA segregated as a physical entity corresponding to either a chromatid or half-chromatid, with rare exceptions.

Incorporation of labelled DNA precursors into nuclear DNA has been accepted by a majority of investigators as evidence for DNA synthesis in cells preparing for division. On the other hand, PELC and his co-workers have described several instances of tissues in which more cells incorporate labelled DNA precursors than eventuelly divide, and they have interpreted these as evidence for "metabolic" DNA. In muscle and brain which never show mitosis similarly labelled cells occur. LIMA-DE-FARIA (1962) has considered in detail these results and contradictory ones of other workers, notably GALL and JOHNSON (1960) and finds the controversy to be unresolved. The results of PELC might equally be evidence of "cell turnover" in that the incorporation of precursors into a normally stable DNA content may represent a breakdown of control preceding eventual cell death.

Reutilization of DNA by one cell type following the partial or complete disintegration of other cells, without prior breakdown of the DNA into small precursor molecules, has been described in several reports. Such reutilization was suggested by LINSKENS (1958) to occur during plant microsporogenesis and investigated by means of autoradiography by TAKATS (1959 and 1961). The most extensive investigations of reutilization have come from the work of HILL and his colleagues, however, but most of their work has involved the injection of radioactively-labelled lymphocytes or thymocytes into animals and then following the uptake of the labelled DNA into the host cells by autoradiography. Thus, HILL and DRASCIL (1960) compared the results on three groups of mice, all of which had previously received lethal doses of X-rays, injected with living homologous P^{32}-labelled thymocytes, UV-killed P^{32}-labelled thymocytes, and a corresponding amount of P^{32}-labelled inorganic phosphate, respectively. The differences between the three groups of animals in the distribution and time of appearance of label in the host cells (which included thymus, bone marrow and spleen cells and endosteal cells of the femur) were interpreted as indicating that large DNA-containing fragments of the donor cells were directly incorporated into the nuclei of the host cells. These results were later confirmed by HILL (1961) for unirradiated animals. The possibility that the phenomenon was the result of auto-irradiation was minimised by HILL and JAKUBICKOVA (1962) in experiments on normal and cortisone-treated animals which received extremely low doses (10^{-3} μc./g. body wt.) of labelled precursors, inorganic phophate or H^3-thymidine, which were at least 500 times less than the dose level required to produce 10 grains above diploid nuclei preparing for mitosis. The significant label found above bone marrow cells suggested that these cells at least, have a tendency to accumulate labelled DNA, perhaps at the expense of the DNA in other cells of the animal. The true significance of this work, together with similar studies by RIEKE (1962) and BRYANT (1962), suggesting as it does that high polymer DNA can be directly incorporated into certain cell nuclei, will be difficult to establish. In the absence of ready explanation it would seem worthwhile to accumulate sound data on this phenomenon, not only at the cytochemical level but also on a biochemical scale.

Stability of DNA both within and between cells is a prerequisite for the use of labelled DNA precursors as a means of following cell turnover and cell migration. The use of H^3-thymidine label for this purpose by, for example, HUGHES et al. (1958) was discussed in detail by CRONKITE, BOND, FLIEDNER and RUBINI (1959). LEBLOND, MESSIER and KOPRIWA (1959) claimed that the method was the most precise available for identifying stable, expanding or renewing cell populations. Instances of "metabolic" DNA turnover or reutilization of DNA will obviously affect the interpretation of the results of this method, but with this reservation it has proved a most useful technique in the study of, for example, proliferation of intestinal epithelia after partial removal of the ileum in rats (Loran and Althausen 1960), the establishment and rate of growth of tumour metastases (Baserga, Kisieleski and Halvorsen 1960), tumour cell turnover rates *in vitro* as compared with *in vivo* (Johnson, Rubini, Cronkite and Bond 1960), lympho-cyte cell lineages in guinea-pig bone marrow (Osmond and Everett 1961), and cell migration during histogenesis of the mouse cerebral cortex (Ange-vine and Sidman 1961). In certain instances at least the labelled DNA is relatively stable under the conditions of tissue culture, as shown by the interesting experiments of Trinkaus and Gross (1961). These workers cul-tured tapetal cells of chicken embryo, previously labelled with H³-thymi-dine, with unlabelled mesonephric cells, and found insignificant transfer of label to the unlabelled cells. In the reciprocal experiment, a small amount of transfer occurred which was considered to result from cytolysis of a few labelled cells, and reutilization of their labelled DNA.

The study of cell proliferation and migration is clearly of importance, not only in natural situations, but also in disease and treatment. From what has been briefly covered here, it is clear that adequate controls must be devised, in all experiments, to test for the possible existence of turnover of DNA or transfer between cells.

The central dogma of molecular biology holds that "DNA makes RNA, RNA makes protein." In general, evidence from cytochemial studies, with certain notable exceptions, supports this useful oversimplification. Excel-lent accounts of the work on nuclear synthesis, RNA and its translocation to the cytoplasm have been given by Prescott (1962), Goldstein (1963) and Prescott (1963) so that no further comment is needed here.

d) Combined use of several methods

The combined use of several cytochemical techniques on the same or similar cells falls into two main categories. The first is for the use of two or more methods which give information on two or more chemical compo-nents of cells, such as the studies on nucleic acids and proteins in plant cells by McLeish (1959) and Rasch and Woodward (1959), in *Paramecium* by Woodward, Gelber and Swift (1961), or in virus infected lung tissue by Sandritter, Muller and Mantz (1960). In the latter, nine different colour producing techniques were used in the same problem. The second category is that of the use of two or more different techniques which all given infor-mation on the same chemical component. An excellent recent example is the work of Jacobson, Swift and Bogorad (1963), who demonstrated the occurrence and distribution of RNA in plastids of maize by means of Azure B staining and incorporation of H³-cytidine, and confirmed the cyto-chemical results by finding RNase-sensitive granules, 170 Å diameter, resembling ribosomes by electron microscopy.

Interference microscope determination of the total dry mass of cells or nuclei has occasionally been combined with cytophotometric estimation of nucleic acid content (in visible or ultra-violet light), for example, on

X-irradiated ascites tumour cells by Caspersson, Klein and Ringertz (1958), on normal mouse liver cells by Gerzeli (1960), on nucleoli and cytoplasm of wheat cells by Longwell and Svihla (1960), and on salivary gland cells in normal and lethal mutants of *Drosophila* by Welch and Resch (1961). As stressed previously, the use of the s a m e cells in all phases of investigation increased enormously the pertinence of the information obtained. This has been demonstrated by the studies of Killander et al. (1962 a and b) and Kimball, Vogt-Köhne and Caspersson (1960), although it is sometimes not fully exploited (e.g. Seed 1962).

A frequently used combination of techniques is that of cytophotometry for estimation of amounts of nucleic acid with autoradiography for demonstrations of synthesis. The Belgian group of Chevremont has made many interesting studies in this way on cytoplasmic DNA moieties (e.g. Chevremont, Chevremont-Comhaire and Baeckeland 1959). It has also proved of value in the study of disturbed situations following surgical intervention (e.g. Frank 1960), chemical treatment (e.g. Patau and Das 1961) or radiation tratment (e.g. Looney, Campbell and Holmes 1960).

Occasionally, the results at the cytochemical level are compared with similar ones obtained at the biochemical level. In this way, Smith, Newton and Wildy (1959) demonstrated clearly the discrepancies which can arise in non-randomly increasing cell populations studied only by biochemical methods.

The ciliated protozoa *Euplotes,* the ribbon-like macronucleus of which shows two "reorganization" bands beginning distally and moving to meet at the center, has been the subject of many combined investigations. Gall (1959) used cytophotometry of Feulgen stain for DNA and alkaline fast green for histone proteins combined with incorporation of H^3-thymidine to show that the bands represented regions of DNA replication, and Prescott, Kimball and Carrier (1962) compared the timing of DNA synthesis in the macro- and micro-nuclei of the same organism. The unique morphological features of nuclear replication in *Euplotes* has stimulated its use in several subsequent studies, but insufficient use is made of the ability of cytochemical techniques to be applied sequentially to the same cells, a procedure which is applicable to any cell type. Hale, Cooper and Milton (1965) have taken full advantage of this ability in extensive studies of DNA synthesis in proliferating normal (phytohaemaglutin-treated) and leukemic leukocytes. Unlike all previous investigations of the timing and rate of DNA synthesis, the results of this group have been interpreted as indicating that the rate of synthesis varies from cell to cell. Instead of a single S-shaped curve of synthesis to which all cells conform, they propose that individual cells in the population fall on one of a family of curves varying in slope and length. Only by direct comparison of grain counts and Feulgen-values for each individual cell could such a result have been obtained.

VI. Conclusion

Cytochemical studies of the nucleic acids represent a major sphere of activity in cell biology in both normal and pathological tissues. The opportunities for error with the consequent production of misleading results are many and varied. In photometry the new sophisticated commercial instruments, properly used, may help to reduce the likelihood of error, but in photometry, as in the use of autoradiographic methods, the key to success is care.

Occasionally, results are produced which represent a departure from reasonable expectations. Such new avenues should be explored energetically, but with as many different types of investigation as possible, and, perhaps, more important, with extrem care in the design of experiments.

References

Alfert, M., 1950: J. Cell Comp. Physiol. 36, 381.
— 1952: Biol. Bull. 103, 145.
Angevine, J. B., and R. L. Sidman, 1961: Nature 192, 767.
Antenen, K. von, 1958: Experientia 14, 190.
Anteunis, A., and S. L. Liu, 1960: Arch. Biol. (Liege), LXXI, 227.
Appleton, T., 1964: J. Roy. Micr. Soc. 83, 277.
Atkin, N. B., and B. M. Richards, 1956: Brit. J. Cancer 10, 769.
Atkins, L., Priscilla Taft, and K. P. Dalal, 1962: J. Cell Biol. 15, 390.

Barka, T., and L. Ornstein, 1959: J. Histochem. Cytochem. 7, 385.
Barnard, E. A., and J. Marbrook, 1961: Nature 189, 412.
Baserga, R., W. E. Kisieleski, and K. Halvorsen, 1960: Canc. Res. 20, 910.
— and Nemeroff, 1962: J. Histochem. Cytochem. 10, 628.
— 1961: J. Histochem. Cytochem. 9, 586.
Becker, N. H., and K. Ogawa, 1959: J. Biophys. Biochem. Cytol. 6, 2, 295.
Belanger, L. F., and C. P. Leblond, 1946: Endocrinology 39, 8.
Bell, L. G. E., 1956: Phys. Techniques in Biol. Res. ed. Oster and Pollister.
Bell, P. R., 1960: Proc. Roy. Soc. B 153, 421.
— and K. Muhlethalez, 1962: J. Ultrastruct. Res. 7, 452.
Bergeron, J. A., and M. Singer, 1958: J. Biophys. Biochem. Cytol. 4, 433.
Biesele, J. J., 1944: Canc. Res. 4, 232.
Bloch, D. P., and H. Y. C. Hew, 1960: J. Biophys. Biochem. Cytol. 7, 515.
— and G. C. Godman, 1955: J. Biophys. Biochem. Cytol. 1, 531.
— 1947: Cold Spring Harb. Symp. Quant. Biol. 7, 7.
Boivin, A., R. Vendrely, and C. Vendrely, 1948: Compt. rend. 226, 1061.
Branez, E., and H. Roels, 1961: Nature 192, 1043.
Branton, D., and F. Ruch, 1964: Exper. Cell Res. 36, 285.
Bryant, B. J., 1962: Exper. Cell Res. 27, 70.
Bucher, Nancy, L. R., and D. Mazia, 1960: J. Biophys. Biochem. Cytol. 7, 65.

Caro, L. G., 1961: Nature 191, 1188.
— 1962: J. Cell Biol. 15, 189.
— and R. P. van Tubergen, 1962: J. Cell Biol. 15, 173.
— — and F. Forrox, 1958: J. Biophys. Biochem. Cytol. 4, 491.
Caspersson, T. O., 1932: Biochem. Z. 253, 97.
— 1936: Skand. Arch. Physiol. 73, Suppl. 8, 1.
— 1950: Cell Growth and Cell Function, Chapman and Hall, London.
— E. Klein, and N. R. Ringertz, 1958: Canc. Res. 18, 857.
— G. Lomakka, and O. Caspersson, 1960: Biochem. Pharm. 4, 113.
Chamberlain, P. J., and P. M. B. Walker, 1965: J. Mol. Biol. 11, 1.
Chargaff, E., and S. Zamenhof, 1948: J. Biol. Chem. 173, 327.
Chayen, J., 1960: Exper. Cell Res. 20, 150.
— and K. P. Norris, 1953: Nature 171, 472.
Chevremont, M., S. Chevremont-Comhaire, and E. Baeckeland, 1959: Arch. de Biol. LXX, 812 and 834.
Cleland, K. W., 1961: Nature 191, 504.
Cronkite, E. P., V. P. Bond, T. M. Fliedner, and J. R. Rubini, 1959: Lab. Invest. 8, 263.
— S. W. Greenhouse, G. Brecher, and V. P. Bond, 1961: Nature 189, 153.
Cummings, J. E., and J. M. Mitchison, 1964: Exper. Cell Res. 34, 406.

DAVIES, H. G., 1958: In: "General Cytochemical Methods," Vol. ed. J. F. DANIELLI. Academic Press.
— M. H. F. WILKINS, and R. G. H. B. BODDY, 1954: Exper. Cell Res. 6, 550.
DEELEY, E. M., 1955: J. Sci. Inst. 32, 263.
— H. G. DAVIES, and J. CHAYEN, 1957: Exper. Cell Res. 12, 582.
— B. M. RICHARDS, P. M. B. WALKER, and H. G. DAVIES, 1954: Exper. Cell Res. 6, 569.
DEITCH, ARLINE, D., 1961: Proc. Amer. Soc. Cell Biol. (Abs.) 45.
DIDERHOLM, H., K.-E. FICHTELIUS, and O. LINDER, 1962: Exper. Cell Res. 27, 431.
DONDERO, N. C., H. I. ADLER, and M. R. ZELLE, 1954: J. Bact. 68, 483.
DONIACH, I., and S. R. PELC, 1949: Progr. in Biophys. 3, 1.
— — 1950: Brit. J. Radiol. 23, 184.
DUIJN, P. van, and J. P. PERSIJN, 1960: Nature 186, 805.
— E. M. den TONKELAAR, and M. J. HARDONK, 1962: J. Histochem. Cytochem. 10, 473.
DUNN, CLARA E., A. D. BASS, and A. HOPE MCARDLE, 1958: Exper. Cell Res. 14, 23.
DUTT, M. K., 1963: J. Histochem. Cytochem. 11, 390.

EBEL, J. P., and S. MULLER, 1958: Exper. Cell Res. 15, 21.
EDWARDS, J. L., A. L. KOCH, P. YOUCIS, H. L. FREEZE, M. B. LAITE, and T. F. DONALSON, 1960: J. Biophys. Biochem. Cytol. 2, 273.
EVANS, T. C., 1947: Proc. Soc. Exper. Biol. Med. 64, 313.

FAUTREZ, J., G. CAVALLI, and E. PISI, 1955: Nature 175, 684.
— E. PISI, and G. CAVALLI, 1955: Nature 176, 311.
FEDER, N., and R. L. SIDMAN, 1958: J. Biophys. Biochem. Cytol. 4, 593.
FEINENDEGEN, L. E., and V. P. BOND, 1962: Exper. Cell Res. 27, 474.
FICO, A., 1959: "Autoradiography," Ch. 3 in "The Cell," ed. BRACHET and MIRSKY. Academic Press.
— and C. PAVAN, 1957: Nature 180, 983.
FIRKET, H., and W. G. VERLY, 1958: Nature 181, 274.
FITZGERALD, P. J., M. L. EIDINOFF, J. E. KNOLL, and E. B. SIMMEL, 1951: Science 114, 494.
— MARGERY G. ORD, and L. A. STOCKEN, 1961: Nature 189, 55.
FLAX, M. H., and M. H. HIMES, 1952: Physiol. Zool. 25, 297.
FRANCK, G., 1960: Arch. Biol. LXXI, 487.
FREED, J. J., J. L. ENGLE, G. T. RUDKIN, and J. SCHULTZ, 1959: J. Biophys. Biochem. Cytol. 5, 205.

GAHAN, P., 1962: Nature 195, 1115.
GALL, J. G., 1959: J. Biophys. Biochem. Cytol. 5, 295.
— and W. W. JOHNSON, 1960: J. Biophys. Biochem. Cytol. 7, 657.
GAULDEN, M. E., and R. P. PERRY, 1958: Proc. Nat. Acad. Sci. 44, 6, 553.
GEORGE, II, L. A., and G. S. VOGT, 1959: Nature 184.
GERZELI, G., 1957: Arch. Biol. 68, 1.
— 1960: Riv. Istoch. norm. pat. 6, 341.
GOLDSTEIN, L., 1963: In: "Cell Growth and Cell Division" (ed. HARRIS). Academic Press.
GOVEART, J., 1957: Arch. Biol. (Liege) 68, 165.
GRISHAM, J. W., 1960: Proc. Soc. Exper. Biol. Med. 105, 555.
GULLBERG, J. E., 1959: Lab. Invest. 8, 94.

HALE, A. J., E. H. COOPER, and J. D. MILTON, 1965: Brit. J. Haematol. 11, 144.
— 1964: J. Path. Bact. 85, 311.
HARTFORD, C. G., and A. HAMLIN, 1961: Lab. Invest. 10, 627.
HAY, ELIZABETH D., and J. P. REVEL, 1963: J. Cell Biol. 16, 29.
HERZ, R. H., 1959: Lab. Invest. 8, 71.
HILL, M., 1961: Nature 189, 916.
— and V. DRASIC, 1960: Exper. Cell Res. 21, 569.
— and J. JAKUBICKOVA, 1962: Exper. Cell Res. 26, 541.
HIMES, M. H., 1961: Proc. Amer. Soc. Cell Biol. (Abs.) 87.
HIRAOKA, T., 1960: J. Biophys. Biochem. Cytol. 8, 286.
HOWARD, ALMA, and D. L. DEWEY, 1961: Exper. Cell Res. 24, 623.
— and S. R. PELC, 1951: Exper. Cell Res. 2, 178.
HOWLING, D. H., and P. J. FITZGERALD, 1959: J. Biophys. Biochem. Cytol. 6, 313.
HUGHES, W. L., V. P. BOND, G. BRECHER, E. P. CRONKITE, R. B. PAINTER, H. QUASTLER, and F. C. SHERMAN, 1958: Proc. Nat. Acad. Sci. 44, 476.
HUTCHINSON, W. C., J. K. BURNS, and A. J. HALE, 1958: Exper. Cell Res. 14, 193.

Inamdar, N. B., and U. V. Wagh, 1959: Nature 183, 1541.
Ishida, M. R., 1959: Int. J. Cytol. 24, 107.
Iversen, S., 1961: Nature 191, 150.

Jacobson, Ann B., H. Swift, and L. Bogorad, 1963: J. Cell Biol. 17, 557.
Jansen, M. T., 1958: Exper. Cell Res. 15, 239.
— 1961: Histochemie 2, 542.
Joftes, D. L., 1959: Lab. Invest. 8, 131.
Johnson, H. A., J. R. Rubini, E. P. Cronkite, and V. P. Bond, 1960: Lab. Invest. 9, 460.

Kaplan, W. D., and J. E. Siskin, 1960: Experientia 16, 67.
Kasten, F. H., 1956: J. Histochem. Cytochem. 4, 462.
— 1957: J. Histochem. Cytochem. 5, 398.
— 1960: Int. Rev. Cytol. 10, 1.
— G. Kiefer, and W. Sandritter, 1962: J. Histochem. Cytochem. 10, 547.
Kelly, J. W., and G. Bloom, 1959: Exper. Cell Res. 16, 538.
Killander, D., C. Ribbing, N. R. Ringertz, and B. M. Richards, 1962 a: Exper. Cell Res. 27, 63.
— B. M. Richards, and N. R. Ringertz, 1962 b: Exper. Cell Res. 27, 321.
Kimball, R. F., Laura Vogt-Kohne, and T. O. Caspersson, 1960: Exper. Cell Res. 20, 368.
Koehler, J. K., K. Mühlethaler and A. Frey-Wyssling, 1963: J. Cell Biol. 16, 73.
Kopriwa, Beatrix M., and C. P. Leblond, 1962: J. Histochem. Cytochem. 10, 269.
Krause, M., and W. Plaut, 1960: Nature 188, 511.
Kurnick, N. B., 1947: Cold Spring Harbor Symp. Quant. Biol. 12, 141.
— 1955: Int. Rev. Cytol. IV, 221.
— and M. Foster, 1950: J. Gen. Physiol. 34, 147.
— and A. E. Mirsky, 1950: J. Gen. Physiol. 33, 265.

La Cour, L., E. Deeley, and J. Chayen, 1956: Nature 177, 272.
Lajtha, L. G., and R. Oliver, 1959: Lab. Invest. 8, 214.
Laquerriere, R., 1957: Compt. Rend. Soc. Biol. CLI, 1272.
Laurila, E., J. Martinen, R. Nihtila, and U. Votla, 1959: Exper. Cell Res. 16, 343.
Leblond, C. P., B. Messier, and Beatrix Kopriwa, 1959: Lab. Invest. 8, 296.
Leeman, L., 1959 a: Nature 183, 1188.
— 1959 b: Exper. Cell Res. 16, 686.
— 1960: Exper. Cell Res. 20, 596.
Lessler, M. A., 1951: Arch. Biochem. Biophys. 32, 42.
— 1953: Int. Rev. Cytol. 2, 231.
Leuchtenberger, Cecilie, 1958: In: "General Cytochemical Methods," Vol. I, 219. Ed. Danielli. Academic Press.
— H. F. Heliveg-Lansen, and L. Murmanis, 1954: Lab. Invest. 3, 245.
— R. Leuchtenberger, and C. Vendreley, 1952: Exper. Cell Res. 3, 240.
— I. Murmanis, L. Murmanis, S. Ito, and D. R. Weir, 1956: Chromosoma 8, 78.
— F. Schrader, 1952: Proc. Nat. Acad. Sci.
— — D. B. Weib and D. P. Gentile, 1953: Chromosoma 6, 61.
Levi, Hilda, 1954: Exper. Cell Res. 7, 44.
— 1957: Exper. Cell Res. Suppl. 4, 207.
— and A. Nielsen, 1959: Lab. Invest. 8, 82.
Liquier-Milward, J., 1956: Nature 177, 619.
Lima-de-Faria, A., 1959: J. Biophys. Biochem. Cytol. 6, 457.
— 1962: Prog. in Biophysics 12, 282.
Lisco, H., R. Baserga, and W. E. Kisieleski, 1961: Nature 192, 571.
Lison, L., 1950: Acta Anat. 10, 333.
Longwell, Arlene C., and G. Svihla, 1960: Exper. Cell Res. 20, 294.
Looney, W. B., R. C. Campbell, and B. E. Holmes, 1960: Proc. Nat. Acad. Sci. 46, 698.
Loran, M. R., and T. L. Althausen, 1960: J. Biophys. Biochem. Cytol. 7, 667.
Love, R., and R. G. Suskind, 1961: Exper. Cell Res. 22, 193.
Lowe, C. K., H. Box, P. R. Venkataraman, and D. S. Sarkaria, 1959: J. Biophys. Biochem. Cytol. 5, 251.

Marmur, J., R. Bownd, and C. L. Schildkraut, 1963: In: "Prog. in N. A. Res. and Mol. Biol." I. Academic Press.
McDonald, Barbara B., 1962: J. Cell Biol. 13, 193.

McLeish, J., 1959: Chromosoma 10, 686.
— and N. Sunderland, 1961: Exper. Cell Res. 24, 527.
McQuade, H. A., and M. Friedkin, 1960: Exper. Cell Res. 21, 118.
Mazia, D., W. S. Plaut, and G. W. Ellis, 1955: Exper. Cell Res. 9, 305.
Meek, G. A., M. J. Moses, and A. Eiring, 1961: Proc. Amer. Soc. Cell Biol. (Abs.) 140.
Mendelsohn, M. L., 1958 a, b: J. Biophys. Biochem. Cytol. 4, 407, 415 and 425.
— 1960: J. Nat. Cancer Inst. 25, 485.
— F. C. Dohan, and H. A. Moore, 1960: J. Nat. Cancer Inst. 25, 477.
— and B. M. Richards, 1958: J. Biophys. Biochem. Cytol. 4, 707.
Merriam, R. W., and H. Ris, 1954: Chromosoma 6, 522.
Meselson, M., and F. W. Stahl, 1958: Proc. Nat. Acad. Sci. 44, 671.
Millhouse, E. W., 1961: J. Histochem. Cytochem. 9, 661.
Monesi, V., 1962: J. Cell Biol. 14, 1.
Moorhead, P. S., and V. Defendi, 1963: J. Cell Biol. 16, 202.
Moses, M. J., A. Agnew, and A. H. Sparrow, 1953: J. Histochem. Cytochem. 1, 383.

Naora, H., 1957: J. Biophys. Biochem. Cytol. 3, 949.
Natarajan, A. T., 1961: Exper. Cell Res. 22, 275.

O'Brien, R. T., and L. A. George II, 1959: Nature 183, 1461.
Ornstein, L., 1952: Lab. Invest. 1, 250.
Osmond, D. G., and N. B. Everett, 1961: Proc. Amer. Soc. Cell Biol. (Abs.) 158.
Ostrowski, K., and W. Sawicki, 1961: Exper. Cell Res. 24, 625.

Painter, R. B., R. M. Drew, and B. G. Giauque, 1960: Exper. Cell Res. 21, 98.
— — and W. L. Hughes, 1958: Science 127, 1244.
— Mara G. Germanis, and J. C. Storts, 1961: Proc. Amer. Soc. Cell Biol. (Abs.) 162.
Patau, K., 1952: Chromosoma 5, 341.
— and N. K. Das, 1961: Chromosoma 11, 553.
— and D. Srinivasachar, 1959: Chromosma 10, 407.
— and H. Swift, 1953: Chromosoma 6, 149.
Patten, S. F., and K. A. Brown, 1958: Lab. Invest. 7, 209.
Pelc, S. R., 1947: Nature 160, 749.
— 1958: In: "General Cytochemical Methods," I, 279. Ed. Danielli. Academic Press.
— J. D. Coombes, and G. C. Budd, 1961: Exper. Cell Res. 24, 192.
— and T. C. Appleton, 1965: Nature 205, 1287.
— and L. F. La Cour, 1959: Experientia 15, 131.
Persijn, J. P., and P. van Duijn, 1961: Histochemie 2, 283.
Plaut, W., 1959: Lab. Invest. 8, 286.
Pollister, A. W., and M. J. Moses, 1949: J. Gen. Physiol. 32, 567.
— and M. A. Ornstein, 1955: In: "Analytical Cytology," 1. Ed. Mellors. McGraw Hill.
Prescott, D. M., 1962: J. Histochem. Cytochem. 10, 145.
— and M. A. Bender, 1962: Exper. Cell Res. 25, 222.
— R. F. Kimball, and R. F. Carrier, 1962: J. Cell Biol. 13, 175.
Przybylski, R. J., 1961: Exper. Cell Res. 24, 181.

Randall, Sir John, 1965: Proc. Roy. Soc. B (in the Press).
Rasch, Ellen, and J. W. Woodard, 1959: J. Biophys. Biochem. Cytol. 6, 263.
Reid, C., 1960: Science 131, 1078.
Reinholz, E., V. Belloch-Zimmerman, and C. Wirth, 1960: Experientia 16, 286.
Revel, J. P., and Elizabeth D. Hay, 1961: Proc. Amer. Soc. Cell Biol. (Abs.) 177.
Richards, B. M., 1957: Ph. D. Thesis, University of London.
— P. M. B. Walker, and E. M. Deeley, 1956: Ann. N. Y. Acad. Sci. 63, 831.
Rieke, W. O., 1962: J. Cell Biol. 13, 205.
Ris, H., and A. E. Mirsky, 1949: J. Gen. Physiol. 32, 489.
Ritter, C., H. S. Di Stefano and A. Farah, 1961: J. Histochem. Cytochem. 9, 97.
Rodkiewicz, B., 1960: Acta Soc. Bot. Poloniae, XXIX, No. 2.
Roels, H., 1954 a: Nature 173, 1039.
— 1954 b: Nature 174, 514.
— 1958: Nature 182, 873.
Ruch, F., 1960: Z. weiss. Mikroskopie 64, 453.
Rudkin, G. T., 1960: IRE Trans. on Med. Electronics, Me-7, 122.
— and Sally L. Corlette, 1957: Proc. Nat. Acad. Sci. 43, 964.
— — 1958: Proc. X Int. Conf. of Genetics, II, 242.
Rogers, A. W., 1961: Exper. Cell Res. 24, 228.

Sagan, L., Y. Ben Shual, J. A. Schiff, and H. I. Epstein, 1964: J. Cell Biol. 23, 814.
Sandritter, W., H. Cramer, and W. Mondorf, 1959: Arch. Gynak. 192, 293.
— H. Fischer, K. Sussenberger, and H. G. Schiemer, 1959: Exper. Cell Res. 17, 197.
— D. Muller, and O. Mantz, 1960: Zeit. Path. 70, 589.
Seed, J., 1962: Proc. Roy. Soc. B, 156, 41.
Sibatani, A., 1950: Nature 165, 356.
— 1953: J. Biochem. (Japan) 49, 119.
— 1954: Biochim. Biophys. Acta 13, 66.
Smith, C. L., A. A. Newton, and P. Wildy, 1959: Nature 184, 107.
Srinivasachar, D., and K. Patau, 1959: Exper. Cell Res. 17, 286.
Stenram, U., 1953: Exper. Cell Res. 4, 383.
Stich, H. F., 1962: Exper. Cell Res. 26, 136.
— and J. M. Naylor, 1958: Exper. Cell Res. 14, 442.
Stockdale, F. E., and H. Holtzer, 1961: Exper. Cell Res. 24, 508.
Stowell, R. E., 1945: Stain Technol. 20, 45.
Sunderland, N., and J. McLeish, 1961: Exper. Cell Res. 24, 541.
Swift, H. H., 1950 a: Physiol. Zool. 23, 167.
— 1950 b: Proc. Nat. Acad. Sci. 36, 643.
— 1953: Internat. Rev. Cytol. 2, 1.
— 1955: In "The Nucleic Acids," Vol. 2, 51. Academic Press.
— N. Kislev, and L. Bogorad, 1964: J. Cell Biol. 23, 91 A.
— L. Rebhun, E. Rasch, and J. Woodward, 1956: In: "Cellular Mechanisms in Diffe-
rentiation and Growth." Ed. D. Rudme. Princeton Univ. Press.

Takats, S. T., 1959: Genetics 44, 541.
— 1961: Proc. 1st Amer. Soc. Cell Biol. (Abs.) 211.
Taylor, J. H., 1960: J. Biophys. Biochem. Cytol. 7, 455.
— A. Morishima, and M. M. Grumbach, 1961: Proc. 1st Amer. Soc. Cell Biol. (Abs.) 214.
— P. S. Woods, and W. L. Hughes, 1957: Proc. Nat. Acad. Sci. 43, 122.
Terner, J. Y., and G. Clark, 1960: Stain Tech. 35, 167.
Thiery, M., 1960: Arch. Biol. LXXI, 389.
Tolles, W. E., 1959: Lab. Invest. 8, 99.
Torre, L., de la, and G. W. Salisbury, 1963: J. Histochem. Cytochem. 10, 39.
Trinkaus, J. P., and M. C. Gross, 1961: Exper. Cell Res. 24, 52.
Tubergen, R. P. van, 1961: J. Biophys. Biochem. Cytol. 9, 219.

Viola, M. P., 1964: Nature 204, 1094.
Vokaer, R., 1951: Gynecologie et Obstetrique, 50.
— C. L. Gompel, and A. Guilain, 1953: Nature 172, 31.

Walker, P. M. B., 1954: J. Exper. Biol. 31, 9.
— 1955: Exper. Cell Res. 8, 567.
— 1956: Phys. Tech. in Biol. Res. Ed. Oster and Pollister.
— 1958: In: "General Cytochemical Methods," I, 163. Ed. Danielli. Academic Press.
— and J. M. Mitchison, 1957: Exper. Cell Res. 13, 167.
— and B. M. Richards, 1957: Exper. Cell Res. Suppl. 4, 86.
— — 1959: Ch. 4 in: "The Cell." Ed. Brachet and Mirsky. Academic Press.
— and H. B. Yates, 1952 a: Symp. Soc. Exper. Biol. No. VI, 265.
— — 1952 b: Proc. Roy. Soc. B, 140, 274.
Welch, R. M., and Kathleen Resch, 1961: Proc. Amer. Soc. Cell Biol. (Abs.) 226.
Wieland, H., and G. Schening, 1921: Ber. dtsch. chem. Ges. 54, 2527.
Wimber, D. E., 1961: Exper. Cell Res. 23, 402.
Woodward, J., Beatrice Gelber, and H. H. Swift, 1961: Exper. Cell Res. 23, 258.
— and H. H. Swift, 1964: Exper. Cell Res. 34. 131.

Protoplasmatologia
  V. Karyoplasma (Nucleus)
    3. Chemistry and Cytochemistry of Nucleic Acids and Nuclear Proteins
    c) Biochemistry of Histones and Protamines

Biochemistry of Histones and Protamines

By

ROGER VENDRELY and COLETTE VENDRELY

Institut de Recherches Scientifiques sur le Cancer, Villejuif (Seine), France

With 14 Figures

Contents

Introduction

Protamines and histones are the most characteristic proteins of the cell nucleus. They are the basic proteins, which are apparently combined with the DNA in the chromatin of the chromosomes. Protamines are never present in somatic nuclei, they occur only in the spermatozoa of some animal species; in contrast, histones are present in all the somatic cells and they occur also in the sperm cells of some species which do not contain a protamine.

The first protamine was isolated in 1872 by Miescher from salmon sperm; later on, Kossel (1928) performed a very detailed study of the protamines in different species of fish. In immature fish testis there is only histone. Miescher (1897) had already noted a difference in constitution between nuclear proteins from mature and immature salmon testis. Bang (1899) isolated a proteic fraction from immature mackerel testis and identified it as a histone, while mature spermatozoa contained a protamine.

In 1947, STEDMAN and STEDMAN reached the same conclusion, that the stemm cells in salmon spermatogenesis contain a histone and mature sperm a protamine.

Protamines are characterized by a high arginine ratio which may sometimes attain 90% of the proteic nitrogen; this explain their high nitrogen ratio which is above 20%. The simplest protamines are composed of a small number of different amino acids: no more than 6 or 7, in addition to arginine. Their molecular weight is low, about 4 to 10 000 (RASMUSSEN et LINDERSTRÖM-LANG 1934, TRISTRAM 1947) so that they pass easily through a cellophane membrane. Protamines form readily salts with a number of acids and they are generally prepared by extraction with diluted sulfuric acid or hydrochloric acid. They are not precipitated by ammonia, they are soluble at high temperature (60° C.) in a divalent mercury salt in the presence of a strong acid (Equal volume of 0.34 M. HgSO₄ and 1.88 M. H₂SO₄) (MIRSKY and POLLISTER 1946 b) and insoluble at 0° C.

Histones are much more complex than protamines. The word histone was originally coined by KOSSEL in 1884 for a basic protein he had extracted from chick erythrocytes. Histones contain practically all the usual amino acids, with a marked predominance of the basic amino acids: arginine, lysine and histidine. They differ from the protamines by a lower nitrogen ratio (about 17%). Their molecular mass is higher than that of the protamines (8000 at 30,000) and they do not diffuse through a cellophane membrane. Like the protamines, they are soluble in diluted acids but are precipitable, at least partially, at pH 9–11 (isoelectric point). In fact, a fraction of the histone (lysine-rich histone) is not precipitated at its isoelectric point by NH³ (DAVISON and BUTLER 1954). They are also precipitated by ethanol acetone, or by salting out with sodium chloride. Lastly, they are soluble at high or low temperature in a solution of sulfuric acid containing mercuric ions (MIRSKY and POLLISTER 1946 b).

KOSSEL estimated first that histones occur only in a few animal tissues, but gradually it was established that the histones are present in all the somatic nuclei throughout the animal and vegetal reign, and they are sometimes encountered in the spermatozoa of some species (PHILLIPS 1962). In the animal species which have protamine-containing spermatozoa, the precursors of the seminal germ cells of the testis contain a typical histone; the transformation into protamine, or the replacement of histone by protamine, occurs apparently at the end of the process of spermatogenesis, in the course of the maturation of the gametes (ALFERT 1956, VENDRELY et al. 1957, ANDO and HASHIMOTO 1958 c).

In fact, histones and protamines form apparently a complex with deoxyribonucleic acid in the cell nucleus. The existence of this complex in the living cell has been questioned a few years ago, because of the fact that it is dissociated largely at high ionic strenght. At the present time, the existence of nucleohistones and nucleoprotamines are generally accepted (DAVISON et al. 1954 a). PEACOCKE (1960), because the constancy of their compositon, whatever the source of nucleohistone and the mode of preparation, suggests that they are a definite complex. More recently, Ts'o and

Bonner (1964) discussed the question of the actual existence of nucleo-histones; though they recognize that we have no proof of the linkage of DNA with histone within the living cell, they point out two a priori reasons for accepting the view that histones provide a structural basis for the functioning of DNA: 1st) histones, by their proximity to DNA are the agents most probably concerned with the orientation and anchoring of the DNA in the chromosomes; 2nd) all data indicate that DNA molecules do not simply float freely inside the nucleoplasm, but have a definite order and are arranged in definite sequences; the histones are the candidates of choice for the position of providing an appropriate structural arrangement.

But it is difficult to give a good definition of these substances, since. as Davison et al. (1954 a) pointed out, we have no satisfactory experimental procedure for the extraction of a non-damaged entire deoxyribonucleo-protein free from foreign proteins of nuclear or cytoplasmic origin. There is, on one hand, a danger of destruction or loss of a part of the basic protein when a thorough purification is wanted, or, on the other hand, a possible contamination by another proteic fraction of nuclear or cytoplasmic origin may occur.

It is possible by methods using high ionic strenght (1 M. NaCl) to extract from the nuclei or from the chromatin practically all the DNA they contain (Mirsky 1947, Vendrely et al. 1957, Zbarsky 1962, Georgiev et al. 1960). Together with this DNA comes an important proteic fraction, part of which is a basic protein (Histone or protamine), but it is difficult to evaluate the exact importance of this component. The extraction of the basic protein with an acidic reagent can be performed [Mirsky and Pollister (1946 b)] with the mercuric reagent; Zbarsky et al. (1959), by repeated treatment of the nucleoprotein by diluted H_2SO_4. It is thus possible to determine the percentage of basic protein in the whole complex. But is the histone totally extracted by a simple acidic treatment? As early as 1914, Steudel declared he was unable to decide wether the protein, which remains attached to the DNA after the acidic extraction, is a non-histone protein, or simply a resistant histone fraction. Steudel was particularly puzzled by the fact that, if he created an artificial complex, mixing pure histone and pure DNA, a part of the histone remained attached to DNA when he treated the syn-thetic nucleohistone with diluted HCl.

As a consequence of this lack of a perfect method for the isolation of histone, we cannot determine very exactly by biochemical techniques the histone content of the nuclei of different tissues and the amount of histone in nucleoproteins. Nevertheless we have now an approximative idea of the place and importance of the histones (or protamines) among the different proteins of the nucleus. We can distinguish in addition to the proteins of the chromosomes, the proteins of the nuclear sap, of the nucleoli, of the nuclear membrane, which are acidic proteins. A part of the chromosomal proteins, the nucleoproteins (DNP) are extracted by 1.5 M. NaCl; another chromo-somal proteic fraction (chromosomal residual proteins) remains with the other nuclear proteins. The proteins attached to DNA to form nucleoproteins soluble in 1.5 M. NaCl are histones and non-histone proteins, histones being the major fraction. Fig. 1, summarizes the different proteic fractions of the nucleus and the methods of separation of these fractions.

It is clear from these data that the histones are the most important proteic fraction associated with DNA and an important component of the chromosomes.

This view was not always held and it is perhaps useful to try to get a clear definition of "chromosomine" which was once considered by the STEDMANS (1947) as the most characteristic component of the cell nucleus. They coined the name "chromosomine" to designate all the protein which was left in the nucleus after extraction of the histone by diluted acid

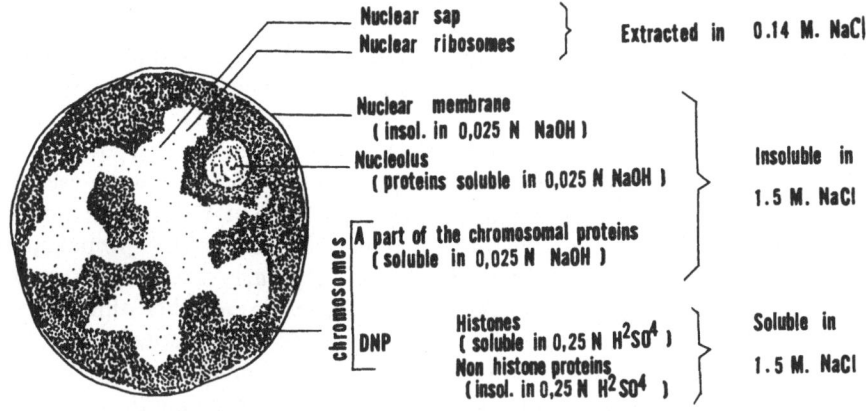

Fig. 1. Diagrammatic view of the fractionation of the various nuclear proteins.

solution and extraction of the DNA by diluted alkaline solution. They claimed that this chromosomine was present in great amounts in all the nuclei, even in Salmon spermatozoa, where they found 20 to 30% of the dry weight of the nucleus as chromosomine. These results are no longer accepted, in fact, as was shown by a number of authors, particularly by FELIX (1952), salmon or trout sperm heads countain practically nothing else than a nucleoprotamine. FELIX thought that the substance called chromosomine by the STEDMANS was a denaturated protamine with a cytoplasmic contaminant precipitated by the acids. STEDMANS views were also criticized by MARKHAM and SMITH (1954), who pointed out that the amount of chromosomine per nucleus has certainly been overestimated. The STEDMANNS used diluted acid solution for the extraction of histone and a part of the histone was certainly left in the residue.

If this chromosomine represents the non-histone protein of the nucleus, its relative amount should vary considerably with the type of nucleus; in erythrocyte nuclei or in sperm nuclei it should be practically absent.

MIRSKY and RIS isolated interphase chromosomes from calf thymus nuclei and obtained a residual protein after removal of the histone with 1 M. NaCl at pH 2.9, and of DNA with DNAase. This residual protein contains tryptophane; a little DNA is also left. It corresponds, at least in part to the "chromosomine" being the non-histone proteic part of the chromosomes.

When the crude chromatin is extracted with 1.5 M. NaCl a part of the chromosomal residual protein is dissolved as a part of the nucleohistone. This non-histone part of the complex is not very important in quantity; Davison and Butler (1956), estimated that it represents no more than 7% of the DNP. But its importance in the structural arrangement of DNA molecule is perhaps very important. Dounce and Sarkar (1960) suggested that this non-histone protein which is linked to the DNA would maintain together end-to-end by covalent bonds the DNA chains, the histone being located in the large groove of the double helix structure of the DNA.

Nevertheless, the histone is always the most important proteic fraction attached to DNA. We do not know exactly the ratio of histone and DNA in the nuclei. Classical reviews report for the whole nucleoprotein a protein/DNA ratio of 1.2 to 1.5, but when the acid extractable histone is considered, the histone/DNA ratio is as low 0.8–0.9 especially in the case of tumour tissue. The figures given by the Stedmans (1947) with histones extracted with SOH_2, 0.5 N, are even lower. But a number of authors quoted by Phillips (1962) in a very complete table, obtained more recently for calf thymus and rat kidney a histone/DNA ratio of 1. Lastly, Laurence et al. (1963), reported that the ratio histone/DNA in routine preparations of calf thymus is 1.1 to 1.2. We obtain also regularly such values in our laboratory. We do not known at the present time whether this ratio is exactly the same in the nuclei of every tissue in all animal species. It seems to be approximately the same in a number of tissus studied so far.

Recently a very low value of 0.48 of the ratio histone/DNA for calf thymus nuclei was reported by Umana et al. (1964) but the discrepancy with the results of other authors was not explained by these authors. It is probable that the low acid concentration, 0.1 N. and the short time of extraction (¾ of an hour) they use may be responsible for this low yield of histone from thymus nuclei. In contrast, they found that interphasic nuclei from rat liver, calf liver, calf pancreas, calf kidney and lamb kidney have a high histone DNA ratio of 1.8 to 2.64. They explain this surprising fact by assuming that the very rapid method of preparation of the nuclei preserves a particularly labile histone fraction. We shall comment further (chapter: metabolism) upon the possible function of this particular and hypothetic histone fraction which would have been ignored by other authors studying the quantitative ratio of histone and DNA in entire nuclei as well as in isolated nucleohistone.

Methods of Preparations

I. Preparation of the Nucleohistones and Nucleoprotamines

A. Nucleohistones

The first methods we have quoted have just a historical interest. They were rather drastic and utilised in most cases diluted alcali or acids; they yielded a mixture of proteic fractions more or less degraded or more or less separated from the nucleic acid moiety. But the extractions done by Lilienfeld (1894) by Bang (1899) and later on by Hammarsten (1924), with pure water, by MIESCHER on pus cells, and by HUISKAMP's (1903) with saline solutions of varied ionic strength, paved the way for modern methods of extraction. HUISKAMP pointed out that the nucleohistone is soluble in water and also in high saline concentration, while it proved insoluble at low saline concentration of NaCl (0.6 to 0.9 per cent) CaCl² or SO₄ Mg. Practically all the present methods of preparation of nucleohistones are based on theses observations. The nucleoproteins are soluble in saline concentrations lower than 0.02 M. NaCl and higher than 0.5 M. NaCl; between these two limits they are insoluble.

Thus the isolation of nucleohistones at the present time is done by two chief procedures: water extraction, or extraction with highly concentrated saline solution.

Nevertheless, whatever the procedure which is applied, it is necessary to eliminate first as much as possible of the cytoplasmic material. The most current method consists in washing repeatedly ground tissue with NaCl (0.14 M.) in the presence of a DNAase inhibitor (arseniate, citrate, sodium fluoride or chelating agents, such as ethylene diamine tetraacetate). All the extraction process must be carried out between 0⁰ C. and 5⁰ C. in order to slow down all possible enzymatic action.

1. Techniques using concentrated saline solutions

The type of these methods is the procedure described by MIRSKY and POLLISTER (1946), POLLISTER and MIRSKY (1946). This method has been extensively used. It can be summarized as follows:

(1) The tissue, first ground in a mixer to obtain a homogenate, is then washed in NaCl (0.14 M.), being several times resuspended in NaCl solution and centrifuged. This treatment eliminates most of the cytoplasmic components.

(2) The extraction of the nucleohistone is done with a solution of 1 M. NaCl. The solution of DNP thus obtained is then precipitated by dilution in 6 volumes of water, which brings the final concentration of NaCl to 0.14 M.

(3) The purification of the DNP is performed by successive dissolution in 1 M. NaCl solution and reprecipitation in 0.14 M. NaCl solution.

The modifications which have been added to this initial method are often of minor importance. SIGNER and SCHWANDER (1949) dissolve the tissue homogenate directly in 1 M. NaCl solution and then precipitate the nucleohistone; the cytoplasmic components are eliminated later by a careful washing of the precipitate. This method presents the advantage of a rapid dissolution and separation of the nucleohistone immediately after the homogeneisation of the tissue, which prevents the risk of degradation by enzymes present in the homogenate. But when the nucleohistone is dissolved in such a complex medium some impurities may be retained by the substance and be difficult to separate from it afterward. This is not very important when the tissue is thymus, which is the most general source of DNP, where the nuclei are predominant and the cytoplasm very reduced. But in tissues like liver, the risk of contamination by cytoplasmic impurities is more important. For that reason some authors have tried to isolated the nuclei first so that no cytoplasmic contaminant is present when the DNP is dissolved. Among the different methods of isolation of nuclei the citric acid methods (DOUNCE 1955, C. VENDRELY 1952) must be rejected because they give nuclei in which the nucleohistone is degraded. The methods using sucrose solution (for instance, CHAUVEAU et al. 1956) on the contrary, preserve the integrity of the complex very well, and it is very easy to get good nucleohistone from nuclei prepared this way. More recent techniques using detergents (ZALTA et al. 1962, HYMER and KUFF 1963, GILBERT and RADLEY 1964) do not permit a further extraction of the nucleohistone; this fact is perhaps due to a denaturation of the histone by the detergent.

It is very easy to get nuclei from fish or bird erythrocytes (C. VENDRELY 1952). A simple treatment with saponine provokes an immediate release of the hemoglobine of the erythrocytes and the residue is an excellent source of nucleohistone.

Isolated nuclei probably give a very pure nucleohistone, but the process of isolation of nuclei is rather long (except for the case of erythrocytes), and since the nuclei probably contain enzymes acting on the DNP (SIEBERT 1963, VENDRELY 1964) the risk of degradation of this substance must not be overlooked.

Whatever the method, it is important to avoid as much as possible enzymatic degradations of the products by nucleases or proteolytic enzymes so that a number of modifications of the initial technique bear on the choice of an enzyme inhibitor: EULER et al. (1945) used fluorides, STERN et al. (1947) arseniates, PETERMAN and LAMB (1848) citrates, BUTLER et al. (1953) mercuribenzoate or ethyliodoacetate, MEDAWAR and ZUBAY (1959), HNILICA and BUSCH (1962) diisopropyl fluorophosphate. Such techniques at high ionic strenght have been employed by a great number of authors: LASKOWSKI (1946), GULLAND et al. (1947), CHARGAFF et al. (1949, 1953), FRICK (1949), GAJDUSECK (1950), LUCK et al. (1953) among others.

The use of glycine as a solvent of the DNP was suggested by STERN and DAVIS 1946 but this procedure has not been used further.

More recently, RASMUSSEN et al. (1962) perform the extraction of the thymus nucleohistone on the gland frozen and ground into a powder in contact with dry ice. The extraction is realised by a boiling solution of guanidinium chloride (3 M. at 105° C.). The solution is filtrated and the nucleohistone is precipitated by dilution and washed with a 0.14 M. NaCl solution.

2. Techniques using low ionic strength solution in the presence of various enzyme inhibitors

In the initial technique described by HAMMARSTEN (1924) the extraction of the DNP was directly performed on the tissue homogenate by contact with distilled water generally; this extraction was preceded by a removal of cytoplasmic components by repeated treatments with 0.14 M. NaCl with an enzyme inhibitor.

This type of extraction has been less employed than the methods at high ionic strength. The following authors, among others used it: STERN

(1949), PETERMAN and LAMB (1948), BERNSTEIN and MAZIA (1953), CRAMPTON et al. (1954 a), SHOOTER et al. (1954).

The procedure used by DOTY and ZUBAY (1956), ZUBAY and DOTY (1959), may be quoted here with some details. It is specially adapted to obtain a DNP of calf thymus on which physical studies must be done. Special care was taken to avoid enzymatic degradation and also molecular interactions, that is to say association between portions of the histones from different DNP molecules which results in the "gelling" of the DNP solution. With this procedure the authors obtained a DNP the molecular weight of which was 19×10^6, thus nearly the double of the molecular weight of the DNA (8×10^6) which can be obtained from it.

The thymus is immediately frozen in liquid nitrogen and then ground in a blendor with 0.075 M. NaCl with versene (0.024 M. sodium ethylene diamine tetraacetate) adjusted at pH 8. At pH 8, the DNAase II cannot work, and the cathepsin cannot act on the histones (MAVER and GRECO 1949). On the other hand, the DNAase I is inhibited by the action of the chelating agent on the divalent cations. The sediment is washed six times by homogeneisations in the 0.075 M. saline solution with versene and centrifugations.

The authors never use in the course of the preparation saline solution more concentrated than 0.075 M., because at about 0.1 M. DNP begins to aggregate and precipitate.

Finally, the sediment is dispersed in distilled water, the DNP is dissolved and the solution is stirred during one hour, in order to prevent the formation of aggregates.

The DNP solution is then dialysed against 0.7 mM. potassium phosphate buffer pH 6.8. The DNP solution thus obtained contains about 15% of DNP in the form of a gel which is eliminated by centrifugation.

This method which seems to be convenient for physical studies is liable to criticism, since the elimination of the gel fraction of the nucleoprotein is quite arbitrary; this gel may represent a significant part, of the substance, as DOUNCE and O'CONNELL (1958) pointed out. SHOOTER et al. (1954) had pointed out that the dispersed material derived from the gel-like material which had been enzymatically degraded.

COMMERFORD et al. (1963) used a very mild and rapid procedure for the extraction of the liver nuclei deoxyribonucleoprotein. The extraction process was performed in 20 minutes. They isolated liver nuclei by mechanical disruption in isotonic phosphate buffer (pH 7.4), wash them with a buffer pH 1.0 consisting of a mixture of EDTA and NaCl, then with a bicarbonate buffer (pH 8.0 ionic strength: 0.15) and at last the nuclei are ground in a mixer. The solution is merely clarified by a centrifugation at 15,000 rpm. during 30 minutes; 90% of the nuclear deoxyribonucleoprotein is thus isolated.

Lastly, WELSH (1960) obtained a non-fibrous DNP from isolated thymus nuclei. Since the method used by this author involves no drastic treatment and all enzymatic degradation seems to be perfectly avoided, this result is not quite clear. Personnally we also used to obtain regularly from bird erythrocyte nuclei at low ionic strenght a non-fibrous nucleohistone.

It is necessary to point out that Luck et al. (1953) and Frick (1956) had already reported that the nucleohistone can be obtained under a non-fibrous form as well as under a fibrous form. The non-fibrous form is obtained after repeated washing of the nuclei with isotonic NaCl. These authors suggest that under these conditions, they remove a factor which might be responsible of the high degree of aggregation of the fibrous form of the nucleoprotein. The relation existing between these two forms is not yet wholly clarified.

The methods using high saline concentration have been more extensively used since the purification of the product is easier than in the low saline concentration method, and the high concentration of salt prevents to a certain extent the action of degradating enzymes. Nevertheless in the presence of salt, the nucleic acid and the protein may be dissociated, so that the native state of the nucleoprotein is not preserved. In contrast the method with low saline concentration (and with inhibitors of enzymes) may yield DNP close to the native state and non-dissociated.

B. Nucleoprotamines

The method with high ionic strength NaCl can be applied for the extraction of the nucleoprotamines from a great number of fish sperms (Mirsky and Pollister 1946 a, Felix 1960) such as salmon, trout, herring, sturgeon and also from mollusk and echinoderm sperms. But this method is not efficient in the case of mammal and bird spermatozoa.

On the other hand, the 1 M. solution is not convenient for the sperm of some species: higher saline concentrations may be necessary, such as is the case for sea urchin, key hole limpet, fresh water clam (Mirsky and Pollister 1946 a), *Patella vulgata* (Mazen 1962) etc.

The method using solutions of low ionic strength cannot be used for the extraction of nucleoprotamines.

II. Preparation of Histones and Protamines

A. Whole Histones

The histones may be extracted from crude chromatin (nuclear material carefully washed in isotonic saline solution), from isolated nuclei, or from isolated nucleohistones.

Whatever the material, the extraction of the total histone is realized most of the time by the initial method of Kossel (1884) using dilute acids: SO_4H_2 or HCl at concentration varying from 0.1 N. to 0.5 N.; the efficiency of the different concentrations will be discussed further. This method was used by a great number of authors (Bakay et al. 1955, Luck et al. 1956, Hirsch-bein and Khouvine 1957, Ui 1957, Belozerski and Uryson 1958, Smellie et al. 1958, Davis and Busch 1959, Hnilica 1959, Holbrook et al. 1959, Klyszejko and Khouvine 1960, Satake et al. 1960, Bijvoet 1957, Davison et al. 1954 b, Zbarsky and Georgiev 1959, Daly and Mirsky 1955, Davison 1957 a, Hamer 1955, Khouvine et al. 1953, Rasmussen et al. 1962, Laurence et al. 1963, Ivai 1964, Neelin 1964, Umana et al. 1964).

Some authors used diluted citric acid in a NaCl solution, but in this case the extraction is not complete and an slow electrophoretically component (lysine-rich histone) is selectively extracted, DAVISON and BUTLER (1954), DALY and MIRSKY (1955). SO_4H_2 1.88 M. containing 0.34 M. SO_4Hg was also used by MIRSKY and POLLISTER (1946 b).

When the acidic solution of histone is thus obtained, the histone can be simply collected by dialysis and freezing-drying, DAVISON (1957 a), DAVIS and BUSCH (1959), KLYSZEJKO and KHOUVINE (1960) or precipitated; different procedures can be employed:

(a) addition of ammonia at pH 10 or 11 and 2 volumes of ethanol or acetone (DAVISON 1957 b, HOLBROOK et al. 1960, BUSCH et al. 1959);

(b) the histone sulfates, which are much less soluble than chlorhydrates, are precipitated by a simple addition of 2 or 2.5 volumes alcohol (STEDMAN and STEDMAN 1951, SATAKE et al. 1960, UI 1957, RASMUSSEN 1962);

(c) the acidic extract is treated with NaCl at saturation, the protein is then precipitated, redissolved in water and collected after dialysis by freezing-drying (DAVISON et al. 1954 b, LUCK et al. 1956);

(d) the acidic histone solution is precipitated by adding trichloroacetic acid to a final concentration of 20% (LAURENCE et al. 1963).

When the extraction is made from the isolated and purified nucleohistone other procedures may be used:

The nucleohistone solution is treated by NaCl (2.6 M.) and after adding 2 volumes alcohol, the DNA is precipitated; the histone which remains in solution, is then dialysed and collected by lyophylisation (CRAMPTON et al. 1954 a). Instead of NaCl, 0.6 M. baryum acetate may be used (CRAMPTON et al. 1957).

Lastly, MIRSKY and POLLISTER (1964 b) proposed a modification of the method of SEVAG in order to separate the histone from DNA: The nucleohistone is dissolved in NaCl 1 M. solution, the pH is adjusted at 11.0 with soda and shaken with chloroform octanol (1 : 1). The histone and a non histone fraction are precipitated and separated at the interface: organic solvent-water; the histone fraction is afterwards extracted with 0.2 N. HCl; HNILICA (1959) used the SEVAG technic without modifications.

All these procedures do not give quite identical products. In fact histone is a mixture of a number of different fractions as we shall see further on. These fractions are more or less readily extracted by the acidic treatment, so that the yield of histone finally obtained is different when different molarities in the acidic solution and different length of time are used for the extraction. Moreover, the nature of the histone thus obtained may vary with the technique employed. Indeed, the experiments of fractionation of the histones show that the arginine-rich fraction of the histone is more soluble in alcohol than the others (STEDMAN and STEDMAN 1950). On the other hand, the arginine-rich fraction precipitates at pH 10.3, while the lysine-rich fraction tends to remain in solution (DAVISON and BUTLER 1954) and it is necessary to use alcohol or acetone to separate it thoroughly.

From these data it is evident that the isolation of all the histone contained in a nucleus, or in a isolated nucleoproteic complex, is not always perfect and the current methods of isolation are not quite satisfactory.

For that reason, we have reconsidered the question of the total extraction of the whole histone. We studied the effect of the acid concentration on the characteristics of the isolated histone fraction. A calf thymus nucleoprotein extracted at high ionic strength was treated repeatedly by 0.1 N. sulfuric acid, then by 0.25 N. sulfuric acid and at last 0.5 N. following the procedure indicated in Table I; the 3 fractions obtained were studied by starch gel electrophoresis at pH 6.86 on one hand, and pH 2.2 on the other hand (Fig. 2). At pH 6.86, the fraction extracted with 0.1 N. sulfuric acid shows 3 main components: 2 migrating very rapidly, the last one being slower and more diffuse. The first extraction (0.1 N. H_2SO_4) removes totally

Table I. *Fractionated extraction of whole histone from a nucleoprotein of calf thymus.*

Acid concentration	Volume of Solvent in ml.	Duration of contact at 2° C. in min.	Amount of histone collected for 10 g. thymus
0.1 N.	100	60	175 mg. (65.5%)
0.1 N.	40	30	
0.1 N.	40	30	
0.25 N.	40	60	60 mg. (22.3%)
0.25 N.	25	15	
0.25 N.	25	1	
0.50 N.	40	60	33 mg. (12.2%)
0.50 N.	25	30	
0.50 N.	25	30	

the most rapid component but not totally the others; the rest of the intermediate fraction is removed at the 0.25 N. concentration with a part of the slowest component.

Finally the 0.5 N. SO_4H_2 dissolves the rest of the slowest component. By comparing these results with the electrophoretic diagrams obtained at pH 2.2 it is possible to identify the rapid component as the lysine-rich histone, the second one as a fraction of the slightly lysine-rich histone, and, lastly, the slow component as arginine-rich histone.

This experiment led us to the following conclusion: (1) The extraction of the whole histone cannot be performed efficiently with a very dilute acid (0.1 N.) which is in agreement with the results of Butler et al. (1954); it is necessary to use a higher concentration of 0.25 N. The use of H_2SO_4 0.5 N. may be useful in spite of the risk of dissolving proteic fractions other than histones and degradating DNA itself.

(2) We do not agree totally with Laurence et al. (1963), when they say that the fractions extracted successively by acidic treatments do not show any marked difference in composition; on the contrary, the differences are rather marked, but they appear only when low concentrations are used at the begining of the experiment. These authors working with 0.25 N.

H_2SO_4 only, did not distinguish between their first product and the following ones. Therefore, it seems that different fractions of the whole histone are detached more or less easily from the complex and the method of preparation must be carefully checked.

ClH N∕100   pH 2,2

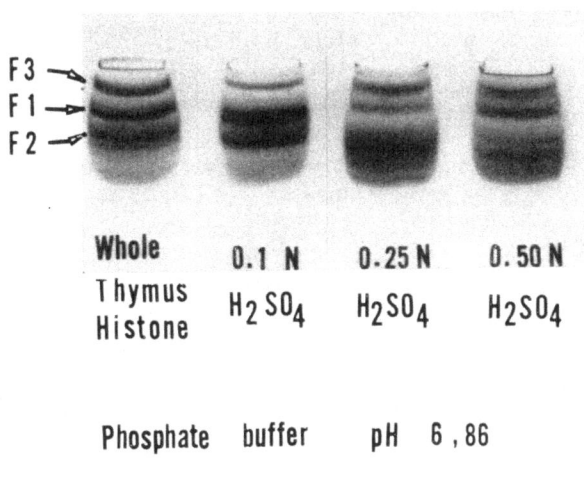

Phosphate buffer   pH 6,86

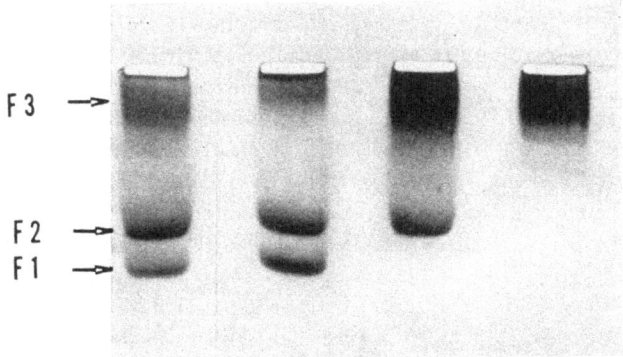

Fig. 2. Starch-gel electrophoresis patterns of thymus histone fractionated by acidic treatment. (R. VENDRELY N. GENTY, Y. COIRAULT, unpublished.)

For the study of histones, and particularly of their quantitative relations to DNA, it is necessary to extract them as thoroughly as possible; for that reason the method described by LAURENCE et al. (1963), seems to be the most satisfactory, provided that more than one extraction with H_2SO_4 0.25 N. is performed. LAURENCE et al. extract the whole histone from the nuclei or nucleoprotein with 0.25 N. H_2SO_4 (it was found that extraction with 0.25 N. HCl gave a more contaminated product). The Dounce homogenizer was found to be convenient for dispersing the nuclear material in the acid, the stainless-steel of the plunger being protected by a plastic sleeve. After dispersal, the material is shaken with the acid (about 1 ml. for every g. wet wt. of original tissue) for 1 hr. and then centrifuged a 1600 g. for 1 hr. A second extraction gives about 5% more histone material, but, except for the experiments in which amounts of histone and DNA were compared, the second extraction was omitted. There is no marked difference in composition between the histones from the first and second extraction. The histones are precipitated by adding 100% (W/V) trichloroacetic acid to the histone solution to a final concentration of 20%. The precipitate collected by centrifugation (1000 g. for 10 min.) is washed with acid acetone (1 drop of conc. HCl to 10 ml. of acetone) and then repeatedly with acetone.

Let us point out that the recent studies of histones concerned chiefly with their possible specificity or metabolism do not bear on the whole histone, but rather on its fractions which are often obtained directly from crude chromatin, isolated nuclei or nucleoprotein. We shall examine this question in a following paragraph.

B. The Protamines

This question has been reviewed very carefully by FELIX (1960).

The first method of extraction of the protamine has been proposed by MIESCHER 1874 for salmon sperm: The sperm was thoroughly defatted and treated with HCl 1 to 2%. The addition of platinium chloride to this solution precipitated the protamine under the form of a double salt.

KOSSEL 1896 replaced the chlorhydric extraction by 3 treatments with sulfuric acid at 1%. The protamine sulfate was generally precipitated from the acidic solution by an excess of alcohol (3 volumes). This procedure allowed KOSSEL to obtain a great number of protamines, but it cannot be applied in every case (mammalian spermatozoa).

FELIX 1960 insists on the efficiency of sodium picrate as a precipitant of the protamines. The high insolubility of protamine picrates makes possible a very complete washing of the precipitate even with diluted soda (0.1 N.). The picrates can be then decomposed by sulfuric acid into protamine sulfate and free picric acid, which is removed by extraction with ether or toluene.

NELSON GERHARDT, a student of KOSSEL's, proposed in 1919 another procedure for the preparation of protamines in which the dry sperm is treated by a solution of copper chloride. The protamine is dissolved and the nucleic acid forms an insoluble salt with copper. The protamine is then precipitated in the form of picrate.

These general principles of extraction are still at the base of the present procedures, the starting material being either sperm heads, or nuclei obtained by cytolysis, or the previously isolated nucleoprotamines.

1. Extraction from sperm heads

BLOCK et al. (1949) isolate first the sperm heads by precipitation with 1 per cent citric acid and then extract the protamine with HCl 0.2%. This solution is heated at 100° C. and NH₃ (pH 8) is added; the precipitate which is formed is eliminated and the protamine is then precipitated under the form of metaphosphate by addition of 33 per cent metaphosphoric acid. The protamine metaphosphate is converted into sulfate, which is at last precipitated by alcohol or acetone. DALY et al. (1951) extracted from fowl sperm a protein similar to protamines. The sperm cells were washed several times with saline, the tails and midpieces were removed by treatment with 0.2 per cent citric acid. The protamine is extracted with 0.25 N. HCl and separated by addition of 3 volumes of 95 per cent alcohol after removal of the precipitate at pH 7.

HAMER (1955) collects the sperms from sea urchins (*Arbacia, Asterias, Loligo*) in sea water, centrifuges and washes them with saline citrate solution. Sperm nuclei are thus obtained. They are treated with 0.2 M. HCl for 24 hours, after centrifugation the pellet is discarded, and the supernatant is dialysed against distilled water. The basic protein is precipitated in the form of a protein picrate; it is then purified by dissolution in acetone. The protein picrate is then converted into sulphate which is precipitated with acetone. The acetone removes all free picric acid which remained in solution.

ANDO and SAWADA (1962) accomplished the plasmolysis of trout sperm cells in distilled water. The nuclei thus obtained were defatted with ethanol and ether and ground with quartz powder in a mortar at low temperature with HCl (0,2 per cent); this extraction is repeated 6 to 8 times. The extracts containing the protamine are neutralized with amberlite (HCO$_3$ form) and lyophilized.

2. Extraction from nucleoprotamines

POLLISTER and MIRSKY (1946) extract the nucleoprotamine from trout sperm with 1 M. NaCl. The protamine is separated from the complex by a mere dialysis against 1 M. NaCl; protamine slowly passes through the cellophane membrane, leaving nucleic acid behind.

ANDO et al. (1957 b) reduce frozen herring sperm into a fine powder, wash it carefully with 0.05 M. sodium citrate solution pH 7. From this material, they extract the nucleoclupeine with 2 M. NaCl containing 0.05 M. sodium citrate. They add to this solution a saturated solution of copper sulfate. The nucleic acid is precipitated and the protamine is then separated under the form of a picrate. The protamine picrate is dissolved in acetone in order to remove impurities and then converted into a protamine sulfate which precipitates in alcohol.

GOPPOLD-KREKELS and LEHMANN (1958) also propose to saturate the solution of nucleoprotamine with NaCl. At 4° C. the protamine gradually settles out. The protamine is then purified by dissolution in 5% trichloroacetic acid and precipitated in 10% trichloroacetic acid; the excess NaCl is thus eliminated.

FELIX (1960) separates first the sperm nuclei from trout or herring by cytolysis in distilled water; this process does not impair the biological potency of these nuclei. The isolated nuclei are treated by a 10% NaCl solution in which they are dissolved with practically no residue. When

9 volumes of water are added to this solution, nucleoprotamine fibers are precipitated. They are then washed with alcohol and dried in ether. The nucleoprotamine is then treated repeatedly with 0.2% HCl (100 ml. for 1 g. nucleoprotamine), 20 volumes acetone are added to this acid extract for 12 hours at 4⁰ C. The protamine chlorhydrate is deposited on the surface of the flask. It is dissolved in a little water, frozen and dried under high vacuum.

The Composition of Histones and Protamines

I. Protamines

The pionneer works of Kossel led him to establish that protamines were composed of very few kinds of amino acids. The basic amino acids are markedly predominant and the most important of them is arginine. But lysine and histidine may be present too. Kossel (1928) therefore divided the protamines into three groups: the m o n o - p r o t a m i n e s which contain only arginine; such are the clupeine, from herring (*Clupea harengus*), the salmine (Salmon, *Salmo Salar*), the truttine (Trout, *Salmo trutta*), the iridine (Trout, *Salmo irideus*), the esocine (pike, *Esox lucius*) etc. The d i p r o t a m i n e s, which contain arginine and lysine such as barbine (from the Barbel, *Barbus vulgaris*), cyprinine (from the Carp, *Cyprinus Carpio*), iridine (which Kossel also quoted in diprotamines since in one preparation he found lysine in addition to arginine) etc. (An other type of diprotamine contains arginine and histidine: the percine [from the perch, *Perca flavescens*] belongs to this group.)

At last come the t r i p r o t a m i n e s which have three basic amino acids: arginine, lysine and histidine: the sturine from the sturgeon, *Acipenser Sturio,* is an example of this type.

The other amino acids found by Kossel and other authors in protamines are alanine and serine in all of them: proline and valine in most and sometimes glycine and isoleucine. In none of them did he find tryptophane, phenylalanine or sulfur-containing amino acids.

Felix (1960) in his review reports numerous analytical results on various fish protamines. Such a table assembles results obtained at different periods and by differents investigators and different techniques, which are difficult to compare.

The progresses in the techniques of isolation and analysis of protamines and, particularly, the use of column or paper chromatography made possible a better knowledge of the protamines.

Some of the results of Kossel are not considered to be quite exact at the present time. The drastic method he used resulted sometimes in the fractionation of the basic protein and a loss of a part of the molecules. For instance, Kossel isolated from carp spermatozoa two types of protamines, cyprinine I and cyprinine II; in fact, as we shall see later, carp sperm contains a typical histone, so that we must conclude that Kossel isolated from this histone two fractions, one of them being particularly rich in arginine, the other in lysine (cyprinine II). A number of his results should thus be checked with modern techniques before being considered as definitive.

The distinction between mono-, di-, or triprotamine was convenient, but, with the progresses of the techniques, it appears that it is not sufficient to characterize the basic proteins of sperms. It may be more appropriate to classify these proteins according to their complexity, depending on the number of amino acids they contain: in some species of fishes, the composition is very simple, they contain only 7 to 8 amino acids (HAMER and WOODHOUSE 1949, FELIX 1953); these are typical protamines. But such proteins may be more complex, for instance in rooster sperm, DALY et al. (1951) demonstrated a basic protein with 12 amino acids, containing 52% arginine, 1.2% histidine and some tyrosine (which is exceptional in protamine). FISCHER and KREUZER (1953) isolated from rooster sperm a protein quite similar, but without any tyrosine and histidine; these differences may be due to the greatest care taken by FISCHER and KREUZER for removing cytoplasmic residue from the sperm. So that this product might be considered as a monoprotamine (FISCHER and KREUZER) or as a diprotamine (DALY et al.) but also certainly as a substance more complex than the typical protamine of salmon sperm nuclei. Such type of protein was encountered by HULTIN and HERNE (1949) by HAMER (1955) and by VENDRELY et al. (1960) (Table II).

But in other species the sperm protein appears to be of the histone type, for instance in cod (STEDMAN and STEDMAN 1951) carp, tench, frog, Nephtys (annelid polychete) (MAZEN 1962). See also Table II.

A great diversity exists among the different animal species and, considering the different amino acid pattern which have been encountered, the distinction between protamines and histones seems to be rather arbitrary whatever the criterion adopted, since a lot of intermediates may be found between typical histones and typical protamines with various amounts of arginine and with various kinds of amino acids. On the other hand, recent progresses showed definitely that histones are a mixture of different fractions which differ from one another especially by their arginine and lysine content, as we shall see in section II (chapter 3). As MURRAY (1964 a) recently pointed out: "... although there are profound differences between histones and protamines, their continued distinction is not readily justified when considered in the context of the recent progress in the fractionation of these proteins. Histones and protamines may be considered as homologous proteins. Their distinction is comparable with the distinction between two histone fractions having widely different properties..." The modern attitude towards these proteins will be to consider them as basic proteins of the cell nucleus, protamines being a particular case of these proteins.

Concerning the problem of species specificity of the protamines, FELIX (1960) considers that since the protamines of different species of fish have different composition, it is not unlikely that the different amino acid pattern may be characteristic of the different species of fish. For instance, ANDO et al. (1959) found that the clupeine from Japanese herring (*Clupea pallasii*) regularly contains glycine, which is not the case for European herring (*Clupea harengus*) investigated by Felix 1958, who found glycine in Baltic sea herring but not in herring from North sea. Nevertheless, it is possible that these differences are due to the different conditions of preservation of the herring gonads and a greater number of results are necessary to solve this question of the species specificity of the protamines.

Table. II. *Composition of*
(Amino acid N in

	Typical protamines			Typical histones			Calf (Thymus histone)[7]	Calf (Thymus histone)[3]
	Salmo salar[1]	Salmo irideus[2]	Salmo irideus[3]	Asterias forbesi[5]	Asterias glacialis[3]	Cyprinus carpio[3]		
Arginine	89.2	89.6	88.2	23.8	22.28	21.51	21.47	23.16
Lysine				19.0	15.11	18.43	19.11	20.76
Histidine				2.9	4.23	1.82	3.87	3.73
Alanine		0.6	0.68	9.85	10.25	8.28	8.75	9.96
Ammonia				3.9			4.06	
Aspartic Acid			0.10	2.9	5.50	3.70	3.10	3.04
Glutamic acid			0.13	4.0	8.77	6.22	5.31	6.06
Glycine		1.9	2.03	4.4	4.40	5.20	5.44	6.12
Isoleucine		0.3	0.31		0.97	3.18	2.82	2.89
Leucine			0.16	5.8	5.68	5.92	5.02	5.67
Phenylalanine				0.9	2.19	1.69	1.36	1.09
Proline	4.3	2.8	2.70	4.2	4.22	4.60	3.23	3.49
Serine	3.25	3.3	3.06	3.8	3.11	4.13	3.77	3.49
Threonine			0.07	3.1	4.03	4.46	3.70	3.84
Tyrosine				1.1	2.14	2.07	1.64	1.68
Valine	1.65	1.7	2.08	3.9	3.94	3.93	3.98	4.55
Methionine				1.4	1.09	0.61	0.63	0.41

[1] Kossel, A., 1928. [2] Ando, T., and C. Hashimoto 1958 b. [3] Vendrely, R., non S. Moore and W. H. Stein 1955. [7] Daly, M. M., A. E. Mirsky, and H. Ris 1951.

On the other hand salmine and clupeine, the proteins of the sperm heads of the salmon and herring, respectively, are extraodinarly similar though these species are rather far from one another. Nevertheless, Stedman and Stedman 1951 found small differences between them, so that the species specificity of the protamines remains a possibility.

Felix (1960) also points out the fact that the protamine composition may be a molecular factor of the ability of two species to be crossbred. For instance, the brook trout (*Salmo trutta*) and the Canadian brook char (*Salmo fontinalis*) have protamine of same composition, they can be crossbred. But the offspring are sterile. It would be worthwhile to study more accurately the protamine composition in relation to the ability of being crossbred in closely related species of fish. But of course other factors, both cellular and molecular and, first of all, differences in DNA molecules, must be involved.

Basic Proteins from Sperm.
per cent of total N)

Intermediate forms											
Protamine like					Histone like						
Patella coerulea [4]	Rooster (Gallin) [6]	Esox lucius (Esocine) [3]	Patella vulgata [4]	Loligo pealii (Squid) [5]	Echinarachnius parma (Sand dollar) [5]	Echinocardium Cordatum [4]	Paracentrotus lividus [3]	Brissopsis lyrifera [4]	Arbacia lixula [4]	Arbacia punctulata [5]	Echinocyanus pusillus [3]
82.9	76.4	73.65	49.3	47.7	34.1	31.0	30.06	28.6	27.8	28.0	27.08
7.6	0.0	5.49	19.6	8.2	13.9	28.7	14.65	26.6	28.6	17.1	15.69
0.0	1.53	0.00	0.93	2.0	1.0	1.3	3.83	1.83	1.52	2.1	3.26
2.4	1.22	2.79	6.1	4.2	9.3	13.7	9.41	13.0	14.8	9.8	8.86
	1.17			4.0	3.7					3.8	
		0.28	0.62	3.4	1.9		5.19			2.7	3.91
		0.45	0.91	3.7	3.0		6.91			3.9	5.75
4.0	2.80	3.05	4.7	3.6	4.8	11.7	4.90	13.3	10.8	5.7	5.03
0.26		0.30	0.59			1.95	2.63	2.4	2.0		2.73
1.6	0.35	0.71	1.75	5.8	6.1	1.36	5.20	3.0	2.4	7.2	4.68
		0.21		1.0	0.9		1.85			1.0	1.54
1.7	1.94	4.38	4.8	2.6	3.4	5.6		5.0	6.1	4.4	2.10
5.7	3.89	4.38	4.6	4.8	5.3	4.7	4.23	11.7	4.3	4.0	3.67
0.94	7.20	0.58	2.1	2.1	4.2	2.3	3.68	2.8	1.39	3.4	3.28
	1.58	0.21		1.9	1.1		2.00			1.1	1.72
1.6	6.80	1.90	2.2	2.8	4.2	2.3	4.45	2.8	1.45	3.7	4.78
	0.00	0.65		x	x		0.98			1.0	0.87

published. [4] HULTIN, T., and R. HERNE, 1948. [5] HAMER, D., 1955. [6] CRAMPTON, C. F.,

II. Occurence of Protamines in the Different Animal Species

KOSSEL's investigations were done on fish sperms, but the occurence of protamines is certainly not limited to the fishes. We have already quoted the example of the galline from rooster sperm which is a protamine containing a rather high number of kinds of amino acids in addition to arginine.

As we have already noted previously, we have no method for the separation of the proteins of sperms in mammals. This fact led us to suggest that the analysis of the arginine ratio on whole sperms may give a clue to the nature of the protein of the sperm (VENDRELY, R., and VENDRELY, C., 1953,

Knobloch and Vendrely 1956). In fact we considered the DNA content of the nucleus together with the arginine content of the nucleus, and we expressed the results in molecules of arginine of the protein per atom phosphorus of the DNA. This way of expressing the results appeared to us to be the best one, since the DNA content of the nucleus is constant per cell and characteristic of the chromosome number (Boivin and Vendrely 1949). The absolute amount of DNA per nucleus varies therefore from one animal species to the other and in the same animal the spermatozoa have half of

Table III. *Individual amounts of Arginine and Deoxyribonucleic acid (in $10^{-6}\,\gamma$) of erythrocyte nuclei and fish sperms.*

Animal Species	Material	DNA per nucleus ($\times 10^{-6}\,\gamma$)		Arginine per nucleus ($\times 10^{-6}\,\gamma$)		Molecule arginine per atom phosphorus	
Trout	erythrocyte	4.90		0.96		0.35	
Salmo Trutta	sperm	2.45			1.50		1.10
Pike	erythrocyte	1.70		0.34		0.36	
Esox Lucius	sperm	0.85			0.35		0.74
Carp	erythrocyte	3.20		0.60		0.34	
Cyprinus Carpio	sperm	1.60			0.35		0.39
Tench	erythrocyte	1.70		0.34		0.36	
Tinca Tinca	sperm	0.85			0.18		0.38
Perch *Perca fluviatilis*	erythrocyte	2.00		0.37		0.33	
Barbel *Barbus barbus*	erythrocyte	3.40		0.64		0.34	
Gardon *Gardonus rutilus*	erythrocyte	1.90		0.36		0.34	
Calf	thymus	6.40		1.49		0.41	
Bull	sperm	3.20			2.16		1.21
Rooster	erythrocyte	2.20		0.45		0.37	
	sperm	1.10			0.78		1.27

the somatic value. The basic protein being constantly linked to the DNA seems to be present in a quantity proportional to the DNA; in all the nucleohistones studied so far the amount of histone is equivalent to that of the DNA so that we could expect to find a definite and characteristic value for the ratio: histone arginine/DNA phosphorus, in all the nucleohistone. This ratio is quite different in the case of a nucleoprotamine. We studied in this way a number of isolated nuclei from erythrocytes, and spermatozoa in different species and compared them to calf thymus nuclei. The erythrocyte nuclei contain chiefly a nucleohistone with a negligible amount of non-histone protein, so that the results obtained on the total nucleus are not very different from that obtained on isolated nucleohistone. In calf thymus, there is a little more non-histone protein so that the ratio must be somewhat adjusted. The results on spermatozoa are also quite similar to that obtained on the nucleoproteins isolated from these spermatozoa.

The Table III (Mazen 1962) shows our results. From these results it appears that an arginine/phosphorus ratio close to 0.34 is characteristic of all the nuclei containing a nucleohistone, that is to say, all the somatic nuclei. In the sperm nuclei the ratio is similar when the sperm contains a histone. In Carp and tench, we found effectively a histone in the sperm. On the contrary, when a typical nucleoprotamine is present, this ratio is about 1 (Trout). We proposed that when this ratio is markedly superior to 0.34 and close to 1, the sperm should be considered as containing a protamine. This was true for all the species studied so far. From these analyses, the rooster sperm appears as containing a protamine, which is in agreement with the data of Daly et al. (1951). The pike case appears to represent an intermediate between typical protamine and histone, but since its ratio is markedly superior to 0.34, we consider it as protamine-like. Bull sperms should be considered protamine-like, from their 1.27 ratio.

This procedure to distinguish histone-containing sperm from protamine-containing sperm may look rather gross, but it allows a rapid study of the sperm of a great number of animal species and is quite useful in the case of sperm from which the protein is impossible to extract, for instance mammalian sperm.

Such studies of compared biochemistry would be rather interesting because we do not know at the present time what type of sperm is the most common in the animal kingdom and what is the significance of these types in the evolution of species or its adaption to the environment.

At any rate it is possible now to adopt a provisional classification of sperm into three types, namely: the histone type, which is represented in sea urchin, starfish, cod, carp, tench, frog; the typical protamine type, which is characteristic of salmon, trout, herring and, at last, the intermediate type found in *Patella cerulea,* pike and some others species (Table II).

III. Histones

The following table (Table IV) includes a number of analytical results on histones of different origins. In addition to the values indicated for these amino acids, cystine was found as traces or less than 0.4%. Tryptophane is generally absent or exists as traces. Klyszejko and Khouvine (1960) reported 0.2% moles tryptophane in calf pancreas histones but these authors point out that their preparation was contaminated by ribonucleoproteins since they found 4% uracil in it. Crampton, Moore and Stein (1955) detected only traces of tryptophane in calf thymus histones.

From the results reported in Table IV, it is not possible to decide wether the histones have different compositions when they come from normal or tumour tissues or from different tissues or different animal species. The slight differences that are observed in these histones may well be due to the method of preparation used. Analytical and chromatographic results obtained by Vendrely et al. (1958 b) on proteic hydrolysates obtained from various preparations of deoxyribonucleoproteins isolated at high ionic strength from different animal tissues failed also to reveal any difference

Table IV. Amino acid composition of whole histones (as moles per 100 moles of all amino acids).

Reference	Calf thymus[1]	Calf thymus[2]	Calf thymus[3]	Calf pancreas[4]	Calf liver[5]	Calf kidney[5]	Rat liver[6]	Guinea pig testis[5]	Fowl erythrocyte[3]	Spontaneous mouse tumour[6]	Crocker tumour[6]	Rat Leukaemia[3]	Wheat germ[7]
Aspartic acid	14.2	4.8	4.37	5.5	5.8	5.4	15.6	6.3	4.63	14.6	15.7	6.36	4.4
Glutamic acid		8.2	8.68	9.7	8.8	8.6		9.1	7.94			10.46	8.6
Glycine	8.5	8.3	8.76	5.7	8.5	9.0	8.1	8.9	7.79	7.8	8.3	8.46	8.1
Alanine	13.8	13.4	14.28	13.9	12.9	13.0	12.2	11.9	12.58	12.2	13.4	11.67	15.5
Valine	6.4	6.2	6.53	8.1	6.0	6.1	6.6	5.8	5.53	6.3	6.1	6.15	5.5
Leucine	10.6	7.7	8.12	14.0	8.5	8.1	11.2	8.0	7.39	12.4	11.8	7.84	12.3
Isoleucine		4.3	4.15		4.6	4.4		4.1	4.24			4.35	
Phenylalanine	1.9	2.2	1.56	3.2	2.0	2.0	2.2	1.9	1.78	2.2	1.6	2.41	2.2
Tyrosine	1.7	2.5	2.41	1.9	2.5	2.6	2.2	0.3	2.48	2.3	2.0	2.49	1.7
Serine	5.2	5.9	5.01	4.8	5.8	5.8	6.4	6.9	6.10	6.2	6.6	5.43	4.9
Threonine	5.8	5.7	5.51	4.9	5.8	5.8	5.8	5.5	5.25	6.1	5.1	5.41	5.4
Proline	5.1	4.9	5.01	5.5	4.4	5.1	5.7	5.8	3.70	4.7	5.3	6.72	5.1
Methionine	—	1.0	0.59	—	1.3	1.1	—	1.2	0.65	—	—	0.60	—
Cystine	—	0.1	—	traces	0.4	0.3	—	0.3	—	—	—	—	—
Arginine	9.2	8.3	9.46	6.5	8.0	8.1	8.2	7.4	9.27	7.4	6.2	8.09	7.3
Histidine	2.0	2.0	1.77	1.9	1.9	1.8	2.6	2.1	1.88	2.5	1.4	2.00	1.5
Lysine	15.6	14.7	14.87	14.3	12.9	12.8	13.1	11.9	13.66	15.2	16.4	11.69	17.6
Lysine/arginine	1.70	1.77	1.57	2.2	1.61	1.58	1.60	1.61	1.47	2.05	2.64	1.44	2.41

[1] Johns et al 1960. [2] Crampton et al 1955. [3] Vendrely et al, impublished data [4] Klyszejko and Khouvine 1960. [5] Crampton et al 1957. [6] Butler 1963. [7] Johns and Butler 1962 a.

in the composition of the different preparations (Table V, Fig. 3) concerning the histones still associated with DNA.

Table V. *DNP extracted from somatic nuclei (basic amino acid residue [gm.] per 100 g. protein).*

	Arginine	Lysine	Histidine
Calf thymus	12.6	15.5	2.4
Calf liver	11.8	15.2	2.5
Calf kidney	12.3	16.1	2.3
Carp erythrocytes	13.5	15.5	2.3
Trout erythrocytes	13.5	15.2	2.5
Pike erythrocytes	11.4	13.5	1.9
Fowl erythrocytes	13.3	16.6	2.5
Duck erythrocytes	13.0	15.7	2.4
Frog erythrocytes	12.8	14.0	2.9

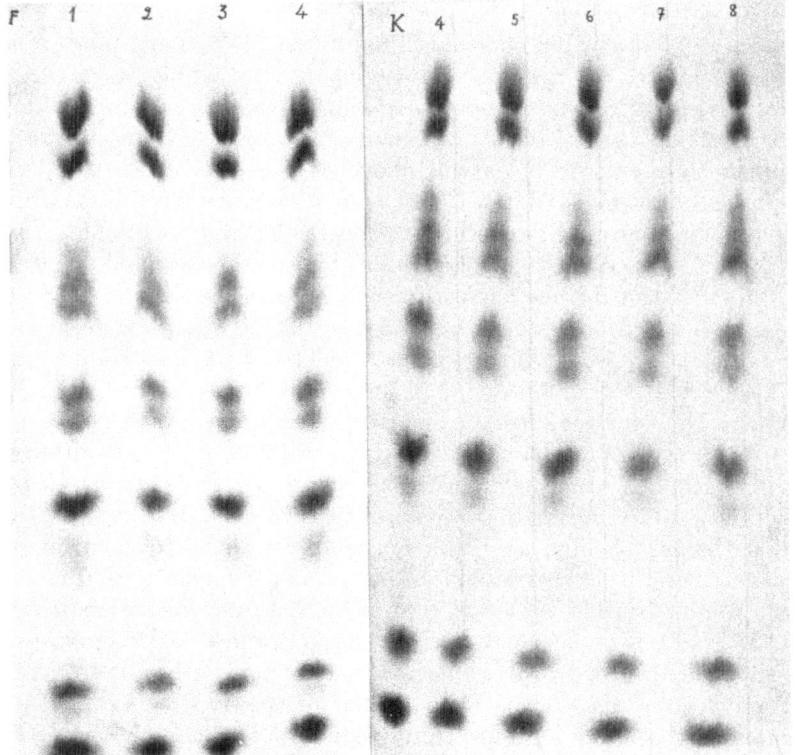

Fig. 3. Chromatograms of nucleohistones from different vertebrates (each volume of hydrolysate corresponds to the same quantity of deoxyribonucleic acid. Nucleohistone from: (1) carp erythrocytes; (2) trout erythrocytes; (3) pike erythrocytes; (4) calf thymus; (5) fowl erythrocytes; (6) duck erythrocytes; (7) tench erythrocytes; (8) frog erythrocytes. Whatman paper No. 1. Solvent: 4 parts, butanol; 1 part, ethanol; 1 part, acetic acid; 2 parts, water. (R. VENDRELY et al. 1958 b).

Nevertheless, it is very important to try to detect differences among different histones since these substances have very likely an important role in the regulation of gene activity and could be determinant factors of cell differentiation, as we shall see in a later chapter.

But the study of the overall composition of different histones is too crude for the detection of differences; more refined techniques are required. The fractionation of histone into its different components and the comparison of the individual composition of each fraction must be done for histones of different origin. Before considering these data we shall consider some analytical results obtained on the histones in whole nucleoprotein complexes.

IV. Analytical Studies on Histones and Protamines in the Nucleoprotein Complex

The analytical results we have reported were obtained on isolated histones or protamines and, as we have pointed out, it is sometimes difficult to tell with certainty wether the histone thus studied is intact and similar to that which was present in the nucleus.

The study of the whole nucleohistone complex allows an other approach to the problem of the histone composition by providing substances in a probably less altered state and by making it possible to use DNA as a reference for the expression of the results. The results on the amino acids could then be refered to the atoms phosphorus of the DNA.

Vendrely et al. (1957), Mazen (1962), made a balance sheet of the different fractions they separated from calf thymus nuclei and sperm nuclei in the course of isolation of the nucleohistone at high ionic strenght. They found that all the DNA of the nucleus was recovered in the nucleoprotein collected at the end of the process of preparation, so that it was possible to use the DNA as a standard of reference for the associated proteins. They were thus able to show that a fraction of the protein probably associated to the DNA was lost in the course of the purification by repeated precipitation of the nucleoprotein: when a nucleohistone from thymus is reprecipitated two times, the ratio mol arginine/atoms P is about 0.32; after eight precipitation it is only 0.24 which means that a part of the protein is lost.

From the basic amino acid analysis of the whole nucleus, that of the non-histone residue left after extraction of the nucleohistone, that of the fraction soluble in 1 M. NaCl and that of the finally collected nucleohistone at 0.14 M. NaCl, it is easy to calculate the importance and composition in basic amino acids of the lost fraction. From the calculations it appears that this fraction represents about 15 to 20% of the basic protein; it has a high percentage of basic amino acids quite similar to that of the histone in the case of erythrocytes. The hypothesis of a lost basic protein fraction was thus very likely and it was confirmed by the isolation from the 0.14 M. supernatant of two fractions, one which is directly precipitated by dialysis against distilled water and the other particularly important is precipitated from the new supernatant by addition of acetone. The first fraction has a high arginine and lysine ratio, the second one is particularly lysine rich. These two lost fractions are thus definitely nucleohistones.

Table VI. *Number of molecules of arginine, lysine and histidine per cent atoms of nucleic phosphorus.*

		Number in the original calculated fraction (DNH or DNP) of the erythrocyte nucleus and of the spermatozoon.				Number in the deoxyribonucleoprotein (DNH or DNP) isolated from the same material.			
		arginine	lysine	histidine	total	arginine	lysine	histidine	total
Carp	Erythrocyte nucleus	33.2	64.2	5.0	102.4	25.8	35.2	5.0	66.0
	Spermatozoon	31.2	58.8	5.0	95.0	24.5	35.4	5.0	64.9
Pike	Erythrocyte nucleus	33.3	68.7	5.0	106.9	23.4	31.5	5.0	59.9
	Spermatozoon	68.7	26.2	2.5	97.4	49.6	14.1	2.5	66.2
Trout	Erythrocyte nucleus	33.9	63.0	5.0	101.9	24.3	32.6	5.0	61.9
	Spermatozoon	103.0	0.0	0.0	103.0	74.9	0.0	0.0	74.9

A. Ratio between the Numbers of Phosphate Groups in DNA and Basic Groups in Histones

When the amounts of arginine, lysine and histidine are expressed as molecules of basic amino acid per 100 atoms nucleic acid phosphorus, it is evident that all the phosphate groups of the DNA are not saturated by the basic amino acids in the DNH precipitated (Table VI); only 60 to 70% of total basic amino acid is found in our experience; BUTLER (1959) using special cares reaches a still higher degree of saturation: 85%. But in the original nucleohistone which may be calculated taking the lost fraction in account, the saturation is practically alway realised (VENDRELY et al. 1959, 1960). It is very likely that the labile fraction is really a part of the original nucleohistone or nucleoprotamine, but its attachement to the DNA is probably very loose or not permanent, so that it is readily separated from the nucleic acid moiety. This fact is perhaps significant in the mechanism of the possible biological action of these proteins.

B. Replacement of Histones by Protamines

Another point must be stressed here: in the basic nucleoprotein from spermatozoa, the saturation of the phosphate groups by the basic amino acids is also realised, but the relative proportion of basic amino acids is different in the different types of sperm: (cf. page 14) in the histone type there is about 1 molecule arginine for 3 atoms phosphorus; in the protamine type 3 for 3, and in the intermediate form 2 for 3. So that, passing from histone to protamine there is a decrease of lysine which is exactly compensated by an increase of arginine. It appears from these results that protamines are not simply derived from histones by the removal of a number of histone fractions: the sperm nucleus contains in its protamine a greater number of arginine residues for the same amount of phosphorus than does the somatic nucleus in its histone. If in fact protamine already exists as a fraction of the histone, the transformation of histone into protamine should include the elimination of more or less lysine-rich fractions and

Table VII.

DNP from	Trout	Pike	Carp
	Mol/100 atoms DNA P		
Gl	0	3.6	30.9
Asp	0	2.4	18.4
Th	0	2.2	22.2
Se	11.0	17.1	20.5
Pr	14.7	17.1	22.9
Al	2.4	10.9	39.6
V	8.3	7.4	19.5
M	0	2.5	3.0
Isol	1.1	1.2	15.8
L	0.4	2.8	29.4
Ty	0	0.8	10.3
Pa	0	0.8	8.4
Ly	0	14.0	42.0
A	74	49	26.0

their replacement by an arginine-rich fraction. But it has not been possible so far to isolate from a histone a fraction similar to a protamine.

This does not mean necessarily that the protamine molecule derives in fact from a complex modification of the histone molecule in the course of spermatogenesis. It is also possible that the histone is destroyed and completely replaced by a new molecule. [Ando and Hashimoto (1958 a, 1958 c)].

It is also possible to perform a complete analysis of the amino acids by the chromatographic technique of Stein and Moore on the whole nucleoprotein (the hydrolysis being performed on the nucleoprotein, the presence of DNA do not modify the analysis, except for glycine because the adenine of the nucleic acid is degradated into glycine which is thus added to the glycine of the protein). The analysis obtained by Vendrely on nucleohistones or nucleoprotamines give a figure quite similar to that obtained by other authors on separated histones or protamines concerning the relative ratio of each amino acid. The Table VII and the diagram (Fig. 4) shows the results obtained in the case of the nucleoproteins of the different types of sperms: carp sperm which contains a typical histone, trout sperm which contains a typical protamine and pike sperm which represent an intermediate case. The results show that the process of transformation of histone into protamine, which is total in trout and partial in pike, is characterized by a decrease of a number of amino acids which parallels the decrease of lysine. Proline, serine, valine and alanine are relatively preserved.

This suggests that in sperm, where the hereditary message is preserved but not expressed, the part of the histone which is involved in the genic

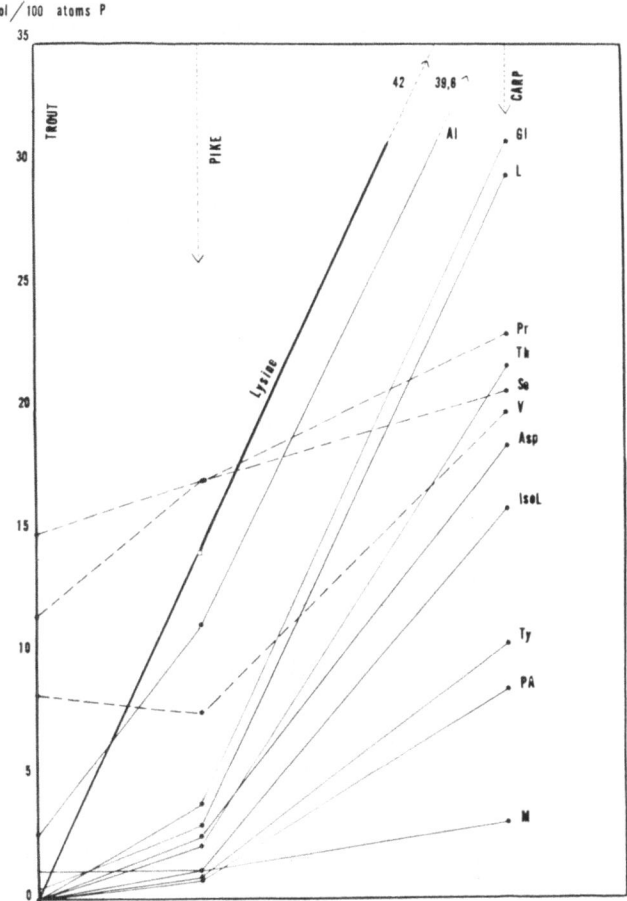

Fig. 4. Comparison of the amino acids ratio (Mol/100 atoms DNA-P) of carp-pike and trout sperm.

activity may be removed without trouble, since it is no longer necessary; arginine replaces the lysine for the saturation of the phosphate groups at these places. The part of the protein which is left on (which replaces the rest of the histone) may be particularly important in the preservation of the gene and in the structural problem of maintaining the DNA fibers together.

BLOCH's cytochemical results (1962) suggest that in the squid the histones which occur in the somatic cells and early spermatogenic cells are replaced successively by different basic proteins, first an arginine-rich histone and finally two protamines. Our biochemical works suggest that a similar step-wise process may exist in some species, stop at midway in others and does not occur in others, as shown in the Fig. 4. The moment of the change of histone into protamine or the replacement of histone by protamine in sperm cells is probably very late in spermatogenesis. Alfert (1956) showed by cytochemical methods that protamine appears at an advanced stage of spermiogenesis in the Salmon. R. Vendrely and C. Vendrely (1953), R. Vendrely et al. (1957), working on bull testis admitted that bull sperm contain a basic protein different from histone and show that this transformation must occur in nearly mature sperm. Das et al. (1964) showed that in *Drosophila melanogaster*, the replacement of lysine-rich histones by arginine-rich histones occurs during the maturation of the sperm rather than at the spermatid stage. Ando and Hashimoto's work (1958 a, 1958 c) on the changes of the proteins in the cell nuclei of the testis during the formation of spermatozoa in the Rainbow trout (*Salmon irideus*) indicate clearly that the basic protein (histone) multiplies with the DNA in the cell nuclei of the whole testis from August to October. Afterward, the multiplication of the cells and thus duplication of DNA is practically stopped and the basic protein changes in composition containing more and more arginine residues. All these results suggest that the changes of histone into protamine may occur at a late stage of spermatogenesis during sperm maturation.

Fractionation of Histones

The first data on the heterogeneity of histones seem to be due to Kossel and Dakin (1904) who studied the basic protein of carp sperm and isolated from it by acidic treatment two types of proteins, one with a high amount of arginine and a low lysine content and the other rich in lysine and very poor in arginine. The separation of these two components was realised by precipitation with hydroferrocyanic acid or sulphosalicylic acid which precipitate the arginine-rich fraction, the lysine-rich fraction being soluble. The authors did not know that carp sperm contains a typical histone and concluded to the presence of two types of protamines. There is no further approach to the problem of the fractionation before 1946, when Stern and Davis observed that it was possible to separate by ultracentrifugation a heavy fraction and a light fraction from a calf thymus histone, but the arginine ratio of both fractions was similar. Ahlström (1947) reported that in the course of dialysis, a great amount of histone passed through the cellophane membrane, the non-dialysable fraction being heterogeneous at electrophoresis and ultracentrifugation.

Systematic assays for the fractionation of histones begin with the investigations of the Stedmans (1950–1951) who showed that a main histone is present in preponderant amount, the sulphate of this main histone being less soluble than that of a subsidiary histone. The main histone is arginine-rich while the subsidiary histone is lysine-rich. These two fractions were characterized in a great number of histones from different animal species (ox, fowl, man, salmon) and in different organs (thymus, spleen, liver). Khouvine et al. (1953) found also that a histone obtained by acid extraction from tumour tissue of rat shows two peaks of slightly different electrophoretic mobility. Daly et al. (1951), Daly and Mirsky (1955) obtained by precipitation at pH 10.6 a arginine-rich histone from a sulfuric solution of histone and a lysine-rich histone through an extraction of nuclei in 0.05 M. citric acid containing 0.5 M. NaCl. This was done from calf thymus, beef liver and turtle erythrocytes.

In addition to these two components, a third one with an intermediate electrophoretic mobility was soon discovered; Davison et al. (1954 b) first distinguished it only at a pH superior to 6.5 and considered it as a mere product of aggregation of the two components separated at lower pH, however Gregoire and Limozin (1954) and Cruft (1953) considered it as a distinct component.

Table VIII. *The Nomenclature of Histone Fractions.*

Type or groups of histones		Nomenclature					
Lysine/arginine molar ratio	Description adopted						
less than 1	Arginine rich	" β		III, IV		f 3	E$_3$
Between 1 and 4	Slightly lysine rich	B γ {0.85, 1.65}	II	{II aa, II a, II b}	contaminants of II b similar to III	f 2 {f 2 (a), f 2 (b)}	E$_1$
Above 4	Very lysine rich	A α {α 1, α 2, α 3}	I	{I a, I b}		f 1	E$_2$
	Reference	(1) (2)	(3)	(4)		(5)	(5)

(1) Crampton et al. (1955). (2) Cruft et al. (1954, 1957 b). (3) Luck et al. (1958). (4) Rasmussen et al. (1962). 5 Johns et al. (1961).

Later, Johns et al. (1960) using carboxymethylcellulose columns and combining sodium acetate chloride and acid elution, obtained from whole thymus histone a very lysine-rich histone possessing a lysine/arginine molar ratio of 6–9/1, a slightly lysine-rich histone (ratio 1, 3/1) and a arginine-rich histone (ratio 0, 8/1). The slightly lysine-rich histone of Johns et al. agrees fairly well with the histone II reported by Luck et al. (1958), the analytical data being similar.

These three principal fractions are presently generally admitted. But the high resolving power of zone electrophoresis on starch gel allowed the separation of a much higher number of fractions in calf thymus histone (Cruft et al. 1957 a) and modern works tend to discover still more fractions. From theoretical point of view, the existence of a great number of histone fractions is important, since histones are supposed to act as gene regulators; this complex function implies a great number of different histones for specific repressions (Dulbecco 1964).

I. Nomenclature

The problem of the fractionation of histones has been so much developped and with so many different methods that it is rather difficult to understand clearly the nomenclature adopted by the different authors, each of them using a different symbol (letter or number) to designate each fraction they separate.

Phillips, in a recent review (1962) has very clearly put together in a table the nomenclature used by various authors trying to compare the terms concerning similar fraction. We reproduce this table with some complements necessitated by the recent development of these researches (Table VIII).

II. Principles of Fractionation

A. Fractionation by Precipitation

1. Fractionation using the differences of solubility in alcohol

The fractions containing more arginine (i.e. arginine-rich and slightly lysine-rich histones) are the less soluble. Cruft et al. (1957 b) used a 33% concentration of alcohol for the specific precipitation of these two fractions. Bijvoet (1957) separates from calf thymus histone by precipitation in 20% alcohol a component he calls I which represents 80% of the whole histone and is essentially an arginine-rich fraction. But this important component contains very probably also the slighly lysine-rich fraction. The component II which is lysine-rich is collected by freezing drying. Ui (1956, 1957) separates with 20% ethanol a fraction I which represents 8/10 of the total histone arginine-rich fraction) and a fraction II by 45% ethanol which represents 2/10 of the whole histone and is a lysine-rich histone with a molecular weight of 8400. He points out that his fractions are identical to the fractions obtained by Daly and Mirsky (1955) (see below).

2. Fractionation methods based on the insolubility in alkaline solution

The most basic histone fractions (arginine-rich and slightly lysine-rich) tend to precipitate first in an ammonia solution at pH 11. The precipitation of the residual fraction (lysine-rich) is then made easier by addition of alcohol or acetone. This procedure has been used by Daly and Mirsky (1955); Davison et al. (1954 b), Davison and Shooter (1956), Busch et al. (1959), Holbrook et al. (1960).

3. Fractionation using trichloracetic acid

Trichloracetic acid as a precipitant of the proteins is liable to produce a fractionated precipitation of the histones following the concentration used. In fact trichloroacetic acid at the final concentration of 5% precipitates 84% of the whole histone, the very-lysine rich histones remaining in the solution. Daly and Mirsky (1955), Luck et al. (1958), Satake et al. (1960), precipitated the arginine-rich and slightly lysine-rich fractions by TCA, 0.45 N. to 0.6 N. (7.5% to 10%); the lysine-rich fraction is precipitated at higher acidic concentrations: 0.8 to 1 N. (10 to 13% TCA).

4. Fractionation by different salt concentration

A more or less satisfactory fractionation may be realised by salting out: a fraction I (arginine-rich and slightly lysine-rich histone) precipitates at

60% saturation in NaCl, a fraction II (lysine-rich) remains soluble even in a saturated NaCl solution and precipitates in a saturated solution of ammonium sulfate (Uı 1957).

B. Fractionation by Differential Extraction

The most general procedure for the preparation of whole histone whatever the starting material (crude chromatin, isolated nuclei, deoxyribonucleoprotein) consists, as we have already noted, in an extraction by dilute acid, HCl or SO_4H_2. Citric acid was sometimes used: Davison and Butler (1954) extracted from calf thymus nucleoprotein a very lysine-rich fraction with N/5 citric acid; Daly and Mirsky (1955) used 0.05 M. citric acid in the presence of 0.5 M. NaCl.

As we have already seen, some authors investigated under what condition the acidic extraction of the histone was thoroughly realised. Butler et al. (1954) report that in the presence of a very dilute acid (N/10 SO_4H_2) only a fraction of the total histone was extracted, it was thus necessary to use N/4 SO_4H_2 or even N/2 SO_4H_2 and perform repeated extractions in order to extract the totality of the histone. They noted that the very lysine-rich histones are more readily removed from the nucleoprotein. We described in a proceeding chapter our results concerning the successive extraction of different fractions by solutions of increasing acidic concentration.

An other procedure of differential extraction was proposed by Johns et al. (1960). These authors studied the effect of varying the concentration of 10 to 98% ethanol (v/v) in the course of extraction of histones by 0.25 N. HCl from calf thymus. The arginine-rich histones were extracted at higher ethanol concentrations (70 to 80% Vol/Vol) and the lysine-rich histones at lower ethanol concentrations (10%). In all cases, the solvent used was 0.25 N. HCl. From these observations, they determined a procedure for a direct preparation of the arginine-rich histone by extraction of calf thymus with acid ethanol. The organ, first washed with a mixture of ethanol and water, is then treated with a solution of ethanol-HCl-water (70 : 25 : 5) and the material extracted is collected by dialysis against ethanol. The arginine-rich histones (representing 24% of the whole calf histone) is thus separated (the ratio lysine/arginine is 0.78).

Johns and Butler (1962 b) and Butler (1964) describe a procedure using 5% perchloric acid which extracts a lysine-rich fraction directly from calf thymus nucleoprotein. This fraction is then precipitated by raising the trichloroacetic acid concentration to 12%.

C. Fractionation by Electrophoresis

A great deal of work has been performed by this method. The technique of free electrophoresis has been first largely used. Cruft et al. (1954) have identified three electrophoretic components in the histone of calf thymus and ox liver and they found that the major component (or β histone) from these two tissues differ in their electrophoretic mobility. This seems to them to be an indication of the tissue specificity of the histones. In addition to that they found that the electrophoretic mobilities of histones of cancer cells

were smaller than those of normal cells. Butler et al. (1954), Davison et al.
(1954) characterized in calf thymus histone 2 components at pH 4.5 and
3 at pH 8.

Luck et al. (1956) working at different pH (4.5, 6.0, 7.7 and 8.6) observed
3 to 5 components in calf thymus histones. Lastly, Cruft et al. (1957 a,
1957 b) studying the histones from at least a dozen different sources showed
that most of them presented 3 electrophoretic groups at pH 6 to 7.6 which
they named respectively α, β and γ in decreasing order of mobility. The
histones they had extracted from mouse liver and spleen, rat kidney, rat
liver and rat hepatoma showed however only one or two of these groups.

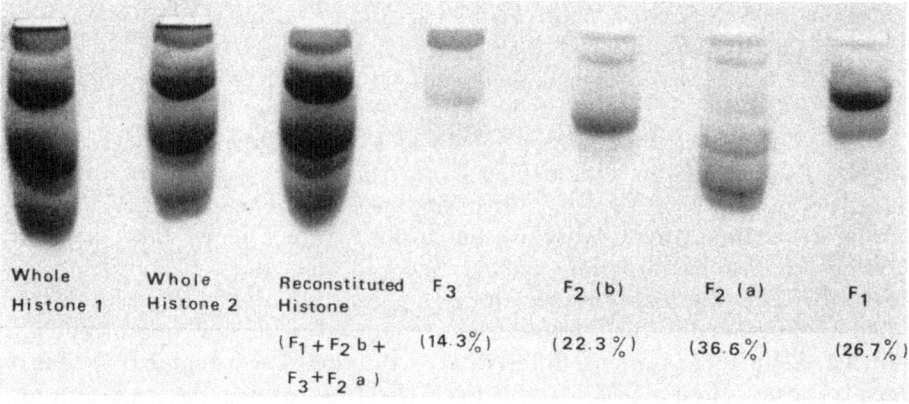

Whole   Whole   Reconstituted F_3 F_2 (b) F_2 (a) F_1
Histone 1   Histone 2   Histone

                                      $(F_1 + F_{2b} +$ (14.3%)   (22.3%)   (36.6%)   (26.7%)

                                      $F_3 + F_{2a})$

Fig. 5. Starch gel electrophoresis pictures of the whole calf thymus histone and histone fractions (R. Vendrely,
Y. Coirault, and M. Picaud, unpublished).

The method of free electrophoresis proved soon much less satisfactory
than the procedure using substrates such as filter paper at the origin (Luck
et al. 1956). Then starch gel was tried and this procedure was soon generally
adopted and has yielded the most interesting results (Cruft et al. 1957 a,
Neelin and Connell 1959). Neelin and Neelin (1960) have succeeded in
characterizing as many as 22 zones for calf thymus histones. Johns et al.
(1960, 1961) performed their electrophoresis on starch gel at pH 2.3 and
obtained from calf thymus histones 10 bands they classified into three
groups: E_1 which is the fastest and contains 5 bands, E_2 intermediate with
3 bands and E_3 the slowest with 2 bands. They showed then by fractionation
on carboxymethylcellulose colums that E_2 corresponded to the very lysine-
rich histone fraction, E_1 to the slightly lysine-rich histone fraction and E_3
to the arginine-rich histone fraction.

Fig. 5 shows starch gel electrophoresis pictures of the whole thymus
histone, of some fractions (F_1, F_{2b}, F_3, F_{2a}) prepared in our laboratory by
the method II of Johns (1964 b) and of the whole histone reconstituted by
recombination of $F_1 + F_{2b} + F_3 + F_{2a}$.

A new substrate was recently introduced for zone electrophoresis of
histones, the polyacrylamide gels (McAllister et al. 1963) which provide a
resolution of the components which is still superior to that obtained with
starch gel, the reproducibility of the results being excellent. Cruft (1964)
used it for the study of calf thymus histone and of its fractions called

a, β and γ. He obtains 9 bands with the whole histone. IRWIN et al. (1963) obtained also a more complex electrophoretic pattern with this gel.

D. Fractionation by Chromatography on Ion Exchange Resins

The first authors who used these methods were CRAMPTON et al. (1955). They chose the resin Amberlite IRC-50 for the fractionation of calf thymus histones. Though they avoided a strong linkage of the material with the resin by using the Ca or Ba form of the resin instead of the Na form and making an elution with calcium and baryum acetate buffers (pH 6.2 or 6.7) of increasing ionic strength, they were not able to obtain more than about 60% of the histone as the sum of 5 peaks.

LUCK et al. (1958) and SATAKE et al. (1960) have been able to realise a nearly complete elution of the sodium form Amberlite column using a guanidinium chloride gradient of increasing concentration (from 7 to 40%). They obtained 4 principal peaks more or less complex. (I, II, III, IV) the first one corresponds to the very lysine-rich histone the others (II, III, IV) corresponding to the slightly lysine-rich histone and the arginine-rich histone. MURRAY (1964 b) also used Amberlite IRC-50 and compared the relative efficiency of guanidinium chloride solution and baryum acetate solution for the elution of the histone fractions. He obtained with guanidinium chloride a better fractionation and assumed that the fractions thus obtained are authentic. NEELIN (1964) used Amberlite for the study of chicken erythrocyte nuclei histones, and performed the elution with increasing concentration gradient of guanidinium chloride buffered in 0.1 M. sodium phosphate. He detects but traces of zones of intermediate mobilities in erythrocytes histones compared to spleen histones and other organs. He found two major components of relatively slow mobilities, instead of a single zone exhibited by extracts from spleen and other tissues.

DAVISON and SHOOTER (1956) were the first authors who used carboxymethylcellulose columns for the study of calf thymus histones. DAVISON (1957 a) describes the working conditions with this substance. The histones are eluted with solution of increasing ionic strengths. With this method he studied fractions of histones soluble at pH 3.0 to 6.5 obtained from nuclei of normal cells in mice, EHRLICH ascites tumours and LANDSCHÜTZ ascites tumours. He also noted that the acidic extracts of nuclear material from calf thymus and other tissues contained non-basic protein contaminants. PHILLIPS and JOHNS (1959 b), in order to reduce the aggregation effects, eluted their carboxymethylcellulose column with diluted acid.

JOHNS et al. (1960) perform their elution first with pH 4.2 sodium acetate buffer for the removal of contaminants then a sodium acetate, sodium chloride buffer pH 4.2 to obtain the F_1 fraction which separates chromatographically into two peaks F_{1a} and F_{1b} (very lysine-rich histone: lysine/arginine 6 to 9) then, with 0.01 N. HCl they obtain F_2 (the slightly lysine-rich, $L/A = 1.3$) and at last 0.02 N. HCl gives F_3 (arginine-rich histone $L/A = 0.8$). The 3 fractions F_1, F_2 and F_3 are approximately in the ratio 20, 50 and 20. JOHNS (1964 a), studying more particularly the lysine-rich histone of calf thymus extracted with 5% perchloric acid, fractionated it on carboxymethylcellulose column at pH 9.0 and eluted it with NaCl gradient from 0 to 1.0 M. He obtained 3 peaks (1, 2 and 3), the third of them being particularly rich in lysine. Another procedure is also indicated by this author using acetone for the fractionation; the yield is better. He obtains thus 3 fractions, A, B and C (A and B corresponding to 3 in the precedent method and C to 1 and 2).

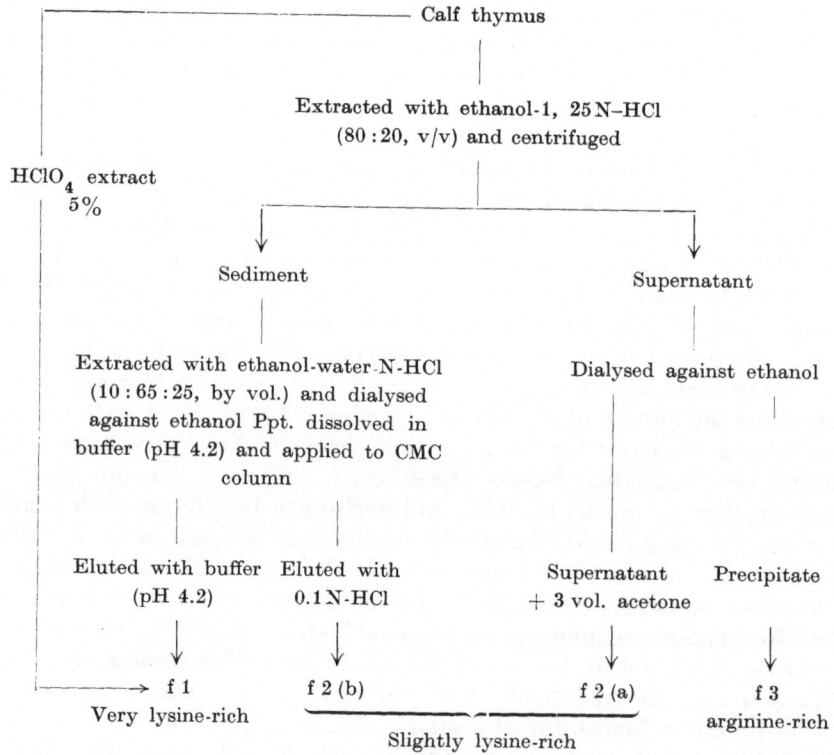

Fig. 6. Summary of method used for the extraction of histone fractions from calf thymus (JOHNS and BUTLER 1962).

CRUFT (1964) used G 75 Sephadex column, the effluent being 0.02 N. HCl. He obtained 4 peaks: 1 (α lysine-rich histone), 2 (β arginine-rich histone), 3 and 4 (γ histone). Due to the use of the Sephadex it is possible to obtain from these experiments some data on the molecular weight of the fractions (more than 40,000 particularly for α histone).

III. General Methods for Obtaining Various Fractions of Histones

We have quoted above in the chapter of fractionation by differencial extraction, the various simple procedures for obtaining directly either the arginine-rich fractions (JOHNS et al. 1960) or the lysine-rich fraction (JOHNS and BUTLER 1962 b); all the details are given in the original publications.

Concerning the separation of the 4 principal fraction as defined by BUTLER and his school, namely F_1 (lysine-rich histone), F_2 (a) and F_2 (b) (slightly lysine-rich histones) and F_3 (arginine-rich histone) the most generally employed method is that indicated by Johns and Butler (1962 b). It is summarized in the following figure (Fig. 6) reproduced from the original publication.

The same procedure slightly modified in the sense of a simplification is described in a more recent paper by Butler (1964) and by Johns (1964 b) (cf. method I, Fig. 7). In a first step, the very lysine-rich histone (F_1) is extracted with 5% perchloric acid which avoids the later use of carboxy-methylcellulose column for the separation of F_1 and F_2 (b).

Method I

Calf thymus (100 g.)

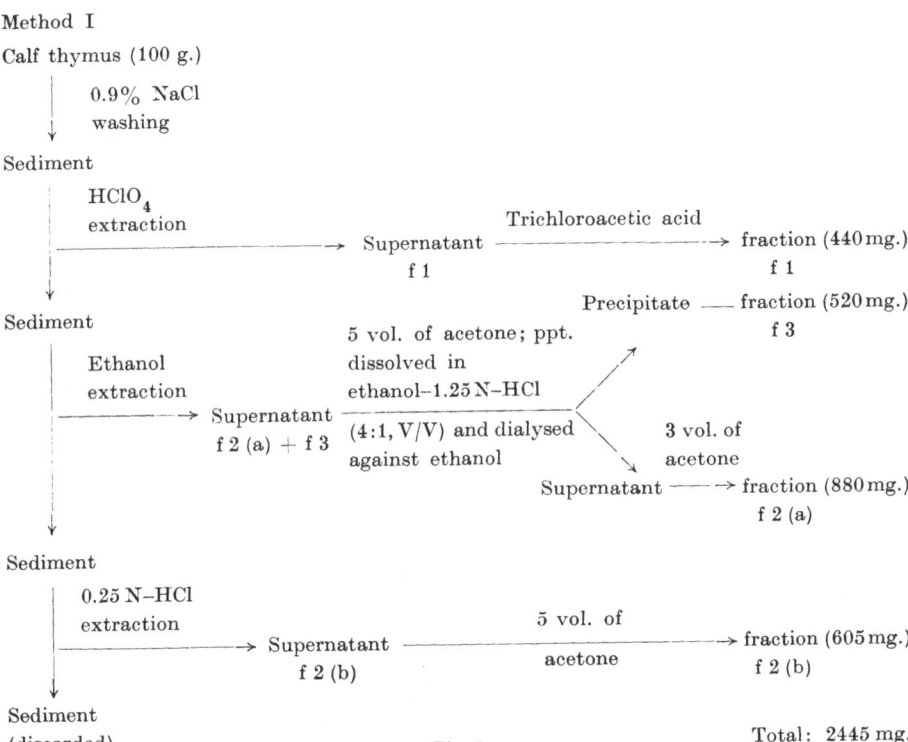

Fig. 7

Total: 2445 mg.

Recently Johns (1964 b) proposed two methods of preparation each of which enables the four main groups of histones F_1, F_2 (a), F_2 (b) and F_3 to be isolated from calf thymus. The combined yields of the four groups amounted to 2.3–2.4 g./100 g. of calf thymus.

These methods are summarized in the following figures (Figs. 7 and 8). The details are given in the publication.

In a recent work, Hnilica (1965) studies the histone F_2 (a) and F_3 obtained by extraction with mixture of absolute ethanol and 1.25 N. HCl (according to the method of Johns et al. 1960), followed by a precipitation in 6 volumes acetone. In contrast with Johns and Butler (1962 b) he considers the fraction F_2 (a) (cf. nomenclature) as arginine-rich histone. He was able to separate F_2 (a) and F_3 by filtration on Sephadex G 75 (elution with 0.01 N. HCl saturated with chloroform). By dialysis against a mixture of n-propanol and ethanol he fractionates the F_2 (a) histones into two components: the

F₂ (a) I which is electrophoretically faster and the F₂ (a) II with slightly lower electrophoretic mobility in starch gel. These two fractions differ significantly in their content of several amino acids (e. g. arginine, lysine, alanine, glycine, etc.).

Similarly, PHILLIPS and JOHNS (1965) have separated the calf thymus histone group F₂ₐ into two subfractions by stepwise precipitation with

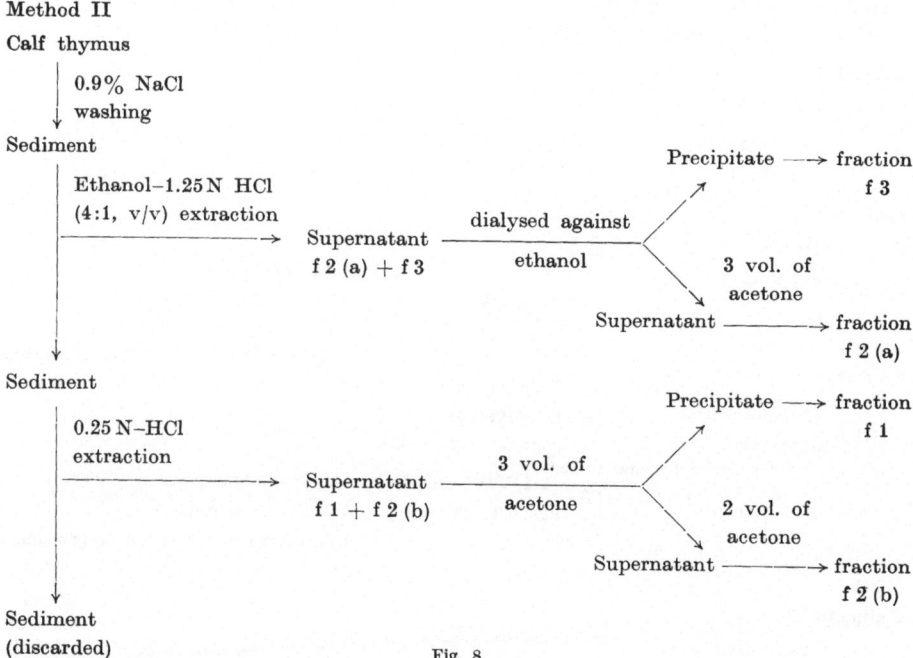

Fig. 8

acetone from acid solution or carboxymethylcellulose and dextran gel column chromatography. The two subfractions differ significantly in their amino acid composition and in the pattern of peptides obtained by tryptic digestion. Both fractions have a very low content of N-terminal amino acids and contain acetyl groups.

SAUTIÈRE et al. (1965), working on the fraction F₃ obtained by the method II of JOHNS (1964 b) have separated by chromatography on carboxymethyl-cellulose two subfractions F₃ CMC I et F₃ CMC II. The last one was then fractionated on Sephadex G 100 and yielded 3 fractions F₃ CMC II SG 100 I, F₃ CMC II SG 100 II, F₃ CMC II SG 100 III. F₃ CMC II SG 100 III was particularly arginine-rich.

IV. Composition of Various Fractions of Histone

In the table (Table IX/X) are reported a number of results chosen from the most recent data on the chemical composition of the three principal fractions of calf thymus histone. A fairly good agreement may be noted between these data from different authors obtained from samples prepared by different methods.

Table IX. *Composition of histone fractions prepared from calf thymus. (Amino acids expressed as moles/100 moles of all amino acids.)*

I—Amino acid analyses of the very lysine-rich histones

Amino acids	A	α 1	α 2	α 3	I	I a	I b	f 1	E 2
Aspartic acid	2.0	0.3	2.4	4.4	1.7	2.5	2.5	3.4	3.5
Glutamic acid	3.2	0.6	3.9	5.1	3.5	4.5	4.3	5.6	5.6
Glycine	6.7	3.7	9.0	10.6	7.0	6.7	7.3	6.4	6.8
Alanine	25.7	31.6	22.8	18.4	25.6	24.0	24.2	23.6	20.5
Valine	4.8	3.1	4.7	5.9	5.8	4.9	4.1	} 9.8	4.8
Leucine	4.15	0.6	4.4	7.5	} 4.8	5.3	5.0		} 7.2
Isoleucine	0.9	0.0	1.0	2.0		1.3	1.2		
Serine	6.6	2.6	5.0	6.9	6.8	6.4	6.5	5.4	6.6
Threonine	5.6	4.3	5.2	4.4	6.0	5.8	5.4	5.1	6.0
Phenylalanine	0.5	0.0	0.5	1.2	0.7	0.6	0.6	1.0	0.9
Tyrosine	0.5	0.0	0.6	1,7	0.5	0.7	0.7	0.2	1.2
Proline	9.5	14.5	9.8	6.6	8.6	8.6	9.1	9.7	8.1
Histidine	0.0	0.0	0.3	1.2	0.07	0.4	0.2	0.3	1.5
Lysine	28.2	38.7	26.2	19.2	27.3	25.3	26.2	27.5	23.1
Arginine	1.8	0.0	4.2	4.4	1.6	3.0	2.6	2.0	4.1
Lysine/Arginine	15.7	∞	6.2	4.4	17.0	8.4	10.1	13.7	5.7
Reference	(1)	(2)	(2)	(2)	(3)	(4)	(4)	(5)	(6)

II—Amino acid analyses of the slightly lysine-rich histones

Amino acids	B	0.8 S γ	1.6 S γ	II	II b	f 2 (a)	f 2 (b)	E 1
Aspartic acid	5.0	5.2	6.2	5.9	5.6	5.8	5.5	6.9
Glutamic acid	8.9	6.8	9.1	9.5	8.7	8.3	9.3	7.4
Glycine	9.3	9.9	8.6	9.7	8.2	12.5	7.4	9.7
Alanine	11.3	18.0	12.2	11.3	11.5	10.0	10.6	11.5
Valine	6.9	6.1	6.5	8.2	6.7	7.7	7.2	6.9
Leucine	8.6	7.9	9.4	} 13.2	8.6	} 14.1	11.2	} 11.6
Isoleucine	4.9	2.1	4.5		4.5			
Serine	5.2	7.5	6.7	4.5	7.0	3.9	7.4	5.9
Threonine	5.8	4.5	5.0	6.6	5.2	5.1	5.7	5.9
Phenylalanine	1.8	1.1	1.0	2.0	1.3	2.0	1.9	1.7
Tyrosine	2.8	1.8	3.2	2.3	3.0	2.7	3.1	2.3
Proline	4.0	4.9	3.7	3.8	4.7	3.0	5.0	4.3
Histidine	2.2	0.6	2.6	3.0	2.8	2.5	2.1	3.0
Lysine	11.8	17.5	11.4	11.7	13.5	11.2	15.8	13.3
Arginine	10.4	5.5	8.6	7.8	7.9	11.2	7.8	9.7
Lysine/Arginine	1.13	3.2	1.32	1.50	1.71	1.0	2.0	1.38
Reference	(1)	(2)	(2)	(3)	(4)	(5)	(5)	(6)

In the next table (Table X) taken from HNILICA et al. (1962) are put together the analytical results concerning, this time, the composition of histone fractions of the same group isolated from different tissues and species.

III—Amino acid analyses of the arginine-rich histones

Amino acids	β	III	IV	f 3	E 3
Aspartic acid	5.8	4.4	4.5	5.0	} 19.2
Glutamic acid	9.7	9.9	10.6	11.2	
Glycine	8.5	8.8	7.9	6.5	6.4
Alanine	11.6	11.8	12.3	12.6	12.5
Valine	5.9	5.8	5.6	6.4	5.3
Leucine	10.1	8.7	8.9	} 13.9	} 10.0
Isoleucine	5.0	5.4	5.4		
Serine	3.7	4.1	4.1	4.7	6.8
Threonine	5.9	7.4	7.4	6.4	6.1
Phenylalanine	2.0	2.5	2.7	2.6	3.2
Tyrosine	2.8	2.4	2.3	2.3	—
Proline	3.7	3.8	4.2	4.8	5.3
Histidine	2.3	1.6	1.6	2.4	4.3
Lysine	9.2	9.4	9.0	9.8	11.0
Arginine	12.4	12.9	12.8	11.1	10.0
Lysine/Arginine	0.74	0.72	0.70	0.88	1.1
Reference	(2)	(4)	(4)	(5)	(6)

(1) CRAMPTON et al. 1957. (2) CRUFT et al. 1957 a. (3) LUCK et al. 1958. (4) RASMUSSEN et al. 1962. (5) JOHNS 1964 b. (6) JOHNS et al. 1961.

V. The N-Terminal Amino Acids

Since the initial work of LUCK et al. (1956), PHILLIPS (1957), BISERTE and SAUTIÈRE (1958), the study of the determination of terminal groups in histones has developed remarkably. These works developed both on unfractionated histones and on various fractions separated from histones.

Concerning the whole histone, the first results by LUCK et al. (1956), which indicated as principal N-terminal groups valine, leucine, alanine and glycine were not confirmed by other authors. PHILLIPS (1957, 1958) reported, on the contrary that proline accounts for 40 to 45% and alanine for 35 to 40%. LUCK et al. (1958) agreed afterwards with these results. As the methods of preparation improved, the terminal groups observed became less numerous. Now it seems that 90% of N-terminal groups are constituted by proline and alanine (PHILLIPS and JOHNS 1959 a). It is necessary to point out that, if the histone preparation is not kept in an acidic medium, it seems that a proteolysis does occur with a progressive liberation of N-terminal groups (alanine, lysine, glycine). Recently, LAURENCE et al. (1963) have compared the whole histone prepared from calf thymus and calf liver nucleoproteins, from Crocker tumour and from a spontaneous mammary tumour. In every case, proline and alanine represent the chief N-terminal amino acids and their sum varies from 61% (liver) to 82% (mammary tumour). The N-terminal equivalent weight is about 30,000.

The N-terminal groups of different histone fractions are now also carefully investigated, which shows certain peculiarities of the 4 principal fractions F₁, F₂ (a), F₂ (b) and F₃. PHILLIPS in 1957 found

TableX. *The amino acid analysis of histone fractions isolated from different sources.*

Origine of histone fraction / Amino acid	Composition of very lysine-rich histones								Composition of slightly lysine-rich histones							
	Calf thymus	Calf liver	Calf spleen	Rat liver	Leukaemic rat liver	Rat spleen	Leukaemic rat spleen	Ehrlich ascites	Calf thymus	Calf liver	Calf spleen	Rat liver	Leukaemic rat liver	Rat spleen	Leukaemic rat spleen	Ehrlich ascites
Aspartic acid ⎫ Acidic	2.9	2.6	2.6	2.5	2.5	2.7	2.4	2.4	5.9	5.7	5.6	6.4	5.2	5.2	5.3	5.8
Glutamic acid ⎭	3.9	3.9	3.8	3.7	3.6	3.6	3.2	3.8	8.0	7.2	7.7	7.1	6.8	6.9	6.8	8.0
Glycine	6.2	6.3	6.8	6.8	6.5	6.7	6.4	7.0	10.4	11.0	10.0	11.1	10.5	10.9	11.0	10.2
Alanine	24.7	24.9	25.0	24.8	25.0	24.6	24.8	24.9	12.0	11.3	12.4	11.5	11.6	11.6	11.8	11.7
Valine	5.6	5.6	5.4	5.2	5.6	5.5	5.6	5.4	7.6	7.9	8.0	7.7	7.6	7.7	8.0	7.6
Leucines	5.8	5.4	5.4	5.1	5.0	5.4	5.4	5.1	13.4	13.1	12.6	14.0	13.0	13.1	13.9	14.0
Phenylalanine	0.8	0.8	0.8	0.8	0.9	0.9	0.8	0.8	1.8	1.8	2.0	1.5	1.8	1.9	2.0	1.9
Tyrosine	0.2	0.3	0.3	0.3	0.4	0.3	0.5	0.2	2.1	2.0	2.0	1.8	2.0	2.0	1.9	1.9
Serine	5.7	5.7	5.8	5.9	5.8	6.0	5.7	5.8	5.4	5.2	5.7	5.4	5.8	5.7	5.1	5.4
Threonine	5.6	5.9	5.9	5.8	6.2	5.9	6.0	5.7	5.4	5.5	5.5	5.4	5.8	5.8	5.5	5.4
Proline	9.0	8.4	8.4	9.0	8.8	8.7	9.0	8.7	3.6	3.3	3.9	3.5	3.9	3.9	3.6	4.0
Histidine ⎫ Basic	0.4	0.4	0.3	0.4	0.4	0.3	0.4	0.4	2.2	1.8	1.7	2.0	2.0	1.8	2.1	1.6
Lysine	26.6	26.9	27.0	26.9	26.6	26.7	27.0	27.0	12.7	13.8	13.4	12.9	13.4	13.7	12.8	12.6
Arginine ⎭	2.6	2.9	2.6	2.8	2.7	2.7	2.8	2.8	9.5	10.4	9.5	9.7	10.5	9.8	10.2	9.9
Lysine: arginine	10.23	9.27	10.38	9.60	9.85	9.88	9.64	9.64	1.33	1.33	1.41	1.32	1.28	1.40	1.25	1.27
Basic: acidic	4.35	4.64	4.67	4.85	4.86	4.71	5.29	4.87	1.75	2.01	1.85	1.82	2.16	2.09	2.07	1.75

Table X (Continued)

Composition of arginine-rich histones

Amino acid		Calf thymus	Calf liver	Calf spleen	Ox liver[1]	Ox spleen[1]	Rat liver	Leukaemic rat liver	Rat spleen	Leukaemic rat spleen	Ehrlich ascites	Fowl erythrocytes[2]	Fowl spleen[2]	Fowl liver[2]
Aspartic acid	Acidic	6.3	5.3	4.9	5.5	5.8	5.2	5.6	5.8	5.4	5.6	5.6	5.9	5.5
Glutamic acid	Acidic	8.4	8.8	8.4	9.5	9.4	8.4	8.1	7.9	7.8	8.7	9.2	9.2	9.0
Glycine		6.6	7.2	7.0	9.5	8.7	6.9	6.7	6.9	6.9	8.6	9.2	9.4	9.2
Alanine		13.4	12.2	12.7	11.0	10.5	12.3	12.6	13.2	12.5	11.3	12.2	11.1	10.7
Valine		6.1	5.9	5.7	6.1	6.4	5.9	6.1	6.8	6.6	6.2	5.8	6.3	6.0
Leucines		13.4	13.5	13.2	14.9	15.6	13.6	13.2	13.5	13.3	13.2	10.2	9.8	9.5
Phenylalanine		2.9	2.9	3.0	2.6	2.7	3.0	2.8	2.6	2.6	2.4	2.2	2.6	2.8
Tyrosine		1.6	2.0	1.8	3.0	3.3	1.8	1.8	1.8	1.5	2.1	2.6	3.0	2.9
Serine		4.9	4.8	4.7	4.2	4.2	4.9	5.1	4.9	5.1	5.2	5.0	4.5	4.8
Threonine		6.1	6.9	6.9	6.0	5.9	7.2	6.5	6.3	6.8	5.9	5.5	6.1	6.3
Proline		4.9	4.5	4.6	4.2	4.5	4.7	5.2	4.8	5.0	4.1	3.3	3.2	3.8
Histidine		2.2	2.6	2.6	2.0	1.8	2.6	2.4	2.4	2.5	2.6	2.0	1.9	2.1
Lysine	Basic	10.2	10.0	10.6	8.8	8.1	10.0	10.4	9.5	10.0	12.8	9.5	8.6	9.4
Arginine	Basic	13.1	13.4	13.9	12.0	11.7	13.5	13.5	13.8	14.0	11.3	11.5	12.1	11.8
Lysine: arginine		0.78	0.75	0.76	0.74	0.69	0.74	0.77	0.69	0.71	1.13	0.82	0.71	0.79
Basic: acidic		1.73	1.84	2.04	1.53	1.42	1.92	1.92	1.87	2.00	1.88	1.55	1.50	1.60

[1] β histones (arginine-rich) described by Mauritzen and Stedman, 1960. [2] β histones (Mauritzen and Stedman, 1959).

that the proline N-terminal group is associated chiefly with the moderately lysine-rich histone, whereas the alanine N-terminal group is associated chiefly with the arginine-rich group. SATAKE et al. (1960) have not identified any N-terminal residue in their lysine-rich fractions (I a, I b) of calf thymus histone obtained by chromatography on Amberlite IRC-50. In slightly lysine-rich fractions II aa and II a, they find alanine (86 and 78% of all N-terminal groups respectively) in II b fraction, proline (97%) and in the arginine-rich fractions, alanine for fraction III (74%) and alanine also

Table XI. *N-terminal groups (molar percentages of all groups found).*

Slightly lysine-rich histones (F_2)

N-terminal groups	Calf thymus	Calf liver	Calf spleen	Rat liver	Rat spleen	Ehrlich Ascites
Alanine	9	8		11	7	
Proline	74	76		73	77	
Others	17	16		16	16	
Wt. (g./mole of N-terminal groups	29.400	32.200		28.200	26.000	

Arginine-rich histones (F_3)

N-terminal groups	Calf thymus	Calf liver	Calf spleen	Rat liver	Rat spleen	Ehrlich Ascites
Alanine	95	84	83	81	86	56
Proline	2	1	2	2	1	32
Others	3	15	15	17	13	12
Wt. (g./mole of N-terminal groups	18.800	13.500	12.000	14.000	17.000	22.700

for fraction IV (82%). JOHNS et al. (1960) studied the N-terminal groups of histone fractions obtained by progressive extraction by means of 0.25 N. HCl with decreasing concentrations of ethanol. They found that the fractions extracted at higher ethanol concentrations (arginine-rich histones) have alanine as predominant N-terminal group. As the ethanol concentration decreases, the fractions obtained (more and more lysine-rich histones) have chiefly proline for their N-terminal groups. HNILICA et al. (1962) studying the slightly lysine-rich histone F_2 and the arginine-rich histone F_3 of various organs in calf and rat and in a tumour, do not find any marked difference in N-terminal groups (Table XI).

LAURENCE et al. (1963) compared the fractions F_1, F_2 and F_3 of histones of spontaneous mouse tumour and of calf thymus; they found very similar results in every case. BUSCH (1962), BUSCH et al. (1963 b) studied the fraction $RP_2 L$ which they considered as characteristic for tumours (see page 63). This fraction seems to be composed of a mixture of the components of fractions F_2 (a) and F_2 (b) (slightly lysine-rich histones); proline (51.5%)

and alanine (29.7%) are the principal N-terminal amino acids of one fraction (F₂ b), the other fraction (F₂ a) has a very low amount of N-terminal amino acids.

Phillips (1963) and Johns (1964 b) point out the low yield of N-terminal amino acids in the fractions F₁ and F₂ (a) due, very likely, to the presence of N-terminal acetyl groups. In the fractions F₂ (b) the predominant N-terminal is in fact proline, in fraction F₃ it is alanine. In the fractions F₁ and F₂ (a) acetyl groups have been found amounting to about one residue in 15,000–18,000 g. (F₁) and 10,000–14,000 g. of protein F₂ (a); F₂ (b), F₃ also contain some acetyl groups amounting to one residue in 32,000–38,000 and 26,000–28,000 g. of protein, respectively.

In the actual state of the investigation, it is possible to conclude that the main N-terminal groups for the 4 fractions are acetyl for F₁ and F₂ (a), proline for F₂ (b) and alanine for F₃.

Protamines:

It is interesting to compare the data obtained on histones with the results obtained on protamines.

Sanger (1945) characterized proline at the NH₂-terminal ends of clupeine from the herring (*Clupea harengus*). Hashimoto (1955) and Ando et al. (1957 a, 1958 b) identified in clupeine of the Japanese herring (*Clupea pallasii*) alanine and proline as terminal residues. Phillips (1955) found that proline represents 4/5 of the terminal amino groups of the salmine; the remaining ones were mainly arginine.

Salmine, Truttine, Iridine and Fontinine have only one N-terminal amino acid, namely proline (Velik and Udenfriend 1951, Felix and Krekels 1953, Ando et al. 1958 a, 1958 b). Finally sturine yields two NH₂-terminal amino acids; alanine and glutamic acid.

Ando and Sawada (1962) compared the Clupeines (*Clupea pallasii*) isolated from single fishes and found in each case the same chromatogram on alumina; these chromatograms are heterogeneous and reveal 3 fractions: W (4%), Y (73%) and Z (23%) and as N termini of mixed molecules in each clupeine specimen: proline (40 to 42%) and alanine (42 to 50%) always in the same ratio. Individual and race difference of clupeine were not detected so far.

In conclusion, typical protamines though markedly different from arginine-rich and slightly lysine-rich histones in their general amino acid composition, are comparable to them concerning the N-terminal groups.

VI. Studies on Peptides from Whole Histone and Histone Fractions

The most promising approach to studies of the differences between histones from various species and cell types and the heterogeneity of histone fractions lies in detailed quantitative works on the peptides furnished by various hydrolytic procedures. By this means, it is possible to obtain, on the other hand, some indications on the arrangement of basic amino acids along the polypeptide chains of the histone (spacing) and some information of the various histone fractions combined with DNA.

A. Repartition of the Basic Amino Acids along the Polypeptide Chains of the Histone

This problem was approached by the study of peptides produced by tryptic digestion of the 4 major fractions of calf thymus histone. It is well known that trypsin frees the carboxyl groups of lysyl and arginyl residues in the peptide chain, though such cleavage is inhibited where the next residue is prolyl, aspartyl or glutamyl. PHILLIPS and SIMSON (1961) studied the spacing (number of residues) between the basic amino acids with tryptic digests of an arginine-rich histone from calf thymus having 1 mole of N-terminal groups per 18,000 weight, of which 95 per cent was alanine. During the proteolysis a precipitate formed amounting 25 to 30% of the histone (molecular weight 3000, 5000, one basic amino acid for every seven other residues). The supernatant was examined by paper electrophoresis and chromatography. In this fraction, basic amino acids occur in pairs in at least eight positions. The result on the arginine-rich histone show, that the spacing between basic amino acids can vary from 0 (juxtaposition) to at least seven residues: PHILLIPS and SIMSON conclude that parts of the deoxyribonucleohistone may have loops of non basic acids lying in a groove of the DNA double helix or, with the longer loops, forming bridges between separate double helix spacing sequence.

After tryptic digestion, PHILLIPS (1964) observed that all the histone fractions, except the lysine-rich fraction, give a precipitate a "core" representing 10 to 27% of the material, the average spacing between basic residue in them, as calculated from the basic amino acid content, is about 6 other residues compared with an average value of 3 for the parent histones before degradation. It can be deduced from the smaller soluble peptides identified in fraction F_3 after tryptic digestion that the spacing between basic amino acids varies from 0 (juxtaposition) to 4 non-basic residues. Including the "core", the spacing varies from 0 to 7 non-basic residues. For F_2 (a), the spacings between basic amino acids vary from 0 to 8 non-basic residues. The lysine-rich histone F_1 does not give an insoluble core with trypsin. The pattern of the tryptic peptides from the lysine-rich histone shows that there are several peptides bearing more than one lysyl residue and one third of the peptide N-terminal groups are lysine. This implies that in 9800 weight (100 residues) there are not less than 7 lysyl-lysyl sequences equivalent to about 50% of all the lysine residues.

In all the calf thymus histone there is thus a marked irregularity in the spacing of the basic amino acid residues and we may already admit that it is the same in histones from several other sources. Thus a salt linkage structure for nucleohistone would be irregular with loops of various lengths of non basic amino acids.

SATAKE et al. (1960) have studied the repartition of the basic groups in the whole histone and in 7 histone fractions (1 a, 1 b, II aa, II a, II b, III, IV) resolved by column chromatography on Amberlite IRC-50. The whole histone mixture and the different fractions were hydrolysed by Streptomyces griseus proteinase. The peptides of the hydrolysates were separated on Amberlite IRC-50, analysed for their amino acid composition,

NH₂-terminal residues; the yields of arginine and arginine peptides derived from whole histone sulfate and from histone fractions were determined. They noted the occurence of the peptides Arg. Arg, Arg. Lys, and Arg. Arg. Arg. in the whole histone. Fraction II b, the principal component of the arginine-rich histones is nearly homogeneous, the minor components of the arginine-rich histones (at least four in number) are still heterogeneous. At last, the lysine-rich histone fraction (I) consists of at least three components, each of which contains 1 and 2 arginine residues per mol.

B. The Problem of the Histone Specificity

In order to compare histones of different origins, CRAMPTON, STEIN and MOORE (1957) performed a chromatographic study on Dowex columns of the peptides liberated by the action of trypsin. The tryptic hydrolysates from samples of fractions B (cf. nomenclature) from different organs of calf (thymus, liver, kidney) and guinea pig (testis), chromatographed on Dowex columns, yielded similar effluent curves (evidence for about 35 chromatographically different substances). The compositions of the peptides responsible for the different peaks have not been determined. They conclude that the similarities among the overall patterns afforded by the four hydrolysates must reflect similarities in the way in which the amino acids are arranged in the original proteins. Nevertheless some differences may be noted; they are perhaps due to alterations appearing in the course of isolation. MURRAY (1964 b) has also examined the heterogeneity of the histone fractions doing analysis of the peptides obtained by tryptic digestion (their number, the relative quantities in which they are found and their amino acid composition). Fraction I a, II a, and IV were thus studied. I a, by the fingerprinting procedure shows 51 peptides while by ion exchange chromatography only 26 components were obtained, but among them 15 only were homogeneous and the others were heterogeneous when examined by the fingerprinting method, paper chromatography or ionophoresis. Fractions II a and IV appeared also heterogeneous and their fingerprints showed 80 and 60 peptides for the respective fractions. MURRAY estimates than in a preparation of histones the number of different histones may be very large.

NEELIN (1964) has studied the fraction 5 of chicken erythrocyte histone obtained by chromatography on Amberlite IRC-50 that would represent a major histone component not found in other nuclei of the same animal. The complex fingerprint of the tryptic peptides of this fraction 5 in no way resembles the fingerprint of chicken globin and also differs significantly from those of all calf thymus histone fractions.

BUSCH and MAVIOGLU (1964) have studied their RP 2-L fraction which they considered as specific of tumours and analysed more particularly one of the components of this fraction, namely the fraction 2 a, which is a slightly lysine-rich fraction. The peptides from fraction 2 a were purified by fingerprinting and analysed. Significant differences were found in the peptides of the Walker tumour and calf thymus: for instance, arginine is present in the Walker tumour peptide n⁰ 7 and is missing from the same peptide in calf thymus; two lysines occur in the calf thymus seventh peptide and only one in the corresponding peptide from Walker tumour. HNILICA et al. (1963, 1964) have already compared the fractions 2 b isolated from calf thymus, rat thymus, rat spleen and Walker tumour, studying the peptides obtained by tryptic hydrolysis. All the fingerprints (obtained by a combination of paper electrophoresis and chromatography) showed 22 major ninhydrin positive spots. The amino acid composition of the corresponding peptides (n⁰ 1, 2, 3, 4, 14, 15, 16) from calf thymus and from Walker tumour were very similar, some differences were found for n⁰ 5 and n⁰ 10 but other analysis are necessary before a definitive conclusion can be reached. The chromatography of peptides on Dowex 50 gives approximately 40 peaks with calf thymus and Walker tumour but some of them were very low and might represent hydrolytic products of a contaminating protein. In Walker tumour, the peak n⁰ 5 is missing. The peptides obtained by column chromatography from calf thymus and Walker tumour contained from 3 to 7 or more residues per molecule. Peptides recovered from Walker tumour fraction 2 b showed some differences between corresponding peptides obtained from calf thymus, only peptide n⁰ 29 was identical in amino acid composition.

VII. Heterogeneity of the Protamines

The heterogeneity of the protamines was suggest primarily by KOSSEL and the question was more recently reviewed by FELIX (1960).

By means of countercurrent distribution, RAUEN et al. (1953) have shown that clupeine, salmine and iridine are not homogeneous. SCANES and TOZER (1956) have found three main components in clupeine and several minor components. FELIX, at last, (1958, 1959) obtained three fractions from clupeine of herring of the Baltic sea. The three fractions had the same arginine ratio, the first fraction was nearly homogeneous and contained only 5 amino acids (arginine, alanine, serine, valine and proline).

ANDO et al. (1957 b) have attempted to fractionate the protamines using the differences of solubility of picrates and sulfates of protamines: clupeine, salmine, iridine. ANDO and SAWADA (1959) have also studied the heterogeneity of iridine and clupeine by chromatography on alumina column and obtained 3 fractions (W, Y and Z); W is very likely a nucleotide-like substance, Y represents 73% of the whole product and Z 24%.

The electrophoretic studies give contradictory results. ZIMMERMANN (1959) using a Tiselius apparatus, with 0.1 M. (pH 7) phosphate buffer, or 0.1 M. veronal-acetate buffer (pH 7) found three fractions with clupeine preparations. UI (1956), using a NH₄Cl—NaCl—NH₃ buffer at pH 8.6, obtained with salmine a simple symmetrical curve. In phosphate buffer pH 7.7, the salmine behaved as a heterogeneous substance. UI estimates that the observed heterogeneity is due to the formation of aggregates in the phosphate buffer, which has been controlled effectively by ultracentrifuge studies. These results appear as preliminary and a further study of the heterogeneity of protamines should be undertaken.

VIII. Essays of Fractionation of the Whole Nucleoproteic Complex

Such a type of work was first undertaken by CHARGAFF et al. (1953), CRAMPTON et al. (1954b). These authors prepared a calf thymus nucleohistone by the low ionic strength technique and submitted it to a denaturation with chloroform, they realised afterwards a series of extractions with increasing concentrations of NaCl. By addition of alcohol to the extracts thus obtained, they separated different nucleoproteic fractions and removed the protein by the SEVAG technique. The study of the different DNA fractions shows that in these fractions the $G+C$ ratio is lower and lower and the $A+T$ ratio is higher and higher with increasing salt concentrations.

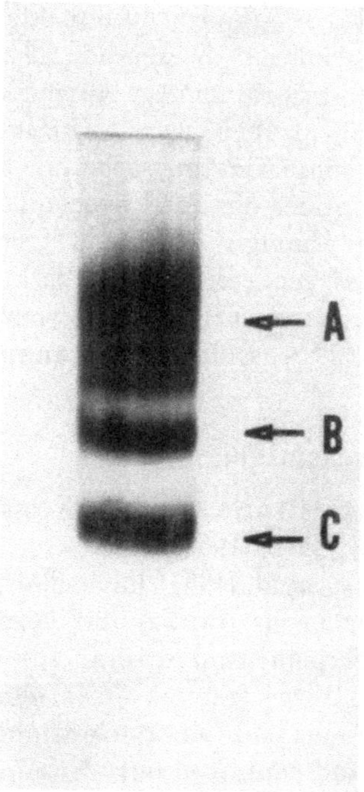

LUCY and BUTLER (1954) showed that the thymus deoxyribonucleoprotein cannot be fractionated by saline solution if it has not been previously submitted to the denaturing action of the chloroform. Moreover, they found that it is not necessary to change the saline concentration to realise the fractionation. They used a constant concentration of 0.6 M. NaCl and found that the fractions extracted successively contain a DNA with increasing $A+T$ ratio. In a further study LUCY and BUTLER (1955) show that a proteic fraction with a high lysine content correspond to a DNA with a predominance of $G+C$ and that a arginine-rich fraction corresponds to a DNA with a predominance of $A+T$.

In 1958, KENT et al. found that the treatment of a preparation of calf thymus or calf spleen deoxyribonucleohistone by sodium heparinate or dextran sulphate causes the liberation of DNA fraction of increasing $A+T$ ratio and decreasing $G+T$ ratio.

Fig. 9. Nucleoproteic fractions (precipitable at pH 4,7) obtained by aseptic autolysis of calf thymus crude chromatin. Electrophoresis on agarosesephadex (G. 200) gel.

Lastly, BAKAY et al. (1960) tried to fractionate the whole calf thymus nucleohistone on DEAE cellulose columns without any very clear result, except some data partly resembling those of LUCY and BUTLER.

In our laboratory such research on a possible structural and analytical relationship between histone and DNA in the nucleohistone has been undertaken with a somewhat different technique. In 1958a, VENDRELY et al. noted that the pycnotic process in calf kidney nuclei was characterized by a depolymerization of the whole nucleoproteic complex. Recently we realised an aseptic autolysis of calf thymus crude chromatin or desoxyribonucleohistone isolated at low ionic strength; the autolysis was made at 37° in the presence of chloroform or thymol and yielded during the first six hours a fraction still precipitable in a 0.14 M. NaCl solution and afterwards a soluble fraction which can be precipitated at pH 4.7 and a very degraded residue. The pH 4.7 fraction is constituted by fragments presenting a molecular weight of 10^5, it was separated by zone electrophoresis on Sephadex-agarose (VENDRELY 1964, VANDERPLANCKE 1964, VENDRELY et al. 1965 b) into three bands: A, B, C (Fig. 9). The analysis of the rapidly migrating fraction (C) indicates an increased arginine ratio and a lower lysine ratio than in whole histone.

Possible Structure of Nucleoprotamines and Nucleohistones

X-ray diffraction studies on DNA fibers provided a basis for the establishment of the famous WATSON and CRICK double helix model for the DNA molecule. It was important to determine what X-ray diffraction pattern was obtained from the macromolecular complexes, nucleoprotamine and nucleohistone. The first study in that field was done by FEUGHELMAN, WILKINS et al. (1955) who showed that deoxyribonucleoprotamine has a fairly well organised structure, as appears from the sharp and well oriented X-ray reflections; the structure of the DNA remains clearly visible, so that these authors proposed for this complex a structure in which the polypeptide chain, which is in a fully extended form, is wrapped helically around the DNA molecule filling the shallower groove of the helix. WILKINS (1956) gave a diagram (Fig. 10) of this model of structure, the side chain reaching out from the polypeptide chain at about right angles, so that on alternate sides the basic end-groups can combine with the phosphate groups of the DNA. The non-basic residues (one-third of the residues in protamine) may occur as folds in the polypeptide chain, so that all the basic groups are able to combine with phosphates. Sequence analysis (FELIX et al. 1956) has indeed shown that the non-basic residues occur in pairs, not singly, so that they can fold. The same X ray diagrams were obtained by WILKINS (1956) on entire Loligo sperm heads, so that this structure may correspond to that of native deoxyribonucleoprotamine.

For the deoxyribonucleohistones the diagrams are far less sharp. In 1956, WILKINS suggested for it a structure similar to that of the deoxyribonucleoprotamine, the arginine and lysine residues being attached to the phosphate groups and the polypeptide chain being wrapped around in both the deep and the shallow groove of DNA. The deep groove was implicated as the "catalytic groove" by which the functions of DNA might be effected.

As PHILLIPS (1962) pointed out, the very lysine-rich fraction which represents 20 per cent of the whole histone of calf thymus, has an extended β-form chain with about 10 moles per cent of proline residues in it like protamine. It is thus possible to imagine that this lysine-rich histone may lie in the shallow groove of the DNA helix as is protamine. An attempt to provide a model for nucleohistone, with localisation of different types of histones, was made by DAVIDSON and BUTLER (1956), BUTLER (1959). These authors proposed for nucleohistones a zigzag structure (Fig. 11), where the basic amino acids are attached to the phosphate groups and the non-basic amino acids are disposed in zigzag between them; the basic residues occur at every fourth position in the histone chain regularly. BUTLER proposed another possibility, in which the basic groups would be linked to the phosphate of one side chain separately. In this model it would be possible that one of the two principal fractions of histone might be attached to one of the chains, and the other to the other chain, so that each DNA chain of the double helix would be distinguished by a different histone. In this model both grooves would probably be filled, but WILKINS et al. (1959) reject this possibility on X-ray evidence.

On the other hand, the relative proportion of the two histones (lysine-rich) are not equivalent, as they should be in this hypothesis.

The zigzag model makes necessary a regular spacing of the basic amino acid residues along the chain. The study of the peptides produced by the degradation of histone with proteolytic enzymes has shown that such a regularity is not encountered (cf. chapter 3, § VI); on the contrary, 25 to 40 per cent of the arginine is paired with another basic amino acid in different histone fractions studied; the basic amino acid may also occur in triplets. Thus an uneven spacing of the basic amino acids seems to be a characteristic of the histones.

PHILLIPS points out that this irregular spacing of the basic amino acids is not compatible with the idea of an uninterrupted α-helical structure for the histone attached to DNA, since the phosphate groups of the DNA chain are about 7 Å part, and the pitch of the protein α-helix is 5.4 Å; each successive amino acid side chain is orientated 100° from its neighbours. The histone chain might be therefore interrupted.

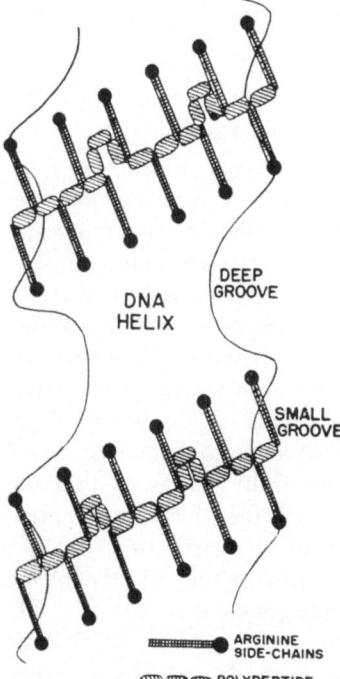

Fig. 10. Diagram showing how protamine binds to DNA. (The phosphate groups are at the black circles. Non basic residues are shown in pairs at fold in the polypeptide chain.) (WILKINS, 1956.)

Further X-ray studies led WILKINS et al. (1959 a, 1959 b) to the idea that the histones might be wrapped around the larger groove of the DNA double helix; they may be in a α-helical form with frequent discontinuities. These authors describe a sharp arc in the X-ray pattern at 35–37Å. More recent studies lay more emphasis on this 35 angström spacing, which is certainly a very significant feature of the native nucleohistones, since it appears in entire erythrocyte nuclei. It is suggested that this spacing, the value of which is about the same as the pitch of the DNA helix, is produced by diffraction from histones bridges formed between roughly parallel DNA molecules.

More recently, ZUBAY (1964), from X-ray diffraction studies and electron microscopic studies, proposed a sheet-like structure for the nucleohistones.

In the gel state the DNA molecules would run in one direction. Histone bridges formed between DNA molecules lie parallel to the large groove of the DNA, with their long axis at an angle of 60⁰ to the long axis of the DNA (Fig. 12). The regular spacing of these bridges measured by X-ray diffraction (35–38 Å) is approximately equal to the pitch of the DNA. How many DNA molecules may be held together by the same histone bridges is unknown.

RICHARDS (1964) also performed X-ray and electron microscope studies on nucleohistones at different degrees of hydration. He finds that the X-ray

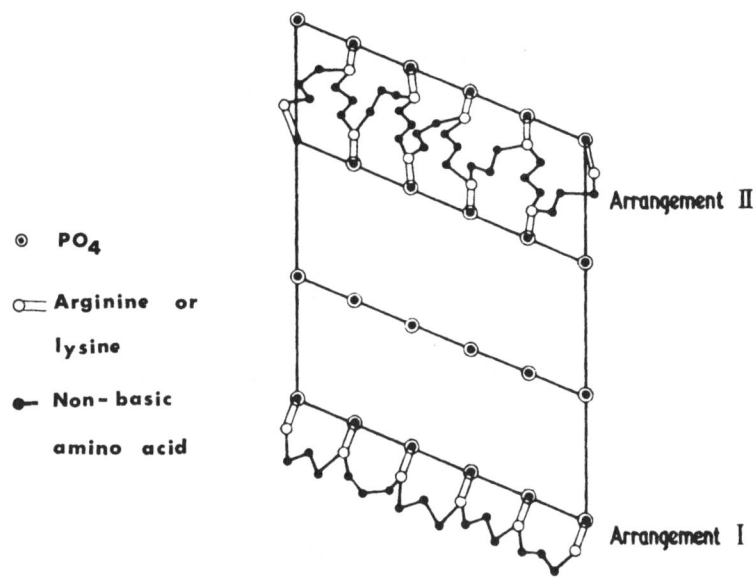

Fig. 11. Two possible arrangements of a regular peptide on the DNA chain (BUTLER 1959).

diffraction pattern of nucleohistone taken at a series of decreasing relative humidities shows a characteristic series of changes. As a nucleohistone preparation dries, the 38 Å ring seen in the wet state is replaced by a 35 Å ring that orients into a meridional arc at lower humidities. When the specimen is almost dry however, the 35 Å arc disappears and a 37 Å ring of weak intensity is found. As complete dryness is approached a meridional arc at 75 Å appears and becomes very strong in fully dried specimens.

The 38 Å reflection seen in the wet nucleohistone may be from the separation between the parallel DNA double helices. In the wet nucleohistone, the protein lies as a gel between the DNA helices and gives no well-defined X-ray diffraction pattern. During the initial stages of drying, the histone molecules shrink and may tend to concentrate at the points of linkage to the DNA; this may be why the relative intensity of the reflection at 35 Å increases. As drying progresses, the relative intensity of this reflection continues to increase. The intense 75 Å reflection in dry nucleoprotein may be due to contraction of histone molecules—each associated with two pitch lenghts of the DNA helix—to the point where successive histone molecules come to lie as dense blobs separated by 75 Å along the long axis of the DNA helix (Fig. 13).

Luzzati and Nicolaieff (1963) performed a different approach to the problem using small angle X-ray diffraction technique. Their results are consistent with the presence of bundles of four DNA molecules surrounded by histones in diluted solutions. At higher concentration the organisation is more complex but always represented by bundles of DNA molecules joined together by histones.

These studies cannot yet permit to draw a definite picture of the arrangement of DNA and histone in the chromosomes of living cells. Nevertheless, they lead us to consider that histone must have (among other functions) a structural function which is important for the constitution of chromosomes. At the present time, the theories on nucleoprotein structure propose that histones hold together two or more DNA double helices. The sheet-like arrangement

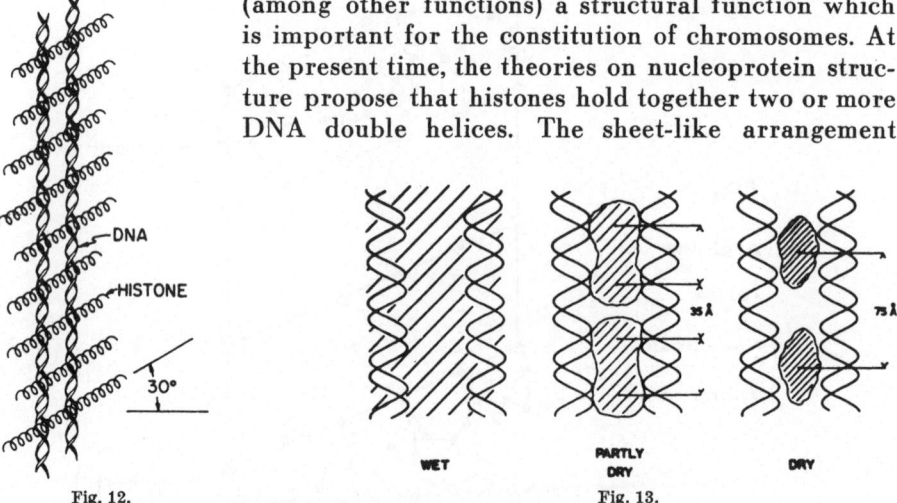

Fig. 12.   Fig. 13.

Fig. 12. Model of oriented nucleohistone in the gel state (Zubay 1964).
Fig. 13. Interpretation of changes in relative intensity in X-ray diffraction patterns of nucleohistone at successive stages of drying (Richards 1964).

proposed by Zubay gives a possible mechanism for the coiling of chromosomes, when they pass from the extended interphase form to the highly condensed stage of mitotic chromosomes. The adjacent coils of the supercoiled DNA would be held together by histones bridges.

Tissue and Species Specificity of the Histones

1. Similarities between Histones of Different Origins

The problem of the specificity of the histones is a most important one, since it appears very likely that the histones may act as regulators of the activity of genes and perhaps also control the duplications of the genes. Thus, differences found in histones from different tissues might be highly significant and provide a clue for the important problem of tissue differenciation. Moreover, the study of histones of cancer cell may show some differences related to a lack of control of the histone on the replication of the DNA molecule and thus on the cellular proliferation. The comparison of histones of different living species may also add some interesting data to problems of evolution.

However, the question is far from being settled at the present time since most studies were done on calf thymus, which has been extensively analysed and fractionated, for it yields readily clean nuclei with very little cytoplasmic contaminants; this is often not the case for other organs. This is probably the reason why only few attempts have been made to compare histones from different sources.

The STEDMANS (1950), from their first results on Salmon erythrocytes and liver, postulated that a specificity existed in histones both with regard to species and tissue, based on analytical differences: number and composition of components. MAURITZEN and STEDMAN (1959, 1960) on the basis of analytic and electrophoretic studies on ox and fowl spleen and liver histones maintained that there is a tissue and species specificity of the arginine-rich β histone, though their results seem to show a similarity between these different histones; the differences they found may be accounted for by the presence of acidic proteins as impurities in the histones. On the other hand, CRAMPTON et al. (1957) studying two main fractions of histones (A and B) prepared from calf thymus, liver, kidney and from Guinea pig testis found them to be quite similar in composition and chromatographic pattern.

DAVISON (1957 a, 1957 b) has indicated that histones obtained from calf thymus, mouse liver and mouse ascites tumour cells, have similar elution patterns at pH 6.5 on carboxymethylcellulose columns. Furthermore, the electrophoretic pattern of the histones of the Walker carcinoma of the rat resembled that of calf thymus histone.

VENDRELY et al. (1958 b) analysed directly the basic amino acids of deoxyribonucleohistones from different tissues of calf and from erythrocytes of carp, trout, pike, tench, frog, fowl and duck and made qualitative chromatographic studies of the hydrolysates. The found in all cases remarquably similar results and similar chromatographic patterns of the amino acids.

DAVIS and BUSCH (1959) performed a systematic analysis of cationic nuclear proteins of tissues of tumour-bearing rats; their chromatographic patterns were quite similar for the histones from different tissues, but some quantitative variations from one tissue to another might suggest some tissue specificity.

HNILICA et al. (1962) developed a very satisfactory method of preparation of the three principals fractions of histones and applied it to the thymus, spleen and liver of calves, and to the liver and spleen of normal and leukaemic rats. Their results on amino acid analysis and N-terminal groups in the various fractions showed no significant differences for the same fractions from different sources. No difference appeared in the patterns of starch gel electrophoresis apart from minor bands probably due to impurities. On the other hand the characteristic analytical differences for each type of fraction were apparent in all the preparations.

These results show a great general similarity between histones from different sources: different tissues in the same animal or different animal species. It is hard to decide wether minor differences are due to impurities or to a real species or tissue specificity.

Some works have been performed on developping tissues of immature animals in order to detect possible differences in the histone, related to differentiation.

Lindsay (1964) studied the electrophoretic pattern of histones from adult and embryonic chicken tissue and did not find any rigorous changes or differences in the histone patterns which correlates with the state ' of differentiation.

Agrell and Christensson (1965) have also examined the nature of the histones in the course of the development of the organs in chick embryo. They have found that in the whole embryo, the ratio of histone to DNA (weight/weight) does not vary during embryogenesis. But, studying the quantitative variations of fractions F_1, F_2 and F_3 they noted the occurence of a gradual transformation of an embryonic pattern in which the amounts of the three fractions are similar, to an adult pattern in which the amounts of F_2 is high and that of F_1 is low. For some organs (eyes, heart, brain) the adult pattern (ratio $F_1 : F_2 : F_3$) is established before the 10th day, for others (erythrocytes, liver) the transformation occurs in a later embryonic period (10 to 16 days).

Neidle and Waelsch (1964) studied the histones of brain, liver, kidney and thymus in different animal species (rat, mouse, guinea pig, rabbit) by disc electrophoresis in polyacrylamide gels. The patterns thus obtained were indistinguishable for the histones of brain, liver and kidney in adult rat. But in the immature rat, the histones of brain and liver seem to be different from the histones of the adult for the same organs. The histones of the thymus were similar to that of the brain and liver in immature rats, and remained the same in adult animals. This suggests that even if some change occurs in histones in the course of maturation in certain organs, the histones of the thymus gland do not participate in this process.

These results of Neidle and Waelsch show once more a similarity of histones from various tissues within a given species. Nevertheless the differences between the thymus histone and the histones of other organs in adults was never reported before, and should be confirmed; the same is true for the differences between the histones of newborn and adults. As the authors point out, it is not possible to decide at the present wether these differences are of a quantitative or qualitative nature; further characterisation of individual bands is necessary. These differences may also be due to differences in extractability and purification of the histones from young and adult organs.

These authors report also species specificity of the histones of rabbit and guinea pig (brain and liver), which would differ from each other and also from rat and mouse histones. But the interpretation of these possible species differences has to be done by a careful characterisation of the bands.

Apart from some results which suggest a species specificity of the histones and a difference in histone between immature and mature animals, most studies show rather a great similarity in composition in histones from different sources and it is not possible to establish clearly with our present techniques the existence of a general species or tissue specificity in the examples studied so far. But in a few particular cases, a quite significant difference from the current type has been noted in histones, namely in the chicken erythrocyte and in vegetal tissues.

II. Histones Different from the General Type

A. The Erythrocyte

From a number of studies performed by different authors, STEDMAN and STEDMAN (1950), MAURITZEN and STEDMAN (1959), NEELIN and CONNELL (1959), CHRISTOMANOS and DIMITRIADIS (1964), NEELIN (1964), the erythrocyte histone appears definitely different from the histones of other tissues and deserves special attention.

NEELIN and BUTLER (1961) and NEELIN (1964), comparing erythrocyte and spleen histone in chicken, found definite differences in their electrophoretic behavior: there were only traces of zones of intermediate mobilities in erythrocytes histones (these components were readily apparent in spleen histone) and two major components of relatively slow mobilities instead of the single zone exhibited by extracts from spleen. The fractionation on Amberlite column and elution with guanidinium chloride under the same conditions as for fraction III and IV of thymus histone (RASMUSSEN, MURRAY and LUCK 1962) yielded a very particular fraction (fraction 5), different from the fraction III and IV obtained from thymus histones: more basic (32 molecules of lysine, histidine and arginine per 100 molecules of total amino acids, instead of 25 for III and IV) with an unusually high serine and relatively low phenylalanine content.

This fraction 5 (NEELIN et al. 1964) appeared to replace the "arginine-rich" histone of other somatic tissues. The other four erythrocyte histone fractions are closely similar in kind to their chromatographic counterparts in calf tissues despite some differences in number and yield.

This observation recalls the earlier studies of the STEDMANS (1950) on the same material, showing that fowl erythrocyte nuclei contains a histone fraction "subsidiary histone" which does not occur in the histones of others tissues. But it is difficult to relate exactly the "subsidiary histone" of the STEDMANS and the "fraction 5" of NEELIN since their analytical characteristics are somewhat different.

HNILICA (1964) also performed electrophoresis of chicken erythrocyte histone and found a band which is absent in mammalian tissues histones. This band was separated on a carboxymethylcellulose column and purified by chromatography on Sephadex G 75. This fraction, labelled F_{2c}, located at the electrophoresis between the two sharp zones of very lysine-rich histone, F_1 one side, F_{2b} of the other, is quite specific regarding its amino acid composition; it has a high proportion of lysine, alanine, serine and arginine and its N-terminal amino acid is threonine, which is quite unknown in other histones, the chief N-terminal of the mammalian

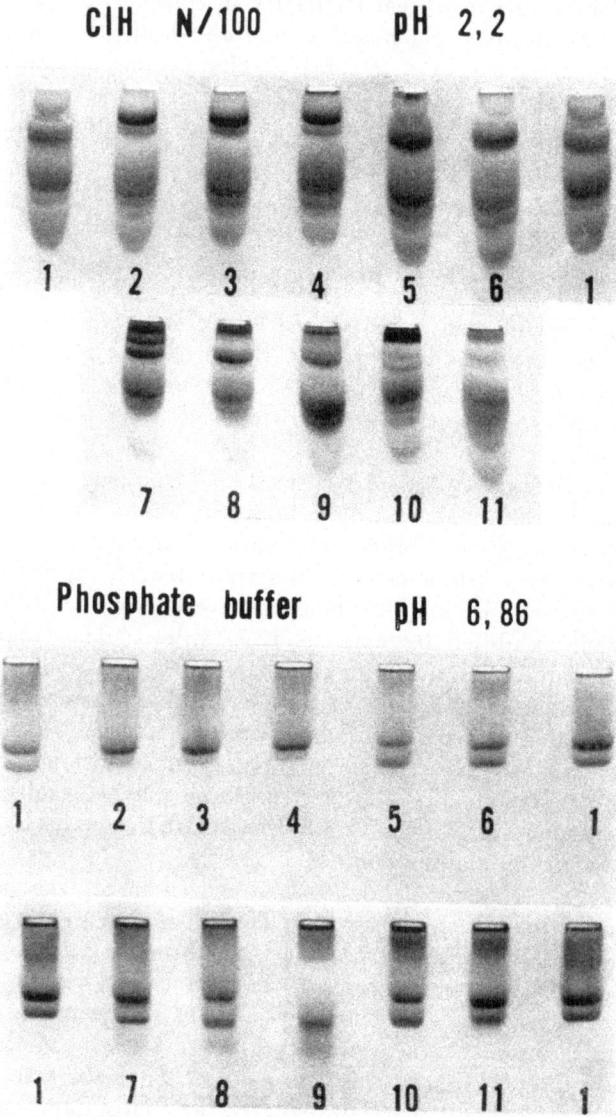

Fig. 14. Starch gel electrophoresis patterns of histone preparations. (Vendrely et. al. 1965.) (1) Calf thymus histone; (2) chicken erythrocyte histone; (3) duck erythrocyte histone; (4) guinea hen erythrocyte histone; (5) turtle erythrocyte histone; (6) viper erythrocyte histone; (7) trout erythrocyte histone; (8) pike erythrocyte histone; (9) carp erythrocyte histone; (10) pleurodele erythrocyte histone; (11) frog erythrocyte histone.

histones being alanine and proline. These results are in agreement with Neelin's findings.

In conclusion: it is important to stress that most of the studies about species or tissue specificity of histones were performed on mammalian cells and in fact no clear-cut differences are found in that case.

The histone of chicken erythrocytes is markedly different from the calf thymus histone, which seems to represent the common type for mammalian histones. The chicken erythrocyte histone is quite different from the histones of other organs in chicken; this point seems to be particularly significant, since erythrocytes are very special cells, highly differentiated, with a very simple cytoplasm, a small condensed nucleus, no more mitotic activity and no functional activity, the synthesis of haemoglobin being stopped at that stage. A careful study of the histone of such a "limit type" of cell is certainly important. The question whether this particular type of histone is characteristic of the erythrocyte in different animal species having such a type of red blood cell is to be studied. Such researches are now in progress in our laboratory on the histones of erythrocytes in different birds, reptiles, amphibians and fishes, using starch gel zone electrophoresis at different pH (VENDRELY et al. 1965 a). We have already found that in different birds the erythrocyte histones are quite similar among themselves, and different from the typical calf thymus histone. In fish erythrocytes the histone is still different from calf thymus histone, but also different from bird erythrocyte histone. This is also true for amphibian and reptile erythrocytes histones, which are different from those of fished, birds and mammals. These results suggest, therefore that there is a tissue specificity for histones in erythrocytes, but also that there are differences when species belonging to different animal groups are considered (Fig. 14). These studies are being extended to other tissues. It appears already that, if species and organs specificity in histones has not yet been proved, the study of further animal groups, may reveal marked differences from the typical calf thymus histone. The problem of vegetal histone is also very interesting and significant.

B. Vegetal Tissues

In 1962 a, JOHNS and BUTLER isolated a nucleohistone from wheat germ by high ionic strength treatment. The histone separated from it was subsequently fractionated with acetone and studied by starch gel electrophoresis. These authors reported that the fraction which would correspond to the arginine-rich histone of the animal tissues seems to be absent in the histone of wheat germ, but they obtained a lysine-rich fraction (A) and a slightly lysine-rich histone (B).

These fractions differ markedly from similar fractions obtained from animal tissues: the predominant N-terminal amino acid in both case is alanine, which is characteristic of the arginine-rich fraction in animal histone. Considering the method of preparation used by the authors, there is no reason to think that the arginine-rich histone would be lost in the course of the isolation. This result is in agreement with previous findings of CRUFT, MAURITZEN and STEDMAN (1957 b), CRUFT, HINDLEY, MAURITZEN and STEDMAN (1957 a), who reported that the main fraction of wheat germ histone was different from the main fraction of animal histone (arginine-rich) and more closely resembled their 1.6 s. γ fraction from ox thymus (slightly lysine-rich). On the other hand, the lysine-rich fraction itself appears different from similar fractions of animal histones: the lysine/arginine ratio is 3 instead of 6. The moderately lysine-rich fraction, very close to the moderately lysine-rich fraction of animal tissues from the analytical results, has nevertheless alanine as terminal N, instead of proline; on starch gel it yields one slow moving band and one main diffuse band, instead of 3 clear bands as in thymus histone. The studies on vegetal histones are not developed enough at the present time to establish that vegetal histones are generally different from animal histones. But MURRAY (1964 c) on pea embryo chromatin made observations similar to those of JOHNS and BUTLER (1962 a) on wheat germ histone in suggesting that this histone is less rich than thymus in components containing a large proportion of arginine. BIRNSTIEL und FLAMM (1964) found a great proportion of lysine-rich histone in chromatin of pea embryo and of tissue cultured tobacco cells.

On the other hand, IWAI (1964) isolated from rice embryos a histone which seems to be similar to calf thymus histone with respect to the amino acid composition of its lysine-rich and arginine-rich fractions isolated on a carboxymethylcellulose column. The same author isolated from the unicellular green alga chlorella a histone with a predominant lysine-rich fraction.

But DE and GHOSH (1965) using cytochemical staining methods conclude that histones are apparently absent from the cells of Cyanophyceae.

So, from these first results some plant histones seems to be different from animal histones, showing a higher proportion of components rich in lysine and poor in arginine than in the calf thymus histone, which is typical of animal histones.

C. Do Microorganisms Contain Histones?

The problem of the presence or absence of histones in bacteria is a very important one on a theoretical level since the microorganisms represent the best material for studies of molecular biology and it is important to know wether the bacterial chromatin differs from the chromatin of metazoans by a lack of histone. If a histone is present in bacteria then, it would be much easier to study the general role of histone in the cell.

When we examine the results reported in the literature, we cannot draw any clear-cut conclusion. Though some authors estimate that bacteria contain a histone associated to DNA, a number of others say that DNA is not associated with histone in microorganisms. This discrepancy is not surprising since bacteria have a very resistant cell wall and the chromosomes in them represent a very small part of the bacterial cell (2 to 8%) so that their isolation from the mass of the bacterial cytoplasm requires rather time consuming and complicated techniques to distroy the cells and remove the cytoplasmic fraction. These techniques: grinding, the use of detergents, the ultrasonic treatment may favour an enzymatic degradation of the hypothetic histone in the course of the preparation. For these reasons, the isolation of a bacterial histone, if this histone exists, is not on easy task.

CHARGAFF and SAIDEL (1949) working on Avian tubercle bacilli and TSUMITA and CHARGAFF (1958) on Bovine Tubercle Bacilli isolated a nucleoprotein which lacked the basic properties of histones or protamines: dicarboxylic amino acids predominated over basic amino acids.

More recently, ZUBAY and WATSON (1959) realized water extraction of E. coli ground with alumina powder, and obtained thus a deoxyribonucleoprotein which they purified several times. The protein of this deoxyribonucleoprotein is not a histone: the amino acid analysis shows that the ratio of basic amino acids is twice lower than that of an ordinary histone. The X-ray diffraction of this product (WILKINS and ZUBAY 1959) shows a A type pattern of crystalline DNA, which suggest that a large part of the DNA is free of protein. But, as the authors point out, it is not impossible that a part of the nucleoprotein might be separated from the DNA by enzymatic processes in the course of the preparation.

Moreover, BUTLER and GODSON (1963) tried to extract a basic protein from nuclear material of protoplasts of Bacillus megatherium. The protoplasts were destroyed by use of detergents (Lubrol W) and ultrasonic desintegration was used afterwards on the crude nuclear material in order to lower the viscosity, which helps in the elimination of ribosomes. The extraction of the basic protein by dilute acid solutions as it is used for animal histones, removes only 0.4% of the total protein. BUTLER and GODSON conclude that the DNA of bacteria is not associated with a basic histone and that the basic proteins extracted by other authors from bacteria would be derived from the basic proteins of bacterial ribosomes. Lastly, BECK and WALKER (1964) showed that sera containing antinucleohistones failed to stain trypanosome nuclei by the fluorescent antibody technique. They conclude that it is probable that the DNA of trypanosome nuclei is not present as DNA-histone complexes of the type found in metazoan cell nuclei.

STEINERT (1965) using cytochemical methods reaches also the conclusion that there is no histone in the kinetoplast of trypanosoma. But HEROIN-DELAUNEY (1965) shows a very marked staining of the kinetoplast with alkaline fast green which suggests the presence of basic proteins in this organelle.

In spite of these works suggesting the non-occurence of histones in microorganisms, other authors estimate that histone might be present in association with DNA in bacteria. As early as 1946 b, MIRSKY and POLLISTER, isolated from Pneumococci (type III) with saline solution chromatin threads which were purified. They obtained a "chromosin" which they estimated to be probably a mixture of histone and non-histone proteins. BRAUN et al. (1957) isolated a nucleohistone from Brucella abortus. CRUFT and LEAVER (1961) extracted from Staphylococcus aureus a basic protein fraction which showed an amino acid composition quite similar to that of a calf thymus histone. PALMADE et al. (1958) used a very mild technique for the isolation of purified nucleoprotein from E. coli. The germs were first transformed into protoplasts by use of penicillin and the nucleoprotein was extracted by a mere shaking with glass powder of the protoplasts in NaCl 1 M. The solution is dialysed during 3 days at 2⁰ C. against a solution of NaCl 4 g. ⁰/₀₀. The precipitate thus obtained is purified three times by redissolution and dialysis. The product shows a N/P ratio of 3.64 and its chromatographic pattern is quite similar to that of a nucleohistone from trout erythrocytes.

More recently, Baghavan and Atchley (1965) using a technique very similar to that of Palmade et al. (1958) obtained a pure fibrous nucleoprotein from protoplasts of *Bacillus subtilis*. This nucleoprotein was obtained by precipitation with $MgCl_2$ and it contained 43% DNA, 52% protein and only 5% RNA. The acid soluble proteins extracted from the purified DNA P fraction are similar to histones; they have a high content of basic amino acids and lack cystine and tryptophane. Their possible ribosomal origin is quite unlikely since the RNA ratio is very low in this nucleoprotein complex.

Hurst and Taylor (1965) in order to study the inhibitory effect of histone fractions on bacterial growth have isolated from *Escherichia Coli* (at the end of the logarithmic phase) a fraction F_3 (using the method of Johns et al. 1960) and a fraction F_1 (method of Johns and Butler 1962). The fraction F_3 showed some inhibitory activity. It would be interesting to perform a precise analysis of these different fractions.

In conclusion, the problem of the histones of bacteria is not solved and further studies are necessary. The use of very mild techniques and the purification of the deoxyribonucleoproteic complex before the acidic extraction of the basic proteins are probably very important.

At the present time it is not impossible to imagine the occurence of histones in association with the bacterial DNA.

D. Cancer Cells

The histones of cancer cells have been investigated by a number of authors in relation to the hypothesis that histones are gene inhibitors: a possible change in the histones of malignant cells may explain the lack of regulation and control during malignant growth.

The earliest research on this subject are due to Stedman and Stedman (1943) who found a much smaller amount of histone in malignant cells than in normal cells. Debov (1951) confirmed these results. However, it is very hazardous to assume that the cancer cell nuclei contain less histone than normal cell nuclei because the yield of histone from the nuclei is very low. A part of the histone may be destroyed enzymatically in the course of isolation, or may be difficult to remove from the nuclei by the ordinary extraction processes. In fact, Davison (1957 a) estimates that the results of Stedman and Stedman may be due to the fact that these authors used too dilute acid for the extraction of the histone.

On the other hand, it is very difficult to isolate tumour nuclei free of cytoplasmic contaminants so that the yield of histone may be apparently lower. Cruft et al. (1954) noted also a very low yield of histone in some tumours (4 per cent of the dry weight of nuclei in a human breast sarcoma, compared with 25 to 30 per cent obtained in many normal cell nuclei). However they showed that such great deficiencies are not characteristic of all cancer tissues since in some cancers the yield of histones from the nuclei approach normal, but never attain really high values.

Davison (1957 a) showed that the histone content from mouse-ascites tumour nuclei and mouse normal nuclei (kidney and spleen) to be the same. ALLFREY et al. (1955) said there is no appreciable difference in the histone content of normal and leukaemic leucocytes. Lastly, KIT and GROSS (1959) found that the ratios of histone to DNA was approximately constant for diploid and tetraploid tumours (carcinoma and lymphoma cells of chinese and syrian hamster), but the histone content of the carcinoma cells was approximately 3 times as great as that of the lymphoma cells.

More precise studies were performed by LAURENCE et al. (1963) who measured the amount of basic amino acid compared to the amount of — PO_4 groups in the DNA. It is thus interesting to know wether in tumours DNA is less "covered" by histone than in normal tissues. LAURENCE et al. have shown that the deficiency of histone arginine + lysine, as compared with DNA phosphorus of about 20% is not significantly greater in Crocker tumour and spontaneous mouse tumours than in normal tissues.

So the problem of the amount of histone associated with DNA in tumour nuclei is not quite resolved, but it seems that cancers may exist without any decrease in the amount of histone per nucleus. The low yield of histone in cancer nuclei which has been reported may be a consequence rather than a cause of this process (a higher loss of easily extractable histone resulting from the particular state of a neoplastic cell). We must, therefore, look for other differences than quantitative ones.

CRUFT et al. (1954) thought that tumour histones were cell specific because they found in them particular physical properties: lower solubilities, smaller electrophoretic mobilities, and isoelectric points of 7.0, smaller diffusion constants and greater ability to aggregate in acid solution. They suggest that these differences are due to an underlying chemical cause, particularly an increased content of acidic amino acids. But until now no marked difference in the overall amino acid composition of histones from normal and neoplastic cells has been reported. HAMER (1951) found a similar composition of histones from calf thymus and from a transplantable rat sarcoma induced by methylcholanthrene. LAURENCE et al. (1963) more recently compared histone from calf thymus, rat liver, a spontaneous mouse tumour and Crocker sarcoma, and found no significant difference in the general amino acid composition. It is possible that the differences which have been reported earlier may be due to the presence of acidic proteic contaminants of the histones. In fact, LAURENCE et al. have shown that the amount of acidic amino acid is highly reduced when the nuclei were isolated with special care.

The progress of the fractionation of the histone made possible a further search for differences in histones from normal and neoplastic cells, namely differences in the composition and relative importance of the different fractions. But the electrophoretic pattern and the chromatographic behaviour of the histones of mouse ascites tumour cells and of the WALKER carcinoma of the rat was found by DAVISON (1957 a and 1957 b) to be similar to that of the calf thymus histones. LAURENCE (1963) obtained also quite similar starch gel electrophoretic patterns for histones of calf thymus, Croker tumour and spontaneous mammary tumour in mice. These results support the suggestion of HNILICA et al. (1962) that any differences between tumour and non-tumour histones are too small to be detected by the methods available.

The comparison of fractions of histones from different sources is difficult, since it is very difficult to obtain perfectly pure fractions. Nevertheless, Hnilica et al. (1963) separated by chromatography on carboxymethylcellulose columns a subfraction of the moderately lysine-rich histone, coded as $F_2 b$, in a state of high purity as shown by electrophoretic mobility and NH_2 terminal amino acid analysis. They prepared this fraction from Walker carcinosarcoma, rat spleen and rat thymus, and from calf thymus. The amino acid composition of this $F_2 b$ fraction was quite similar in each case and no specific differences appeared. The chief NH_2-terminal amino acid was proline in all $F_2 b$ fractions. Nevertheless it was possible to think that in spite of this overall similarity the distribution of the peptides after tryptic hydrolysis might reveal a specificity. In fact this distribution was the same in the histones from all four tissues thus studied. The analysis of the peptides after fingerprinting showed a great similarity: the distribution of the spots as well as the distribution of arginine, histidine and tyrosine were essentially the same for all the tissues studied.

On the other hand, peptides were separated by column chromatography and analysed for their amino acid composition. In that case many differences were noted between fraction $F_2 b$ of the Walker tumour and of calf thymus (Hnilica et al. 1964). This result, different from that obtained from analysis of peptides after fingerprinting, may be explained by the fact that with column chromatography the peaks are still heterogeneous and the mixture of peptides obtained include those from the minor protein contaminants.

A preliminary study of the peptides of another subfraction of the moderately lysine-rich histone, 2 a fraction in calf thymus and Walker tumour was performed by Busch and Mavioglu (1964). Differences were noted in the composition of corresponding peptides, but the 2 a fraction was not very pure.

Thus, a specificity in the peptide composition of some fractions of histones from tumours has not yet been demonstrated quite clearly in these preliminary studies. More extensive studies are necessary to throw some light on the important question of the specificity of histones in cancer.

We must quote here the studies concerning the $R P_2 L$ fraction which Busch and his coworker reported to be present in the histones of tumours and absent in any normal tissue, even tissues showing marked mitotic activity (such as regenerating rat liver, or embryonic tissue). This fraction was isolated from the acid soluble proteins of neoplastic cells, chromatographed on carboxymethylcellulose columns and eluted with formic acid (Davis and Busch 1959, 1960, Busch 1962, Busch et al. 1962). They showed that this fraction was composed chiefly of fractions 2 a and 2 b (according to the fractionation of Johns et al. 1962 b) and a very little amount of fraction F_1. They concluded that it represented a qualitatively specific cancer histone fraction: this fraction $R P_2 L$ was characterized by an intense uptake of labelled amino acid. But further researches (Busch et al. 1963 b) showed that, when acid-soluble proteins were extracted from isolated deoxyribonucleoproteins, there was no longer any difference between tumours and normal tissues in the histone chromatograms. Thus, there is no specific cancer cell histone fraction. Nevertheless, the existence of a special R P_2 L peak in chromatograms of the histone extracted from whole cancer nuclei must be explained. BUSCH et al. admitted that the early emergence of the "slightly lysine-rich complex" (R P_2 L) on the chromatogram, which reflects a decrease of its overall basicity, may be due to the fact that these proteins were linked to acidic proteins present in higher concentration in neoplastic cells than in normal cells, or qualitatively different; so that the difference between tumour cell nuclei and normal cell nuclei would lay rather in the acidic proteins of the nuclear sap, than in the histones.

III. Conclusion

The search for specific differences in histones from different sources in often disappointing since very careful analyses tend to show a general similarity of the different histones studied; even an impressive developmental change such as the remplacement of mesonephros by metaphros in chicken embryo LINDSAY (1964), or the cancerous state of the cells, BUTLER (1963), are not correlated with marked changes in the histone.

The interpretation of minor differences in electrophoretic patterns, or in amino acid composition of some fraction, is difficult since a very small amount of impurities may be responsible for these differences. There are two possibilities for such a study:

(1) The use of more refined analytical studies, such as fingerprinting on the histone fractions, which allows the determination of the amino acid arrangement in a histone molecule; such studies are now in progress, but they must be done on very pure fractions of histone and these are difficult to obtain.

(2) The comparison of unrelated living species where the differences among the histones may be so important that they can be readily detected with current techniques, such as starch gel electrophoresis and amino acid analysis of the different fractions; chicken erythrocytes, fish, reptile or amphibian erythrocytes (VENDRELY et al. 1965) and wheat germ seem to represent such favourable cases.

Concerning the cancer histones, no particular histone characteristic of the neoplasic state of the cell has been found. The only true difference between neoplasic and normal histone would be found in their metabolism.

Histone Metabolism

I. General Results on Histone Metabolism in Normal and Tumour Tissues

The first studies on the metabolism of histones together with other nuclear proteins are those of DALY et al. (1952) who injected ^{15}N-labeled glycine intraperitoneally into mice over a period of hours. They showed that the uptake of isotope was very great in the cytoplasmic proteins of pancreas, liver and kidney. The residual proteins of the nuclei of these tissues were labelled nearly as much as the cytoplasmic proteins. In comparison, the specific activity of the histones of each tissue was very low. BRUNISH and LUCK (1952) studied the incorporation of alanine-1-^{14}C and phenylalanine-2-^{14}C into nuclear proteins of rat liver *in vivo* and reported that the histones were the least radioactive of the proteins isolated, whereas the proteins of the nucleus corresponding to the "residual proteins" of MIRSKY and RIS were the most radioactive. Other authors (SMELLIE et al. 1953), working on the uptake of ^{15}N glycine into histone and other proteins of rat liver, and ALLFREY et al. (1955) on pancreas, liver and kidney of mice, came to the same conclusion.

From these early results on non-proliferative tissues, the histones thus appeared as relatively inert in the general metabolic activity of the cell.

Concerning the metabolism of tumour histones, ROTHERHAM et al. (1957) compared the labelling of hepatoma and liver proteins 20 hours after administration of radioactive glycine. They reported that, while the specific activity of the cytoplasmic proteins and the overall of labelling of the nuclear proteins of the liver was twice that of the tumour (hepatoma), the specific activity of the histones of the tumour was twice the specific activity of the histones of the liver. Such studies have been done extensively on a great number of normal tissues and tumours by BUSCH and his coworkers (1958). They reported that, 1 hour after injection of lysine-U-^{14}C into tissues of tumour bearing rats, the specific activity of the histones of the Walker and Jensen tumours was greater than those of the proteins of any other cellular fraction (microsomes, mitochondria, cytoplasmic sap, HCl insoluble nuclear proteins and 0.20 M. NaCl soluble nuclear proteins). In normal tissues, including brain, heart, intestine, liver, kidney, muscle, pancreas, spleen, testis and thymus, the specific activity of the histones and of the other nuclear proteins were lower than in the tumour tissues. CAMPBELL et al. (1957) obtained similar results, indicating that the nuclear proteins of tumours had a relatively higher level of protein labeling than did the liver.

More recently, STEELE and BUSCH (1963) compared the incorporation of L-lysine-U-^{14}C into nuclear proteins of liver and Walker 256 carcinosarcoma of rats. They found that in liver the specific activities of the acidic nuclear protein fractions, other than those of the nuclear sap, were highest and the specific activity of the histone fraction was the lowest. In the tumour, on the contrary, the specific activity of the histones was higher than that of the acidic proteins.

ZBARSKY et al. (1964) have also shown that in cancer nuclei in vitro (Ehrlich ascites tumour) the biosynthesis of histone is higher than in normal nuclei (rat liver) and the biosynthesis of acid proteins is lower in spite of their particularly important ratio in these nuclei.

All these results suggested that the incorporation of amino acids in the histones is relatively low in normal tissues and rather high in tumours. These facts may be interpreted in two ways: (1) The tumour histones contain a specific fraction with a high turnover which is not represented in normal tissues, (2) the incorporation of amino acid into histones is related to the duplication of chromosomes and is higher when the cells are actively dividing (embryonic cells, tumour cells).

The first possibility received some support from works of DAVIS and BUSCH (1959–1960), indicating the existence in neoplastic cells of a proteic fraction, which they coded RP_2L and which was the second radioactive peak eluted by formic acid in carboxymethylcellulose columns. 30 to 55 per cent of the isotope, radioactive lysine, which was incorporated into nuclear proteins was found in this fraction. They isolated this fraction from Walker tumour, Jensen sarcoma, Flexner Jobling carcinoma, Ehrlich ascites tumour, Sarcoma 180 and a human malignant melanoma. Since it was not found in growing non-tumour tissues, such as embryonic rat tissues and regenerating rat liver, they concluded that it was specific for cancer cells. BYVOET and BUSCH (1961) indicated that this fraction was bound to DNA and so it was considered as a histone fraction. HIDVEGI et al. (1964) found it in Lettré Ehrlich ascites carcinoma cells and Hela cells but they found it also in normal bone marrow of various animals (rabbit, rat, etc.).

But, as was shown by BUSCH et al. (1963 a, 1963 b), the difference between normal and cancer cells no longer appears if the fractionation is performed not on whole nuclei or cells, but on isolated deoxyribonucleoprotein. Therefore the RP_2L fraction cannot be considered as a pure histone fraction and its high specific activity is due to acidic proteins combined with it. The authors suggest that acidic proteins may compete with DNA to combine with histone. This process may represent a possible mechanism of release of histone from DNA, which would be necessary before DNA nucleotidyl transferase or polymerase could proceed with the DNA synthesis.

II. Histone Metabolism and Cell Division

Let us now examine the question of the relation of histone synthesis and DNA synthesis.

A. Correlation between Histone and DNA Synthesis

A number of works, especially with cytochemical techniques, suggested that the two syntheses may be necessarily linked and that no histone synthesis occurs when DNA does not duplicate (ALFERT 1955). BLOCH and GODMAN (1955) showed that the synthesis of DNA is correlated with a synthesis of histone, the histone/DNA ratio being constant in different tissues in the course of normal growth, in pathological growth (RASCH and WOODWARD 1959), or even after cell irradiation (DAS and ALFERT 1961, RINGERTZ 1963). MEEK (1964), showed by microdensitometry of the amount of FEULGEN and fast green dyes bound that this ratio histone/DNA was constant in interstitial cells of mice testis and in spermatogonia undergoing mitosis as well as in spermatocytes in interphase. All the preceding results were obtained by cytochemical techniques but some biochemical works indicate also a relationship between the synthesis of histone and the rate of division of the cells. OKUDA et al.)1963) show that the *in vitro* incorporation of lysine-^{14}C is 6 times higher in the deoxyribonucleoprotein from regenerating liver than from normal liver. FLAMM and BIRNSTIEL (1964), in tobacco cells cultures, showed by biochemical measurements that the histone/DNA ratio was about 1 in this type of cells, and that the rate of incorporation of amino acids into histone was a function of the culture age, suggesting a correlation between the synthesis of histones and the number of cell division.

There is certainly a synthesis of histone when the chromosomes duplicate prior to cell division. But it is important to know wether DNA and histone synthesis occur simultaneously or not and whether histone synthesis may exist without DNA synthesis.

B. The Moment of the Histone Synthesis

The regenerating liver after partial hepatectomy offers a good system for the study of the different syntheses related to cell divisions, namely DNA and histone synthesis. HOLBROOK et al. (1962) showed that the synthesis of histone seems to occur during a longer period of the pre-mitotic cellular cycle than does DNA synthesis. The peak of incorporation of precursors into histone (glycine-1-^{14}C) appeared before that of the incorporation of adenine-8-^{14}C into DNA. The increase of histone seems to be relatively greater than the increase of DNA before mitosis, so that the ratio histone/DNA is higher before mitosis (IRVIN et al. 1963). The sequence of events is first an increase in RNA synthesis, then an increase in histone synthesis, then DNA synthesis. The authors suggest that the initial increase of histone concentration in the nucleus would inhibit the RNA synthesis and somehow stimulates directly or indirectly DNA synthesis; when the histone concentration attains a maximum, the DNA synthesis stops. They showed indeed, by adding histones to Ehrlich ascites carcinoma cells, that low concentration of histones inhibited RNA synthesis and stimulated DNA synthesis while high concentration inhibited both. BUTLER and COHN (1963) showed also that the histone synthesis begins before DNA synthesis in liver after hepatectomy. UMANA et al. (1964), studying the metabolic behavior of the histones in relation to the mitotic cycle in regenerating rat liver, found also that, while the DNA synthesis began 18 hours after hepatectomy and culminated at 25 hours, the histone synthesis began after 12 hours and reached a maximum at 19 hours.

So, it seems that in the sequence of events preceding mitosis, the synthesis of histone may precede DNA replication. Autoradiographic studies of PLAUT (1963) on *Drosophila* polytene chromosomes and photometric and autoradiographic works of BRYAN (1964) on liver or kidney cells of new born mouse reach the same conclusion.

C. Possible Dissociation between Histone and DNA Synthesis

Recent developments suggest that there is no necessary correlation between DNA and histone metabolism. Though the turnover of histone is very low in non-dividing cells for instance in normal liver, BUSCH et al. (1964) have shown that in rat liver 3 to 6% of the whole histone is synthesized each 24 hours, while no DNA is synthesized in the same time. In Walker tumour 50% of the histone might be synthesized in 24 hours.

BYVOET (1964), studying the rate of synthesis of histones and DNA by use of labelled lysine and ^{32}P injected in rats, found that the ratio of the two rates of synthesis in 3 different transplantable tumours was 2.5 times greater than in normal thymus, spleen and kidney but lower than in liver. All these results suggest that in tumours and in liver, the rate of synthesis of histones is higher than required for the replication of chromosomal material.

On the other hand, it is also possible to dissociate DNA and histone synthesis experimentally. LINDNER et al. (1963) showed that fluorouracil in Ehrlich ascites tumour cells stops DNA synthesis without effect on histone synthesis, which results in a considerable increase of the histone/DNA ratio. FLAMM and BIRNSTIEL (1964), in tobacco cells cultivated *in vitro,* found also that 5-fluorodeoxyuridine inhibited DNA synthesis and subsequent mitosis, while RNA and histone continued their synthesis normally. Histone synthesis thus appears to be independant of DNA synthesis.

ONTKO and MOOREHEAD (1964) showed that when the DNA synthesis is stopped after irradiation of Ehrlich ascites cells, the histone synthesis demonstrated by the incorporation of 1-^{14}-C-glycine still proceeds readily. So, in Ehrlich ascites tumour cells as well as in tobacco cells, the histone synthesis is not dependant on active DNA replication or cell division.

D. Different Metabolic Behaviours of Histones

We have seen that there is certainly a relationship between the synthesis of histone and the synthesis of DNA but these two syntheses are probably not quite synchronous and moreover, histone can be synthesized when no DNA is formed. In fact autoradiographic researches of PRESCOTT and BENDER (1963) suggest that chromosomal proteins do not duplicate in a semiconservative pattern as DNA, but have a continuous turnover. This leads us to imagine two types of histones regarding their metabolic behaviour. UMANA et al. (1964) suggest that a metabolically stable histone would be intimately attached to DNA occupying the large groove of the double helix, and a labile histone would be more loosely associated with the DNA molecule and subject to an active turnover. The metabolic, loosely bound histone is probably lost in the course of the histological techniques for tissue sections and microspectrophotometric measurements would concern only the constant part of the histone firmly bound to DNA. This hypothesis seems quite interesting but the large variations of the histone/DNA ratio found by UMANA et al. (1964) in regenerating liver and in different organs of normal and starved animals should be confirmed for they have never been reported by other authors so far.

III. Where is Histone Synthesized?

These studies of histone metabolism lead us to ask where histones are synthesized. If this synthesis is not necessarily linked to DNA synthesis, as recent works have shown, it is possible to imagine this synthesis as a normal proteic synthesis on ribosomic particles. Since it was not possible to demonstrate the presence of true histones in the cytoplasm (BUTLER 1963), we must assume a synthesis of histones within the nucleus.

In fact, Allfrey et al. (1964) described a technique for purification of calf thymus nuclei. These nuclei were able to incorporate amino acids *in vitro,* high resolution autoradiography on thin sections of isolated nuclei and whole cells showed that the amino acid incorporation occurs within the nucleus and is not due to cytoplasmic contamination. Reid and Cole (1964) realised *in vitro* the incorporation of labelled amino acid into histones of isolated calf thymus nuclei. This incorporation was shown to be true synthesis by tryptic digestion and fractionation of the resulting redioactive peptides. This suggest that histones are synthesized in the nucleus.

Flamm and Birnstiel (1964) showed that after 20 minutes incubation with radioactive D,L-lysine-1-^{14}C, D,L-arginine-^{14}C or protein hydrolysate, the specific activity of nucleoli was double that of the chromatin in a culture of tobacco cells. The initial place of synthesis of the basic protein of the nucleus is thus probably the nucleolus.

In opposition to these results Bloch and Brack (1964), studying the incorporation of arginine into histone during spermiogenesis in the Grasshoper Chortophaga viridifasciata, conclude that the histone would be synthesized in association with the RNA granules in the cytoplasm and then migrate into the nucleus.

IV. Conclusions

The problem of the metabolism of histones is a very important one: if histones act upon the DNA as regulators of genic activity, the metabolism of the histones may give a clue to the mechanisms of this action. The high turnover of the histones in cancer cells may be highly significant in this respect. Different interpretations have been proposed for this phenomenon. The turnover of different fractions of the histone have been studied (Busch et al. 1964); after injection of lysine the specific activities are different for the different fractions, however, the interpretation of these results is not evident: if the result is expressed in counts/minute/mg. of protein, the most lysine-rich fraction is the most labelled after one hour; but if it is expressed in counts/minute/μmole lysine, the most lysine-rich fraction has the lowest specific activity. Such researches are just begining, and more results are necessary to have an idea of the mechanism of the action of histones in the cell nucleus through its metabolic behaviour. The points which must be retained are the dissociation between DNA and histone synthesis, the fact that histone synthesis seems to precede DNA synthesis, the possibility of the occurence of a "metabolic" histone fraction and the high turn-over of histones in cancer cells.

The Function of the Histones

In very old publications, the proteins of the chromosomes were considered as the only possible support of the hereditary characters, the proteic nature of the genes being unquestioned. However with the development of the researches on nucleic acids, it appeared that DNA was in fact the only substance bearing the hereditary message, so that the proteins associated to it were considered as structural components of chromosomes with no partic-

ular significance except that of protective substances for the DNA molecule, preventing it from combination with other proteins and also maintaining these molecules at the right place in the chromosomes. The fact that
the histones are constantly associated with DNA and of universal occurrence
in somatic cells favours the idea of a structural function. It is evident that
DNA does not float freely in the nuclear sap even in interphase. It must be
arranged in a difinite order and then agregated and oriented in the mitotic
chromosomes. The histone are perfectly able to assume this function. Physical studies by ZUBAY, RICHARDS (see chapter Structure) suggest that an impor-
tent function of the histone is to connect DNA molecules in a proper order.

Besides such a purely structural role, a number of possible functions
have been proposed for these substances. One of the first hypotheses was
formulated by CASPERSSON (1941) and reconsidered by BUTLER in 1959.
These authors considered that the histones might represent a step in the
transfer of the information from DNA to the system synthesizing proteins
but this concept is no longer considered valid, since the new theories of
molecular biology attribute to the messenger RNA the function of transmission of genetic information from the DNA to the site of protein synthesis.
But the fact that histones show a marked turnover and a synthesis which is
not restricted to the replication of chromosomes makes it possible that the
histones intervene somehow in the biological activity of the cell.

Since the DNA strands in the WATSON and CRICK structure are complementary to one another in their base sequence, the messenger RNA
produced by each strand should have quite the opposite information for
the synthesis of proteins, so that it is important to have a mechanism to
prevent this ambiguity. DOUNCE (1959) suggested that the histones might
block one of the DNA strands at the moment of the production of RNA,
so that only the other strand should act as a template for the RNA synthesis. A more elaborate theory was advanced by LESLIE (1962), who thinks
that the histones might act on the RNA synthesized on the DNA template
by stabilising the specific RNA which will be necessary for the protein
synthesis and act as ribonucleases toward any other RNA. The combined
histone should thus remain linked with the specific RNA until it reaches
the ribosomes, which means that histone should be found in the cytoplasm
and in the ribosomes. In fact, as we have already noted, the ribosomes contain basic proteins but they do not seem to be actual histones. Concerning
the ribonuclease activity, LESLIE found that the histones or basic proteins
obtained from nuclei and ribosomes of HLM cells and from similar fractions
of guinea pig liver showed depolymerase activity at pH 7.8, but it cannot
be excluded that the ribonuclease activity might be due to contamination of
histone by cellular ribonuclease, so that this theory has no sufficient
experimental basis.

Another function was proposed by BLOCH and GODMAN (1955) concerning
the possible role of histones in DNA synthesis. These authors suggested that
histones may act as templates for DNA synthesis and vice versa. But we
now have some reason to think that histones may be synthesized in the
nucleolus and not directly copied on the DNA within the chromosomes.

Lastly, the function of gene inhibitors for histone was proposed by
STEDMAN and STEDMAN in 1950. It has received some support from recent
experimental results of HUANG and BONNER (1962) and other authors. The
STEDMANS postulated that, since all types of cells in an organism have the
same genic complement, there must be in the nucleus a genic regulator
responsible for cell differentiation. This regulator may be histone and they
suggested also that the histones might possibly regulate mitosis (STEDMAN
and STEDMAN 1943). It is worth pondering on this idea, since it concerns
one of the most intriguing problems of cell biology: gene regulation. The
only system which has been well studied in respect to gene regulation is
the bacterial system, and it is legitimate to take it as a basis for general
speculations, assuming that similar mechanisms exist in other cells. There
are two types of genes: regulator genes and structural genes. The structural
genes provide for the synthesis of messenger RNA, which in turn codes
the synthesis of a protein; the regulator genes control this process through
a repressor that operates over some restricted area of the gene. As pointed
out by DULBECCO (1964), the histones may be considered as candidate for
genetic repression.

In bacteria there probably exist two types of regulatory systems: one
concerning the regulation of messenger RNA, the other controling DNA
multiplication. We shall therefore examine the facts now available on the
action of histone upon the synthesis of messenger RNA and then the action
of histone on DNA synthesis.

I. Action of Histones on Messenger RNA Synthesis

In 1962, HUANG and BONNER showed that histone was able to suppress
chromosomal RNA synthesis of pea embryo chromatin. This chromatin was
extracted from the embryonic axes of germinating pea seeds by homogenization in sucrose solution and successive filtrations and centrifugations.
Each kilogram of fresh tissue thus yields 400 to 500 mg. of crude chromatin,
containing proteins, DNA and RNA. This material is then purified by
removal of non-chromosomal protein by sucrose gradient centrifugation.
The purified chromatin consists of 31% DNA, 17.5% RNA, 33% histone
protein and 18% non-histone protein, which includes the chromosomal RNA
polymerase. This material was shown to possess the ability to carry out
extended DNA-dependent synthesis of RNA from the four riboside triphosphates (HUANG, MAHESHWARI and BONNER 1960, BONNER, HUANG and
MAHESHWARI 1961). In fact, DNA in the form in which it is present in chromatin is remarkably inefficient in the support of DNA-dependent RNA
synthesis. It is less effective in supporting this synthesis than an equal
amount of pure DNA in the presence of soluble RNA polymerase. The
authors concluded that the chromosomes contain a factor which renders
chromosomal DNA not fully effective in RNA synthesis. The fact that this
factor is the chromosomal histone has been demonstrated by separating
protein and nucleic acid by means of a centrifugation in 4 M. cesium
chloride, the total proteic fraction thus obtained (histone and non-histone
protein) was extracted with tris buffer (0.05 M., pH 8.0) and the solubilized

non-histone protein recombined with the nucleic acid from which it had previously been removed. The new reaction mixture is thus a chromatin without histone. This removal of histone increases the RNA synthesizing activity of the preparation by approximately fivefold (Huang and Bonner 1962).

Now, if the nucleohistone is reconstituted from its purified components, DNA isolated by the Sevag procedure and histone purified by repeated solubilisation in 2.5 M. NaCl and precipitation at low ionic strenght, such nucleohistone is inactive, or nearly so, in the support of DNA-dependent RNA synthesis in the presence of RNA polymerase. Calf thymus histone is as effective as pea embryo histone in the suppression of the activity of pea embryo DNA and vice versa. The RNA polymerase is not inactivated under these conditions since it proved fully effective when free DNA was added to the reaction mixture in the presence of nucleohistone. The action of histone is therefore to prevent the support of RNA synthesis by DNA. The fact that whole chromatin as obtained from the pea embryo possesses some ability to support DNA-dependent RNA synthesis, while the nucleohistone is totally inactive, may be explained by the presence in pea chromatin not only of DNA in the form we isolate as the native nucleohistone, but also some further form which is not complexed with any stabilizing histone and thus remains active in RNA synthesis (Bonner and Huang 1964).

DNA when it is complexed with histone in the native nucleohistone is altered not only in regard to its biological properties but also to its physical properties. The temperature of half melting of the DNA (T m) in the form of nucleohistone is 14⁰ higher than that of deproteinized DNA in the same medium (Huang and Bonner 1962).

The next step of these studies was to investigate the biological role of the different histone components of the total histone. Huang and coll. (1964) used for such investigations histone of calf thymus which can be separated into a variety of fractions with a number of methods. They have performed this separation by gradient elution with guanidinium chloride from an Amberlite IRC-50 column. They separated the four main histone fractions with different arginine/lysine ratio. In the reconstituted nucleohistones, from each fraction, histone and DNA were present in equivalent amount. They first noted that the half melting point, T m, increased with increasing lysine and decreasing arginine content of the histones; thus, combination with histones I b or II b confers upon DNA a marked increase in stability to thermal denaturation. Concerning the support of DNA-dependent RNA synthesis, the nucleohistone with the lowest arginine/lysine ratio (fraction I b lysine-rich) is nearly inactive, the nucleohistone with the highest arginine/lysine ratio (fraction IV arginine-rich) is rather active, the two others fractions (II b and III) being intermediate. So, the lysine-rich histones form with DNA highly stabilized, high melting nucleohistones, which are inactive in support of RNA synthesis (Table XII, quoted from Bonner and Huang 1964).

We may therefore imagine that in the living cells the active sites of DNA may be free or associated with arginine-rich histone, while inactive sites may be associated with lysine-rich histone.

ALLFREY and MIRSKY (1962), using a quite different test system, also were able to show an inverse correlation between the presence of histone and nuclear RNA synthesis. They used isolated thymus nuclei (ALLFREY et al. 1957) which are able to carry out a wide range of biosynthetic activities, including the synthesis of RNA (ALLFREY and MIRSKY 1957, 1962). This synthesis of RNA seems to concern largely that of the "messenger type" RNA (ALLFREY and MIRSKY 1962). These authors noted that the addition of histone to such isolated nuclei resulted in the repression of a number of their biosynthetic activities (ALLFREY 1961), which is not surprising, considering that the histones have a high ability to form complexes with enzymes or

Table XII. *Effectiveness of varied nucleohistones in support of DNA-dependent RNA synthesis (120 µg. pea DNA per 0.5 ml. reaction mixture in presence of chromosomal RNA polymerase).*

DNA provided as:	RNA synthesized/10 min ($\mu\mu$ moles nucleotide[1])
Nucleohistone I b	0
Nucleohistone II b	24
Nucleohistone III	80
Nucleohistone IV	216
Whole thymus nucleohistone	0
DNA alone	320

[1] Incorporation by enzyme alone subtracted.

their substrates or cofactors. In order to determine wether the repression of RNA synthesis of the nuclei *in vitro* is another instance of their capacity to combine with complex enzyme systems and inhibit them, or whether it is an indication of its true biological function, ALLFREY et al. (1963) removed the histones from the nuclei by action of trypsin which performs a preferential hydrolysis of peptide bonds involving arginine and lysine. They incubated the nuclei with 0.5 mg. per ml. of trypsin (1 mg. enzyme per 40 mg. nuclei). After 30 minute incubation at 37°, 70% of the total histone is removed, the loss of non-histone protein being comparatively small. In such nuclei they noted a stimulation of nuclear RNA synthesis of 200 to 400%. When the histones were added back to trypsin treated nuclei the synthesis of RNA was again inhibited. A preliminary study of the distribution of ^{32}P in separated nucleotides of the newly synthesized RNA of histone depleted nuclei, suggest that its composition would be different from that made in the normal thymus nuclei. LIAU et al. (1965), working on isolated nucleoli of Nowikoff ascites tumours, showed that there was in fact a regulation of RNA synthesis by histones in these nucleoli. The addition or removal of histones affects the amount of RNA synthesized by the nucleoli and also its composition. The RNA synthesized in the nucleoli stripped of basic proteins approaches the composition of DNA while the RNA formed in the presence of added histones approaches the composition of ribosomal RNA. These facts suggest that nuclear histones are involved in the regulation of the production of ribosomal RNA by nucleolar DNA. Thus the removal of histone would result in an activation of normally repressed DNA primers of different average base composition. If this is true, the nuclei should be capable of new synthetic function after removal of the histone by trypsin. But this is still to be demonstrated.

ALLFREY et al. (1963) have also studied the effect of different added histones on the DNA-dependent RNA synthesis in thymus nuclei. The lysine-rich histones and the arginine-rich histones were prepared by the method of JOHNS and BUTLER (1962 b), using isolated nuclei rather than whole tissue; the lysine-rich histone fraction II was isolated by the method of DALY and MIRSKY (1955). The arginine-rich histone fraction I was isolated by isoelectric precipitation at pH 10.6; some arginine-rich and lysine-rich histones were prepared by chromatography on carboxymethylcellulose. The histone fractions are added here in a system containing histone with DNA, so that the DNA is already largely repressed. Nevertheless, they found that F 1 histone fraction (lysine-rich histone) is a less effective inhibitor than the F 3 histone fraction (arginine-rich fraction). HINDLEY (1963), on the other hand, studied the relative ability of reconstituted calf thymus nucleo-histones to support DNA-dependent RNA synthesis. The different fractions were isolated by chromatography on carboxymethylcellulose (JOHNS et al. 1960). He noted that DNA complexed with arginine-rich histone or with slightly lysine-rich histones is almost completely inactive in the support of DNA-dependent RNA synthesis, while DNA complexed with the very lysine-rich histone allows RNA synthesis at a rate which is 70% of that obtained using an equivalent amount of free DNA. ORD et al. (1965) showed that the proteins extracted from rat thymus with 250 mM. hydrochloric acid (which can be considered as arginine-rich) is more active to depress DNA-dependent RNA synthesis that the proteins extracted in 50 mM. hydrochloric acid (lysine-rich), which is in agreement with the results of ALLFREY et al.

These results were in contradiction with the data obtained by HUANG et al. (1964), who showed, as already noted, that a lysine-rich fraction was a very strong inhibitor of RNA polymerase activity in preparations from pea seedlings. But, more recently, ALLFREY and MIRSKY (1964 a) tested the histones used by HUANG et al., prepared by chromatography on Amberlite IRC 50, on their thymus RNA polymerase and found results in agreement with those of these authors. Therefore MIRSKY suggested that the inhibitory capacities of the different histone fractions towards RNA synthesizing systems cannot be simply attributed to their lysine or arginine content; he thinks that other factors may intervene, such as the mode of isolation, the surface charge distribution, the secondary structural modifications induced by acids or by drying, aggregation between different histones in solution, differences in coiling due to differences in proline content.

It is evident that reconstituted nucleohistones might be quite different from nucleohistones as they exist in the living cell and these experiments cannot give us at the present time an exact idea of what occurs in the living nucleus between DNA and histone. BARR and BUTLER (1963), using DNA

11*

from calf thymus and variable amounts of histones and histone fractions, have studied the effect of the added histones on the activity of a RNA polymerase of *Bacillus megatherium* in the presence of calf thymus DNA; the ratio histone/DNA in the reconstituted complex was considered. They showed that $F 2_a$, $F 2_b$ slightly lysine-rich histone and $F 3$ (arginine-rich histone) representing 35, 25 and 20% of the whole histone are able to cause an inhibition of RNA synthesis, varying with the histone/DNA ratio (80 per cent inhibition if the complex contain 15 μg. of DNA and 55 per cent inhibition with 30 μg. DNA). $F 1$ (lysine-rich histone), which is about 20% of the whole histone, causes total inhibition of RNA synthesis at all concentrations used; it must be concluded that in that experiment, $F 1$ inhibits all DNA molecules and not only merely those with which it has a special relationship in the native nucleohistone. Thus, the inhibition of RNA synthesis in reconstituted nucleohistone does not seem to be specific, but as Barr and Butler (1963) point out, we have no evidence that the reconstitution of the original material can be achieved by the present methods of reconstitution.

The mode of action of histones in the living chromosomes can be imagined from experiments on the lampbrush chromosomes of amphibian oocytes. The lateral loops of these chromosomes are active sites of RNA synthesis (Gall and Callan 1962). Isawa et al. (1963), Allfrey and Mirsky (1964 a) noted that RNA synthesis in the loops is completely suppressed by addition of actinomycin D to suspensions of isolated *Triturus viridescens* oocytes and, at the same time, the loops become completely retracted. When puromycin is added to oocytes, the chromosomal protein synthesis is stopped but the loops retain their extended morphology. The authors postulated by analogy that the histone by combining with the DNA would yield a condensed nucleohistone coil in which the primer activity of DNA for RNA synthesis would be impossible. To test this hypothesis, they added different histones to the isolated lamphrush chromosomes. The $F 3$ (arginine-rich fraction which proved highly inhibitory for the RNA synthesis) caused a very rapid retraction of the loops. In contrast $F 1$ (lysine-rich histone, less inhibitory) has a much slower and less marked action on the retraction of the loops. Nevertheless it is necessary to point out that the retraction of the loops can be obtained by other polycations (D and L polylysine for instance).

Other types of experiments permit also to imagine the nature of the inhibitory effect of the histones on DNA-dependent RNA synthesis: Allfrey et al. (1964 b) showed that some modification of histone structure such as acetylation and methylation may affect the capacity of the histones to inhibit ribonucleic acid synthesis *in vivo*. In fact, when isolated arginine-rich histones are subjected to a limited acetylation they lose a great part of their effectiveness as inhibitors of DNA-dependent RNA synthesis. Yet such modified histones are still strongly basic and have an affinity for DNA similar to the non-modified histone. In the cell nucleus, the acetylation and methylation of the histones occur very probably after completion of the polypeptide chain (Allfrey and Mirsky 1964 b). So, as Allfrey et al. points out, the effect of histone on nuclear RNA metabolism is not a mere result of the combination of this basic protein to DNA, it must include more subtle mechanisms for the inhibition or reactivation of RNA production at different loci along the chromosome.

From all these results, it appears that the histone has a definite inhibitory effect on DNA-dependent RNA synthesis of chromatin in *in vitro* systems. This fact is a strong support for the theory of the STEDMANS of histones considered as gene inhibitors. Cell differentiation may thus be realized by a differential inhibition of a number of genes. Recent experiments of BONNER et al. (1963) on chromosomally directed protein synthesis in "*in vitro* systems", using various tissues of pea plants are in agreement with this idea. These authors have coupled a chromatin-dependent RNA synthetic system to a messenger RNA-dependent ribosomal protein system derived from *Escherichia coli*. The ribosomal system under the direction of chromatin derived from developing pea cotyledons, synthesizes proteins, including the pea seed reserve globulin which is characteristic of the cotyledons. On the contrary, chromatin of pea buds, which do not synthesize globulin *in vivo*, does not support the synthesis of globulin by the isolated ribosomal system. The control of the genetic activity characteristic of this or that living tissue seems to be preserved in the isolated chromatin. The removal of the histone from pea bud chromatin, in which the genes for globulin synthesis are normally repressed, gives a DNA which supports globulin synthesis. This experiment is very elegant, and if it is reproduced on other materials would prove definitely the control by the histones of gene activity.

Concerning the possible mechanism of gene recognition by the histones, HUANG and BONNER (1965) suggest that this recognition would be due not to the histones but to a special type of RNA attached to them. This RNA, they have found associated with the histone, has an average chain length of 40 nucleotides and is distinguished by its high content of dihydrouridylic acid. The "represseron" as they call it, would be formed of 15 or more histone molecules bound together by hydrogen bonds with an RNA molecule covalently attached to it. The RNA would detect the nucleotide sequence and hence which gene is to be repressed and the histone effects the repression.

All these experiments seems to be quite significant and suggest a gene regulation by the histones; nevertheless some authors do not agree with this interpretation of the results. SONNENBERG and ZUBAY (1965) showed that native calf thymus DNP forms an impenetrable gel-like structure which has a poor primer ability for RNA synthesis *in vitro;* sonicated DNP forms a dispersed precipitate with one third the priming efficiency of the equivalent amount of DNA. They conclude that the *in vitro* reduction by histone of DNA-primed RNA synthesis is due to this precipitation effect rather to a genetic repression.

II. Action of Histones on DNA Synthesis

A few recent works demonstrate that the histones may have a regulatory effect on the synthesis of DNA. We have already quoted the *in vivo* experiments of IRVIN et al. (see page 64) in regenerating liver, suggesting that the histone/DNA ratio may vary in the different stages of the cell life; the beginning of the mitosis being correlated with a marked lowering of this ratio.

In vitro experiments were performed more recently and all of them show the inhibitory action of histones on DNA synthesis.

Billen and Hnilica (1964), Hnilica and Billen (1964) used partially purified DNA polymerase preparations from *E. coli* (Billen 1963) in their *in vitro* system for the synthesis of DNA. They incubated this system with whole calf thymus histone and the differents histone fractions F 1, F 2$_a$, F 2$_b$ and F 3 obtained as described by Johns and Butler (1962 b) and by Hnilica and Busch (1963). They found that whole histone at sufficiently high concentrations is able to suppress DNA synthesis totally. More than 50% inhibition was observed when the ratio of the weight of added histone to the weight of DNA primer was 2 : 1.

The four fractions of the histone were able to prevent DNA replication when present at a ratio histone/DNA superior to 4 : 1, but at lower concentrations, these different histone fractions show different activities. The F 1 (very lysine-rich histone) is definitely the most active, inhibiting the DNA replication at a histone/DNA ratio of 1 : 1. Higher ratio of F 2$_b$ (lysine-rich histones) and F 2$_a$ and F 3 (arginine-rich histone) were required for a complete inhibition. In all these cases the inhibitory effect was concentration dependent (a rising ratio of histone to DNA increases the inhibition).

Bazill and Philpot (1963) in thymus nuclei, Schwimmer and Bonner (1964), Gurley et al. (1964) confirmed the inhibitory effect of histones on the biosynthesis of DNA *in vitro*. Gurley et al. (1964) studied this inhibition on two stages of DNA synthesis: the kinase stage and the DNA polymerase stage, using enzyme preparation from regenerating rat liver (Behke and Schneider 1963). They noted but a small or no effect at all of the histones at the kinase stage. At the polymerase stage as in the experiments of Billen and Hnilica, they found that the F 1 (very lysine-rich fraction) was most inhibitory but they needed a very high ratio of histone to DNA (4.0) to obtain a total inhibition.

Hnilica and Billen (1964) showed also that a mild enzymatic digestion of histones readily destroys their inhibitory activity on DNA replication. Thus, a minimal size is required for the histone molecule to be active. A high lysine content seems to be very important, the very lysine-rich fraction being the most effective; only polylysine shows approximately the same degree of inhibition, but free lysine is ineffective. Thus a definite sequence of lysine residues seems to be essential for the maximal inhibitory effect. Sluyser et al. (1965) reported also that the intraperitoneal injection of histone components into partially hepatectomized rats led to a decrease of the DNA synthesis in the regenerating liver. No significant difference appeared between the specific inhibitory activities *in vivo* of the different histone fractions but *in vitro* the fraction F 1 seems the more active.

The results of Bollum (1963) on *Euplotes* nuclei seems to indicate also an inhibitory action of histone on DNA replication. These nuclei showed no significant priming activity when DNA polymerase was added, but after treatment with 0.01 N. HCl (which removes the histone), the priming activity is revealed.

LEHNERT (1964), using nucleoproteins from nuclei of normal or regenerating rat liver, showed that the non-histone, acid-insoluble part of the nucleoprotein inhibits also the incorporation of ^{3}H-thymidine into DNA.

III. Action of Histones on Living Cells and Organisms

The action of histones or protamines on living cells or living organisms are less significant than assays on cell free systems because of their toxicity; in fact they display a strong bactericidal and a cytotoxic effect.

SANDRITTER et al. (1959), BECKER and GREEN (1960) showed that histones and protamines are incorporated very rapidly into cells grown *in vitro*. It is clear that the presence of these highly positively charged proteins in the cells have a deep and harmful effect on their metabolism. A depressive effect of histone on tumour growth was noted by ZBARSKI and PEREBOTCHIKOVA (1954). These authors mixed histone (from calf thymus, mouse liver etc.) to homogenates of Crooker sarcoma before implantation. The tumours appeared after a twice longer time in mice injected with sarcoma + histone, than in mice injected with sarcoma alone. The fraction of histone soluble in ammonia showed a higher toxicity for the mouse and a more pronounced inhibitory activity on the tumour growth.

More recently, VOROBYEV and BRESLER (1963) added histone (from rat liver, rat thymus or calf thymus) to tumour tissue (liver cancer) and let the mixture stand for 16 h. at 4⁰ C. before inoculation into rats. They showed that the tumour cells were not damaged by this treatment, but there was a very strong inhibition of the tumour growth. BLAZSEK and GYERGYAY (1965) showed that the *in vitro* incubation of Ehrlich ascites cells with histones before inoculation resulted in a strong inhibition of the tumour growth independent of the kind of histone used. They think that the inhibitory effect of histone on tumour growth is probably due to a mere electrostatic interaction between the histone and the surface of the tumour cell which bears a negative charge and may not be attributed to a specific inhibition of the genes.

HNILICA and HOLOUBEK (1961) injected calf thymus histones into rats bearing transplantable acute myeloid leukaemia one hour before transplantation in a first group, 3 hours after transplantation in a second group and then on the third and fifth day. Under these conditions, they noted no cytostatic effect on the leukaemic cells; on the contrary, probably as a result of histone toxicity, the survival of animal treated with histone was less than 30—40 per cent of controls and of transplantation experiments. It appears from this experiment that the histones cannot be considered as possible therapeutic agents against cancer.

Let us quote here some works concerning other effects of histones: HOLOUBEK (1962) studied the stimulation of DNA synthesis in different organs of the rat after injection of tumour histones. His data suggest that the stimulation of DNA synthesis, which was demonstrated in tumour bearing animals after the implantation of a tumour, was due to tumour histones released in the animal by partial lysis of the graft.

CAFFERY et al. (1964) found an inhibitory effect by injected histones on the induction by hydrocortisone of two enzymes in mouse liver: tryptophane pyrrolase and tyrosine α ketoglutarate transaminase. MARKERT and URSPRUNG (1963) injected histones from frog liver to frog eggs and noted that the developments of these eggs were arrested in blastula, but globulin and albumin are more active than histone in this respect. GOODWIN and SIZER (1964) found that histone added at low concentration 100 μg./ml. to a culture of chick embryo brain causes an inductive response in lactic dehydrogenase activity. At higher concentrations of histone, the response is repressive. Some of these results may be interpreted as indicating a possible regulatory function of the histones.

The histones and protamines seem also to stabilize nucleic acids against enzymatic actions. The addition of histone to poliovirus ribonucleic acid preparations (LUDWIG and SMULL 1963) increases the infectivity of this RNA. SZYBALSKI and SZYBALSKA (1963) showed also that basic protamines increase the transforming action of DNA on human cells *in vitro*, and WILZOK and MENDECKI (1963) noted that protamines and histones increase the incorporation of polymerized DNA in living cells. This stabilizing action reflects probably the protective action of histones towards DNA in the chromosomes of the living cells.

IV. Conclusion

A great deal of work has been performed recently on histones and the hypothesis of their possible role of gene regulators has received some support from recent experiments showing their inhibitory action on DNA-dependent RNA synthesis and on DNA replication. Though this role of regulation appears very likely, it is not well understood and some authors still think that histone have only a structural role in the constitution of the chromosomes. The chief problem is now to understand how the histones may perform the important task they are supposed to have in the cell nucleus, and any result can throw some light on the mechanism of their action. Studies on the structure of nucleohistones and interaction between histone and DNA are very important in this respect.

The search for histone specificity in different tissues and species may give also significant results. The heterogeneity of the histones and the different activities of the different fractions in gene inhibition is also a very promising field for the understanding of the histone action. The fact that acetylation or methylation of histones modify their activity shows perhaps one of the mechanisms realised in the cell for gene regulation.

Bibliography

AGRELL, I. P. S., and E. G. CHRISTENSSON, 1965: Changes of histone composition in the developing chick embryo. Nature 207, 638—640.
AHLSTROM, L., 1947: Nucleoprotein in normal and cancer cells. V. The histone from the nuclei of thymus cells. Ark. Kem. Min. Geol. (A) 24, 31.
ALFERT, M., 1955: Quantitative cytochemical studies on patterns of nuclear growth. In "Fine structure of cells", Noordhoff, Groningen, 157—163.
— 1956: Chemical differentiation of nuclear proteins during spermatogenesis in the Salmon. Biophys. Biochem. Cytol. 2, 109—114.

ALLFREY, V. G., 1961: Amino acid transport and early stages in protein synthesis in isolated cell nuclei, in "Biological structure and function", Academic Press, Vol. I, 261—280.
— M. M. DALY, and A. E. MIRSKY, 1955: Some observations on protein metabolism in chromosomes of non dividing cells. J. Gen. Physiol. 38, 415—424.
— R. FAULKNER, and A. E. MIRSKY, 1964: Acetylation and regulation of histones and their possible role in the regulation of RNA synthesis. Proc. Nat. Acad. Sci., U. S. A., 51, 786—793.
— V. C. LITTAU, and A. E. MIRSKY, 1963: On the role of histones in regulating ribonucleic acid synthesis in the cell nucleus. Proc. Nat. Acad. Sci., U. S. A., 49, 414—421.
— — — 1964: Methods for the purification of thymus nuclei and their application to studies of nuclear protein synthesis. J. of Cell Biol. 21, 213—231.
— and A. E. MIRSKY, 1957: Some aspects of ribonucleic acid synthesis in isolated cell nuclei. Proc. Nat. Acad. Sci., U. S. A., 43, 821—826.
— — 1962: Evidence for the complete DNA dependence of RNA synthesis in isolated thymus nuclei. Proc. Nat. Acad. Sci., U. S. A., 48, 1590—1596.
— — 1964 a: Role of histone in nuclear function. — In "The nucleohistones", J. BONNER and P. Ts'o, ed., Holden-Day, inc., San Francisco, London, Amsterdam; 267—288.
— — 1964 b: Structural modifications of histones and their possible role in the regulation of RNA synthesis. Science 144, 559.
— — and S. OSAWA, 1957: Protein synthesis in isolated cell nuclei. J. Gen. Physiol. 40, 451—490.
— — and H. STERN, 1955: The chemistry of the cell nucleus. Adv. Enzymol. 16, 411—500.
ANDO, T., E. ABUKUMAGAWA, Y. NAGAI, and M. YAMASAKI, 1957 a: On the N-terminal residues and sequence of clupeine from Clupea pallasii: the occurence of proline in addition to alanine as the N-terminus. J. Biochem. 44, 191—194.
— and C. HASHIMOTO, 1958 a: Changes of the protein in the cell nuclei of the testis during the formation of spermatozoa of the Rainbow-trout (salmo irideus). Proc. Intern. Symposium on Enzyme Chem. Tokyo and Kyoto 1957, 380.
— — 1958 b: Studies on protamines IV. The basic proteins of the testis cell nuclei of adult and immature rainbow-trouts in breeding season. J. Biochem. 45, 453—460.
— — 1958 c: Studies on protamines V. Changes of the proteins in the cell nuclei of the testis during the formation of spermatozoa of the rainbow-trout (Salmo irideus). J. Biochem. (Tokyo) 45, 529—540.
— S. ISHII, and M. SATO, 1959: Studies on protamines VI. Amino acid composition of clupeine and Salmine. J. Biochem. (Tokyo) 46, 933—940.
— — M. YAMASAKI, K. IWAI, C. HASHIMOTO, and F. SAWADA, 1957 b: Studies on protamines. I. Amino acid composition and homogeneity of clupeine, salmine and iridine. J. Biochem. (Tokyo) 44, 275—288.
— K. IWAI, and M. KIMURA, 1958 a: Studies on protamines. II. Determination of the end groups of clupeine and salmine and their molecular weight. J. Biochem. (Tokyo) 45, 27—39.
— and F. SAWADA, 1962: Studies on protamines. X. Heterogeneity of clupeine from single herring. J. Biochem. (Tokyo) 51, 416—421.
— M. YAMASAKI, E. ABUKUMAGAWA, S. ISHII, and Y. NAGAI, 1958 b: Studies on protamines. III. N-terminal residues of salmine and clupeine. J. Biochem. 45, 429—451.

BAKAY, B., L. B. KIRSCHNER, and G. TOENNIES, 1960: Fractionation of calf thymus deoxyribonucleoprotein on ion exchange columns. Biochem. Biophys. Res. Comm. 2, 459—463.
— J. J. KOLB, and G. TOENNIES, 1955: On the component proteins of calf thymus nucleoprotein. Arch. Biochem. 58, 144—168.
BANG, I., 1899: Studien über histon. Z. Physiol. Chem. 27, 463—486.
BARR, G. C., and J. A. V. BUTLER, 1963: Histones and gene function. Nature 199, 1170—1172.
BAZILL, G. W., and J. S. L. PHILPOT, 1963: Studies on the assay of primer DNA in the presence of histone and in isolated nuclei. Biochim. Biophys. Acta. 76, 223—233.
BECK, J. S., and P. J. WALKER, 1964: Antigenicity of trypanosoma nuclei. Evidence that DNA is not coupled to histones in these protozoa. Nature 204, 194—195.

Becker, F. F., and H. Green, 1960: The effects of protamines and histones on the nucleic acids of ascites tumor cells. Exper. Cell. Res. 19, 361—375.

Behke, R. M., and W. C. Schneider, 1963: Incorporation of tritiated thymidine into deoxyribonucleic acid by isolated nuclei. Biochim. Biophys. Acta. 68, 34—44.

Belozerski, A. N., and S. O. Uryson, 1958: The composition of the nuclear nucleo-proteins of some plants. Biokhimiia, USSR (English translation) 23, 532—537.

Berstein, M., and D. Mazia, 1953: The deoxyribonucleoprotein of sea urchin sperm. I. Isolation and analysis. Biochim. Biophys. Acta. 10, 600—606.

Bhagavan, N. V., and W. A. Atchley, 1965: Properties of a deoxyribonucleoprotein complex derived from Bacillus subtilis. Biochem. 4, 234—239.

Bijvoet, P., 1957: Ethanol precipitation analysis of thymus histone. Biochim. Biophys. Acta. 25, 502—512.

Billen, D., 1963: On the nature of a deoxynucleotide polymerase. Deoxyribonucleic acid complex. Biochim. Biophys. Acta. 68, 342—353.

— and L. S. Hnilica, 1964: Inhibition of DNA synthesis by histones in "The nucleo-histones", J. Bonner and P. Ts'o ed. Holden-Day, Inc. San Francisco, London, Amsterdam; 289—297.

Birnstiel, M. L., and W. G. Flamm, 1964: In comment of article of Johns, E. W., Studies on lysine-rich histones, in "The Nucleohistones", J. Bonner and P. Ts'o, ed. Holden-Day, inc., San Francisco, London, Amsterdam; 57.

Biserte, G., and P. Sautiere, 1958: Groupes a aminés terminaux des histones du thymus de veau et de tumeurs expérimentales du rat. C. R. Acad. Sci. (Paris) 246, 1764.

Blazsek, V. A., and F. Gyergyay, 1965: A note on inhibition of the Ehrlich ascites tumour growth with histones. Exptl. Cell Res. 38, 424—437.

Bloch, D. P., 1962: Histones synthesis in non replicating chromosomes. J. Histochem. Cytochem. 10, 137—144.

— and G. C. Godman, 1955: A microphotometric study of the synthesis of deoxyribonucleic acid and nuclear histone. J. Biophys. Biochem. Cytol. 1, 17—28.

Block, R. I., D. Bolling, H. Gershon, and H. A. Sober, 1949: Preparation and amino acid composition of salmine and clupeine. Proc. Soc. Exper. Biol. Med. 70, 494—496.

Block, D. P., S. D. Brack, 1964: Evidence for the cytoplasmic synthesis of nuclear histone during spermiogenesis in the Grasshopper Chortophaga Viridifasciata (De Geer). The J. of Cell Biology 22, 327—340.

Boivin, A., R. Vendrely, and C. Vendrely, 1948: L'acide deoxyribonucléique du noyau cellulaire dépositaire des caractères héréditaires. Arguments d'ordre analytique. C. R. Acad. Sci. (Paris) 226, 1061—1063.

Bollum, F. J., 1963: Intermediate states in enzymatic DNA synthesis. J. Cell. Comp. Physiol. 62, Suppl. I, 61—71.

Bonner, J., and R. C. Huang, 1964: Role of histone on chromosomal RNA synthesis in "The Nucleohistones", J. Bonner and P. Ts'o, ed., Holden-Day, Inc., San Francisco, London, Amsterdam; 251—261.

— — and R. V. Gilden, 1963: Chromosomally directed protein synthesis. Proc. Nat. Acad. Sci., U. S. A., 50, 893—900.

— — and N. Maheshwari, 1961: The physical state of newly synthesized RNA. Proc. Nat. Acad. Sci., U. S. A., 47, 1548—1560.

Braun, W., J. W. Burrous, and J. H. Phillips, 1957: A phenol extracted bacterial deoxyribonucleic acid. Nature 180, 1356—1357.

Brunish, R., and J. M. Luck, 1952: Amino acid incorporation in vivo into liver fractions. J. Biol. Chem. 198, 621—628.

Bryan, J. H. D., 1964: Nucleoprotein synthesis in the new-born mouse. Nature 204, 574.

Busch, H., 1962: An introduction to the biochemistry of the cancer cell. Academic Press, New York, London.

— P. Byvoet, and H. R. Adams, 1963 a: Chromatographic analysis of acid soluble proteins obtained from the nuclear sap and deoxyribonucleoproteins. Exper. Cell Res. Suppl. 9, 376—386.

— J. R. Davis, and D. C. Anderson, 1958: Labeling of histones and other nuclear proteins with L lysine U-C[14] in tissues of tumor-bearing rats. Cancer Res. 18, 916—926.

— — G. R. Honig, D. C. Anderson, P. V. Nair, and W. L. Nyhan, 1959: The uptake of a variety of amino acids into nuclear proteins of tumors and other tissues. Cancer Res. 19, 1030—1039.

BUSCH, H., L. S. HNILICA, S. C. CHIEN, J. R. DAVIS, and C. W. TAYLOR, 1962: Isolation and purification of RP 2-L, a nuclear protein fraction of the Walker 256 carcinosarcoma. Cancer Res. 22, 637—645.
— and H. MAVIOGLU, 1964: Peptides of histone fraction 2 A. In "The Nucleohistones", J. BONNER and P. Ts'o, ed., Holden-Day, inc., San Francisco, London, Amsterdam; 79—83.
— W. J. STEELE, L. S. HNILICA, and C. TAYLOR, 1964: Metabolism of histones. In "The Nucleohistones", J. BONNER and P. Ts'o, ed., Holden-Day, inc., San Francisco, London, Amsterdam; 242—245.
— — — C. W. TAYLOR, and H. MAVIOGLU, 1963 b: Biochemistry of histones and the cell cycle. J. Cell. and Comp. Physiol.. Suppl. I, 62, 95—110.
BUTLER, J. A. V., 1959: Les histones. Leurs structures, variations et rôle dans les organismes animaux. Exposés ann. de Biochimie Médicale, 21st series, 41—55.
— 1963: The nuclear proteins of normal and cancer cells. Exper. Cell Res., Suppl. 9, 349—358.
— 1964: Fractionation and characteristics of histones, in "The Nucleohistones", J. BONNER and P. Ts'o, ed., Holden-Day, inc., San Francisco, London, Amsterdam; 36—45.
— and P. COHN, 1963: Studies on histones. 6-Observations on the biosynthesis of histones and other proteins in regenerating rat livers. Biochem. J. 87, 330—333.
— P. F. DAVISON, and D. W. F. JAMES, 1953: Isolation and properties of histones from thymus nucleoproteins. Proc. Biochem. Soc. Biochem. J. 54, XXI.
— — — and K. V. SHOOTER. 1954: The histones of calf thymus deoxyribonucleoprotein. I. Preparation and homogeneity. Biochim. Biophys. Acta. 13, 224—232.
— and G. N. GODSON, 1963: Biosynthesis of nucleic acids in Bacillus megatherium. I. The isolation of a nuclear material. Biochem. J. 88, 177—182.
BYVOET, P., 1964: Histone synthesis and mitotic activity in normal and neoplastic tissue of the rat. Proceed. of the Amer. Ass. for Cancer Res. (55th Ann. Meeting Chicago) 5, 9, Nr. 34.
— and H. BUSCH, 1961: DNA binding of RP 2 L, a nuclear protein of neoplastic tissues. Nature 192, 870—871.

CAFFERY, J. M., L. WHICHARD, and J. L. IRVIN, 1964: Effects of histones on the induction of two liver enzymes by hydrocortisone. Arch. Biochem. Biophys. 108, 364—365.
CAMPBELL, P. N., O. GREENGARD, and H. E. H. JONES, 1957: The intracellular distribution of amino acid incorporation by slices of liver and liver tumors and by ascites cells. Exper. Cell Res. 12, 689—692.
CASPERSSON, T., 1941: Studien über den Eiweißumsatz der Zelle. Naturwiss. 29, 33—43.
CHARGAFF, E., C. F. CRAMPTON, R. LIPSHITZ, 1953: Separation of calf thymus deoxyribonucleic acid into fractions of different composition. Nature 172, 289—292.
— and R. LIPSHITZ, 1953: Composition of mammalian desoxyribonucleic acids. J. Amer. Chem. Soc. 75, 3658—3661.
— and H. F. SAIDEL, 1949: On the nucleoproteins of avian Tubercle Bacilli. J. Biol. Chem. 177, 417—428.
— E. VISCHER, R. DONIGER, C. GREEN, and F. MISANI, 1949: The composition of the desoxypentose nucleic acids of thymus and spleen. J. Biol. Chem. 177, 405—416.
CHAUVEAU, J., Y. MOULE, and C. ROUILLER, 1956: Isolation of pure and unaltered liver nuclei, morphology and biochemical composition. Exper. Cell Res. 11, 317—321.
CHRISTOMANOS, A. A., and A. DIMITRIADIS, 1964: Zur Kenntnis der Histone der roten Blutkörperchen des Huhnes I., Enzymologia 27, 23—29.
COMMERFORD, S. L., M. J. HUNTER, and J. L. ONCLEY, 1963: The preparation and properties of calf liver deoxyribonucleoprotein. J. Biol. Chem. 238, 2123—2134.
CRAMPTON, C. F., R. LIPSHITZ, and E. CHARGAFF, 1954 a: Studies on nucleoproteins. I. Dissociation and reassociation of the deoxyribonucleohistone of calf thymus. J. Biol. Chem. 206, 499—510.
— — — 1954 b: Studies on nucleoproteins. II. Fractionation of DNA through fractional dissociation of their complexes with basic proteins. J. Biol. Chem. 211, 125—142.
— S. MOORE, and W. H. STEIN, 1955: Chromatographic fractionation of calf thymus histone. J. Biol. Chem. 215, 787—801.
— W. H. STEIN, and S. MOORE, 1957: Comparative studies on chromatographically purified histones. J. Biol. Chem. 225, 363—386.

CRUFT, H. J., 1964: Electrophoresis and gel filtration of histones, in "The nucleo-histones", J. BONNER and P. Ts'o, ed., Holden-Day, inc., San Francisco, London, Amsterdam; 72—78.
— J. HINDLEY, C. M. MAURITZEN, and E. STEDMAN, 1957 a: Amino acid composition of the six histones of calf thymocytes. Nature 180, 1107—1109.
— and J. L. LEAVER, 1961: Isolation of histones from staphyloccus aureus. Nature 192, 556—557.
— C. M. MAURITZEN, and E. STEDMAN, 1954: Abnormal properties of histones from malignant cells. Nature 174, 580—582.
— — — 1957 b: The nature and physicochemical properties of histones. Phil. Trans. Roy. Soc. London, Ser. B. 241, 93—145.

DALY, M. M., V. G. ALFREY, and A. E. MIRSKY, 1952: Uptake of N^{15} glycine by com-ponents of cell nuclei. J. Gen. Physiol. 36, 173—179.
— and A. E. MIRSKY, 1955: Histone with high lysine content. J. Gen. Physiol. 38, 405—413.
— — and H. RIS, 1951: The amino acid composition and some properties of histones. J. Gen. Physiol. 34, 439—450.
DAS, N. K., and M. ALFERT, 1961: Accelerated DNA synthesis in onion root meristem during X-irradiation. Proc. Nat. Acad. Sci., U. S. A., 47, 1—6.
DAS, C. C., B. P. KAUFMANN, and H. GAY, 1964: Histone protein transition in Droso-phila Melanogaster I. Changes during spermatogenesis. Exptl. Cell Res. 35, 507—514.
DAVIS, J. R., and H. BUSCH, 1959: Chromatographic analysis of radioactive cationic nuclear proteins of tissues of tumor-bearing rats. Cancer Res. 19, 1157—1166.
— — 1960: Chromatographic analysis of cationic nuclear proteins of a number of neoplastic tissues. Cancer Res. 20, 1208—1213.
DAVISON, P. F., 1957 a: Histones from normal and malignant cells. Biochem. J. 66, 703—707.
— 1957 b: Chromatography of histones. Biochem. J. 66, 708—712.
— and J. A. V. BUTLER, 1954: The fractionation and composition of histones from thymus nucleoprotein. Biochim. Biophys. Acta. 15, 439—440.
— — 1956: The chemical composition of calf thymus and nucleoprotein. Biochim. Biophys. Acta. 21, 568—573.
— B. E. CONWAY, and J. A. V. BUTLER, 1954 a: The nucleoprotein complex of the cell nucleus and its relations. Progress in Biophysics and Biophys. Chem., Perga-mon Press, London, 148—194.
— D. W. F. JAMES, K. V. SHOOTER, and J. A. V. BUTLER, 1954 b: The histones of calf thymus deoxyribonucleoprotein. II. Electrophoretic and sedimentation behaviour and a partial fractionation of histones of DNP. Biochem. Biophys. Acta. 15, 415—424.
— and K. V. SHOOTER. 1956: Sedimentation, electrophoretic and chromatographic studies of whole and fractionated calf thymus histones. Bull. Soc. Chim. Belg. 65, 85—96.
DE, D. N., and S. N. GHOSH, 1965: Cytochemical evidence for the apparent absence of histone in the cells of Cyanophyceae. The J. of Histo- and Cytochem. 13, 298.
DEBOV, S. S., 1951: Amount of protein fractions in the cell nuclei of normal and malignant tissues. Biokhimiya 16, 314—320.
DOTY, P., and G. ZUBAY, 1956: The preparation and properties of completely disper-sed thymus nucleoprotein. J. Amer. Chem. Soc. 78, 6207—6208.
DOUNCE, A. L., 1955: The isolation and composition of cell nuclei and nucleoli, in "The nucleic acids, Chemistry and Biology", ed. E. CHARGAFF and J. N. DAVIDSON. Acad. Press (New York) 2, 93—154.
— 1959: The state of DNA in the resting-cell nucleus. Ann. N. Y. Acad. Sci. 81, 794—799.
— and M. O'CONNELL, 1958: Composition and properties of the thymus desoxyribo-nucleoprotein of Doty and Zubay. J. Amer. Chem. Soc. 80, 2013—2015.
— and N. K. SARKAR, 1960: Nucleoprotein organization in cell nuclei and its relation-ship to chromosomal structure. Faraday Soc. Meet. on "The cell Nucleus", Cam-bridge August-September 1959, Butterworths, London, 206 210.
DULBECCO, R., 1964: The histones as candidates for a role in genetic repression. — In "The Nucleohistones", J. BONNER and P. Ts'o, ed., Holden-Day, inc., San Francisco, London, Amsterdam; 362—366.

EULER, H., L. HAHN, H. HASSELQUIST, M. JAARMA, and M. LUNDIN, 1945: Studies on the isolation and characterization of cell nuclei from calf thymus, dog liver and Jensen rat sarcoma. Svensk. Kem. Tid. **57**, 217—227.

FELIX, K., 1952: Zur Chemie des Zellkerns. Experientia VIII, 312—317.
— 1953: Protamines and nucleoprotamines. In "Chemical structure of proteins", J. and A. CHURCHILL, London.
— 1958: Formation et fonction des nucléoprotamines. Bull. Soc. Chim. Biol. **40**, 17—33.
— 1959: Some principles of molecular biology "Symposium on Molecular Biology". Chicago University Press, 1—15.
— 1960: Protamines, Adv., in protein Chem., Academic Press., New York, London; 1—56.
— H. FISCHER, and A. KREKELS, 1956: Protamines and nucleoprotamines. Progress in Biophys. London, Pergamon Press **6**, 1—23.
— and A. KREKELS, 1953: Die Endaminosäuren einiger Protamine. Z. Physiol. Chem. Hoppe-Seyler's **295**, 107—109.
FEUGHELMAN, M., R. LANDGRIDGE, W. E. SEEDS, A. R. STOCKES, H. R. WILSON, C. W. HOOPER, M. H. F. WILKINS, R. K. BARCLAY, and L. D. HAMILTON, 1955: Molecular structure of deoxyribose nucleic acid and nucleoprotein. Nature **175**, 834—838.
FISCHER, H., and L. KREUZER, 1953: Über Gallin. Z. Physiol. Chem. Hoppe-Seyler's **293**, 176—182.
FLAMM, W. G., and M. L. BIRNSTIEL, 1964: Studies on the metabolism of nuclear basic proteins. — In "The Nucleohistones", J. BONNER and P. Ts'o, ed., Holden-Day. inc., San Francisco, London, Amsterdam; 230—240.
FRICK, G., 1949: Some physico-chemical properties of thymo-nucleoprotein prepared according to MIRSKY and POLLISTER. Biochim. Biophys. Acta. **3**, 103—116.
— 1956: A separation in weak magnesium chloride solutions of nuclear material into two fractions with different purine and pyrimidine content. Biochim. Biophys. Acta. **19**, 352—365.

GAJDUSEK, D. C., 1950: Desoxyribonucleoprotein from bovine spleen. Biochim. Biophys. Acta. **5**, 397—403.
GALL, J. G., and H. G. CALLAN, 1962: H^3 uridine incorporation in lampbrush chromosomes. Proc. Nat. Acad. Sci., U. S. A., **48**, 562—570.
GENTY, N., Y. COIRAULT, and R. VENDRELY, 1966: Effect of the acid concentration on the characteristics of the isolated histone fractions. C. R. Acad. Sci., Paris, in Press.
GEORGIEV, G. P., L. P. YERMOLAYEVA, and I. B. ZBARSKY, 1960: Les relations quantitatives des fractions proteiques et nucleoproteiques des noyaux cellulaires de differents tissus. Biokhimiya **25**, 318—322.
GILBERT, I. G. F., and J. M. RADLEY, 1964: A procedure for the isolation of cell nuclei for trace metal studies. Biochem. Biophys. Acta. (Amsterdam) **82**, 618—621.
GOODWIN, B. C., and I. W. SIZER, 1964: Histone regulation of lactic dehydrogenase in embryonic chick brain tissue. Science **148**, 242—244.
GOPPOLD-KREKELS, A., and H. LEHMANN, 1958: The paper-chromatographic distribution of protamines. Z. Physiol. Chem. Hoppe-Seyler's **313**, 147—151.
GREGOIRE, J., and M. LIMOZIN, 1954: Recherches sur l'hétérogeneité et la composition des preparations de thymohistone. Bull. Soc. Chim. Biol. (Paris) **36**, 15—30.
GULLAND, J. M., D. O. JORDAN, and C. J. THRELFALL, 1947: Deoxypentose nucleic acids. Part 1. Preparation of the tetrasodium salt of the deoxypentose nucleic acid of calf thymus. J. Chem. Soc. 1129—1130.
GURLEY, L. R., J. IRVIN, and D. J. HOLBROOK, 1964: Inhibition of DNA polymerase by histones. Biochem. Biophys. Res. Comm. **14**, 527—532.

HAMER, D., 1951: Aspects of the chemistry of the proteins in the nucleus. A short review and some experimental results. Brit. J. Cancer **5**, 130—139.
— 1955: The composition of the basic proteins of Echinoderm sperm. Biol. Bull. **108**, 35—39.
— and D. L. WOODHOUSE, 1949: Amino acid composition of salmine. Nature **163**, 689—690.
HAMMARSTEN, E., 1924: Zur Kenntnis der biologischen Bedeutung der Nukleinsäure-Verbindungen. Biochem. Z. **144**, 383—466.

Hashimoto, C., 1955: Studies on DNP-Protamines by means of the absorption spectra. Bull. Chem. Soc. (Japan) 28, 385—389.

Heroin-Delauney, Y., 1965: Quelques données autohistoradiographiques sur le kineto-plaste du Trypanosoma Cruzi. Ann. Histochim. 10, 25—34.

Hidveji, E. J., F. Antoni, I. Arky, and V. Varteresz, 1964: Studies on histone metabolism and heterogeneity in tumours and bone marrow. — In "Cellular control mechanisms and cancer". Elsevier pub., 226—229.

Hindley, J., 1963: The relative ability of reconstituted nucleohistones to allow DNA dependent RNA synthesis. Biochem. Biophys. Res. Comm. 12, 175—179.

Hirschbein, L., and Y. Khouvine, 1957: Ultracentrifugation et electrophorèse d'histones de placentas humains. C. R. Acad. Sci. (Paris) 244, 517—520.

Hnilica, L. S., 1959: The extractibility of calf thymus histone fractions. Experientia 15, 139—140.

— 1964: The specificity of histones in chicken erythrocytes. Experientia 20, 13—14.

— 1965: The fractionation of arginine rich histones from calf thymus. Experientia 21, 124—126.

— and D. Billen, 1964: The effect of DNA-histone interactions on the biosynthesis of DNA in vitro. Biochim. Biophys. Acta. 91, 271—280.

— and H. Busch, 1963: Fractionation of the histones of the Walker 256 carcinosarcoma by combined chemical and chromatographic techniques. J. Biol. Chem. 238, 918—924.

— and V. Holoubek, 1961: Effect of calf thymus histone on rats bearing transplantable acute myeloid leukaemia. Nature 191, 922—923.

— E. W. Johns, and J. A. V. Butler, 1962: Observations on the species and tissues specificity of histones. Biochem. J. 82, 123—129.

— C. W. Taylor, and H. Busch, 1963: Analysis of peptides of the moderatly lysine-rich histone fraction, F 2 B of the Walker tumor and other tissues. Exper. Cell Res., Suppl. 9, 367—375.

— — — 1964: Peptides of histone fraction 2 B. In "The Nucleohistones", J. Bonner and P. Ts'o, ed., Holden-Day, inc., San Francisco, London, Amsterdam; 84—91.

Holbrook, D. J., J. H. Evans, and J. L. Irvin, 1962: Change in content of nuclear proteins and nucleic acids in regenerating liver. Exper. Cell Res. 28, 120—125.

— J. L. Irvin, and E. M. Irvin, 1959: Relative rates of synthesis of various proteins in nuclei. Fed. Proceed. 18, Nr. 1, Part. I, 248.

— — and J. Rotherham, 1960: Incorporation of glycine into protein and nucleic acid fractions of nuclei of liver and hepatoma. Cancer Res. 20, 1329—1337.

Holoubek, V., 1962: Stimulation of DNA synthesis with histones from tumor tissues. Proceed. of the Soc. for Exper. Biol. and Med. 110, 759—761.

Huang, R. C., and J. Bonner, 1962: Histone, a suppressor of chromosomal RNA synthesis. Proc. Nat. Acad. Sci. (U. S. A.) 48, 1216—1222.

— — 1965: Histone-bound RNA, a component of native nucleohistone. Proc. Nat. Acad. Sci. (U. S. A.) 54, 960—967.

— — and K. Murray, 1964: Physical and biological properties of soluble nucleohistones. J. Molec. Biol. 8, 54—64.

— N. Maheshwari, and J. Bonner, 1960: Enzymatic synthesis of RNA. Biochem. Biophys. Res. Comm. 3, 689—694.

Huiskamp, W., 1903: Beiträge zur Kenntnis des Thymusnucleohistons. Z. physiol. Chem. 39, 55—96.

Hultin, T., and R. Herne, 1948: Amino acid analysis of a basic protein fraction from sperm nuclei of some different invertebrates. Ark. f. Kemi. Mineral, Geol., 26 A, Nr. 20, 1—8.

Hurst, A., and D. R. Taylor, 1965: Growth inhibition of Escherichia Coli by some basic proteins prepared from the same strain. Nature 207, 438—439.

Hymer, W. C., and E. L. Kuff, 1964: Isolation of nuclei from mammalian tissues through the use of Triton X-100. The Journ. of Histochem. and Cytochem. 12, 359—363.

Irvin, J. L., D. J. Holbrook, J. H. Evans, H. C. McAllister, and E. P. Stiles, 1963: Possible role of histones in regulation of nucleic acid synthesis. Exper. Cell Res., Suppl. 9, 359—366.

Ivai, K., 1964: Histones of rice embryos and of Chlorella. In "The Nucleohistones", J. Bonner and P. Ts'o, ed., Holden-Day, inc., San Francisco, London, Amsterdam; 59—65.

IZAWA, M., V. G. ALLFREY, and A. E. MIRSKY, 1963: The relationship between RNA synthesis and loop structure in Lampbrush chromosomes. Proc. Nat. Acad. Sci. (U. S. A.) **49**, 544—551.

JOHNS, E. W., 1964 a: Studies on lysine rich histones, in "The Nucleohistones", J. BONNER and P. Ts'o, ed., Holden-Day, inc., San Francisco, London, Amsterdam; 52—58.
— 1964 b: Studies on histone fraction from calf-thymus. Biochem. J. **92**, 55—59.
— and J. A. V. BUTLER, 1962 a: Studies on histones. 4. The histones of wheat germ. Biochem. J. **84**, 436—439.
JOHNS, E. W., and J. A. V. BUTLER, 1962 b: Further fractionation of histones from calf thymus. Biochem. J. **82**, 15—18.
— D. M. P. PHILLIPS, P. SIMSON, and J. A. V. BUTLER, 1960: Improved fractionations of arginine-rich histone from calf thymus. Biochem. J. **77**, 631—636.
— — — — 1961: The electrophoresis of histones and histones fractions on starch gel. Biochem. J. **80**, 189—193.

KENT, P. W., M. HICHENS, and P. F. V. WARD, 1958: Displacement, fractionation of deoxyribonucleoproteins by heparin and dextran sulphate. Biochem. J. **68**, 568—572.
KHOUVINE, Y., J. GREGOIRE, and J. P. ZALTA, 1953: Desoxyribonucleoproteides de l'epithelioma atypique du rat. I. Histones et proteines non basiques. Bull. Soc. Chim. Biol. (Paris) **35**, 244—256.
KIT, S., and A. L. GROSS, 1959: Quantitative relationships between DNA content and glycolysis or histones of diploid and tetraploid cells. Biochem. Biophys. Acta. **36**, 185—191.
KLYSZEJKO, L., and Y. KHOUVINE, 1960: Desoxyribonucleoproteides du pancréas de boeuf. I. Obtention et composition des histones, des acides desoxyribonucleiques et des proteines non basiques. Bull. Soc. Chim. Biol. (Paris) **42**, 761—773.
KNOBLOCH, A., and R. VENDRELY, 1956: An estimation of the nature of nucleoproteins by an analytical method. Nature **178**, 261—262.
KOSSEL, A., 1884: Über einen peptonartigen Bestandteil des Zellkerns. Z. Physiol. Chem. **8**, 511—515.
— 1896: Über die basischen Stoffe des Zellkerns. Z. physiol. Chem. **22**, 176.
— 1904: Beiträge zum System der einfachsten Eiweißkörper. Z. Physiol. Chem. **40**, 565.
— 1928: The protamines and histones. Ed. Longmans, Green and co. London.

LASKOWSKI, M., 1946: Studies on thymonucleodepolymerase. Arch. Biochem. **11**, 41—48.
LAURENCE, D. J. R., P. SIMSON, and J. A. V. BUTLER, 1963: Studies on histones. 5. The histones of the Crocker sarcoma and spontaneous mammary tumours of mice. Biochem. J. **87**, 200—205.
LEHNER, S. M., 1964: The inhibition of DNA synthesis by nuclear proteins. Biochem. Biophys. Acta. **80**, 338—339.
LESLIE, I., 1962: Nucleoproteins studies on a strain of human liver cells and their significance in relation to genetic codes for template ribonucleic acid. "New developments in tissue culture", J. W. GREEN, ed., Rutgers University Press. 39—62.
LIAU, M. C., L. S. HNILICA, and R. B. HURLBERT, 1965: Regulation of RNA synthesis in isolated nucleoli by histones and nucleolar proteins. Proc. Nat. Acad. Sci. (U. S. A.) **53**, 626—633.
LILIENFELD, L., 1894: Zur Chemie der Leucocyten. Z. Physiol. Chem. **18**, 473—486.
LINDNER, A., T. KUTKAM, K. SANKARANARAYAN, R. RUCKER, and J. ARRADONDO, 1963: Inhibition of Ehrlich ascites tumor with 5-fluorouracil and other agents. Exper. Cell Res., Suppl. **9**, 485—508.
LINDSAY, D. T., 1964: Histones from developing tissues of the chicken: Heterogeneity. Science **144**, 420—422.
LUCK, J. M., H. A. COOK, N. T. ELDREDGE, M. I. HALEY, D. E. KUPKE, and P. S. RASMUSSEN, 1956: On the fractionation of thymus histone. Arch. Biochem. Biophys. **65**, 449—467.
— D. W. KUPKE, A. RHEIN, and M. HURD, 1953: Non fibrous deoxypentose nucleohistone from liver. I. Préparation and composition. J. Biol. Chem. **205**, 235—243.
— P. S. RASMUSSEN, K. SATAKE, and A. N. TSVETIKOV, 1958: Further studies on the fractionation of calf thymus histone. J. Biol. Chem. **233**, 1407—1414.

Lucy, J. A., and J. A. V. Butler, 1954: Fractionation of deoxyribonucleoprotein by successive extraction with constant salt concentration. Nature 174, 32—33.
— — 1955: Fractionation of deoxyribonucleoprotein. Biochem. Biophys. Acta. 16, 431—432.
Ludwig, E. H., and E. Smull, 1963: Infectivity of histone poliovirus ribonucleic acid preparations. J. of Bacteriol. 85, 1334—1338.
Luzzati, V., and A. Nicolaieff, 1963: The structure of nucleohistones and nucleo-protamines. J. Mol. Biol. 7, 142—163.

Markert, C. L., and H. Ursprung, 1963: Production of replicable persistent changes in zygote chromosomes of Rana pipiens by injected proteins from adult liver nuclei. Develop. Biol. 7, 560—577.
Markham, R., and J. D. Smith, 1954: The proteins. Ed. H. Neurath, K. Bailey, Academic Press. 2, 1.
Mauritzen, C. M., and E. Stedman, 1959: Cell specificity of β-histone in the domestic fowl. Proc. Roy. Soc. B. 150, 299—311.
— — 1960: Cell specificity of β-histones from the ox. Proc. Roy. Soc. B. 153, 80—89.
Maver, M. E., and A. E. Greco, 1949: The Nuclease activities of cathepsin preparations from calf spleen and thymus. J. Biol. Chem. 181, 853—870.
Mazen, A., 1962: Contribution a l'étude biochimique des nucleohistones et nucleo-protamines du noyau cellulaire. Bull. Biol. de la France et de la Belgique 96, 305—361.
McAllister, H. C., Y. C. Wan, and J. L. Irvin, 1963: Electrophoresis of histones and histones fractions on polycrylamide gels. Analyt. Biochem. 5, 321—329.
Medawar, P. B., and G. Zubay, 1959: Preparation of nucleoprotein and deoxyribo-nucleic acid from small quantities of lymphoïd tissue desintegrated by ultra-sound. Biochem. Biophys. Acta. 33, 244—246.
Meek, E. S., 1964: A quantitative cytochemical study of chromosomal basic protein in static and proliferative cell populations. Exper. Cell Res. 33, 355—359.
Miescher, F., 1871: Über die chemische Zusammensetzung der Eiterzellen. Hoppe. Seyler's Med. chem. Unters. 441. Die histochemischen und physiologischen Arbeiten von Friedrich Miescher, Leipzig, 1897, 1.
— 1874: Die Spermatozoen einiger Wirbelthiere. Verh. naturforsch. Ges. in Basel, 6, 138. Publishe in: Die histochemischen und physiologischen Arbeiten von F. Miescher, Leipzig, 1897, 55.
— 1897: Die histochemischen und physiologischen Arbeiten von Friedrich Miescher, 2, F. C. W. Vogel, Leipzig.
Mirsky, A. E., 1947: Chemical properties of isolated chromosomes Cold Spring Harbor Symp. on Quant. Biol. 12, 143—146.
— and A. W. Pollister, 1946 a: The nucleoprotamine of trout sperm. J. of gen. Physiol. 30, 101—116.
— — 1946 b: Chromosin, a deoxyribose nucleoprotein complex of the cell nucleus. J. Gen. Physiol. 30, 117—149.
— and H. Ris, 1947: Isolated chromosomes. J. Gen. Physiol. 31, 1—6.
Murray, K., 1964 a: Histone nomenclature in "The Nucleohistones", J. Bonner and P. Ts'o, ed., Holden-Day, inc., San Francisco, London, Amsterdam; 15—20.
— 1964 b: The heterogeneity of histones in "The Nucleohistones", J. Bonner and P. Ts'o, ed., Holden-Day, inc., San Francisco, London, Amsterdam; 21—35.
— 1964 c: In comment of article of Johns, E. W., Studies on lysine-rich histones. In "The Nucleohistones", J. Bonner and P. Ts'o, ed., Holden-Day, inc., San Francisco, London, Amsterdam; 57.

Neelin, J. M., 1964: Histones from chicken erythrocyte nuclei. In "The Nucleo-histones", J. Bonner and P. Ts'o, ed., Holden-Day, inc., San Francisco, London, Amsterdam; 66—71.
— and G. C. Butler, 1961: Comparison of histones from chicken tissues by zone electrophoresis in starch gel. Canad. J. Biochem. Physiol. 39, 485—491.
— P. X. Cailahan, D. C. Lamb, and K. Murray, 1964: The histones of chicken erythrocyte nuclei. Canadian J. of. Biochem. 42, 1743—1752.
— and G. E. Connell, 1959: Zone electrophoresis of Chicken-erythrocyte histone in starch gel. Biochem. Biophys. Acta. 31, 539—541.
— and E. M. Neelin, 1960: Zone electrophoresis of calf thymus histone in starch gel. Canad. J. Biochem. and Physiol. 38, 355—363.

NEIDLE, A., and H. WAELSCH, 1964: Histone: Species and tissue specificity. Science 145, 1059—1061.

NELSON-GERHARDT, M., 1919: Untersuchungen über Salmin. Z. physiol. Chem. 105, 265.

OKUDA, J., D. SZAFARZ, and Y. KHOUVINE, 1963: Comparaison entre les proteines basiques et acides d'une fraction extraite de noyaux isolés de foie de rat normal et en régénération. C. R. Acad. Sci. (Paris) 257, 2904—2905.

ONTKO, J. A., and W. R. MOOREHEAD, 1964: Histone synthesis after inhibition of deoxyribonucleic acid replication. Biochem. Biophys. Acta. 91, 658—660.

ORD, M. G., J. H. RAAF, J. A. SMIT, and L. A. STOCKEN, 1964: Metabolic and chemical properties of basic proteins isolated from nuclei of rat liver and thymus gland. Biochem. J. 95, 321—331.

PALMADE, C., M. R. CHEVALLIER, A. KNOBLOCH, and R. VENDRELY, 1946: Isolement d'une desoxyribonucleohistone à partir d'*Escherichia coli*. C. R. Acad. Sci. 246, 2534—2537.

PEACOCKE, A. R., 1960: The structure and physical chemistry of nucleic acids and nucleoproteins. Progress in Biophys. and Biophys. Chem. 10, 55—114.

PETERMANN, M. L., and C. M. LAMB, 1948: The nucleohistone of beef spleen. J. Biol. Chem. 176, 685—693.

PHILLIPS, D. M. P., 1955: N-terminal groups in salmine. Biochem. J. 60, 403—409.

— 1957: The N-terminal groups and partial fractionation of thymus histones. Biochem. J. 67, 9 P.

— 1958: The N-terminal groups of calf thymus histones. Biochem. J. 68, 35—40.

— 1962: "The histones", Progress in Biophysics and Biophys. Chem. 12, 213—280.

— 1963: The presence of acetyl groups in histones. Biochem. J. 87, 258—263.

— 1964: Studies on peptides from calf thymus histone, in "The Nucleohistones", J. BONNER and P. Ts'o, ed., Holden-Day, inc., San Francisco, London, Amsterdam; 46—51.

— and E. W. JOHNS, 1959 a: The chromatography of thymus histones and the demonstration of proteinase activity in the unfractionated preparation. Biochem. J. 71, 17 P.

— — 1959 b: A study of the proteinase content and the chromatography of thymus histones. Biochem. J. 72, 538—544.

— — 1965: A fractionation of the histones of group F₂ a from calf thymus. Biochem. J. 94, 127—130.

— and P. SIMSON, 1961: Identification of some peptides from an arginine-rich histone and their bearing on the structure of deoxyribonucleoprotein. Biochem. J. 78, 32 P.

PLAUT, W., 1963: On the replicative organization of DNA in the polyténe chromosome of *Drosophila melanogaster*. J. Mol. Biol. 7, 632—635.

POLLISTER, A. W., and A. E. MIRSKY, 1946: The nucleoprotamine of trout sperm. The Jour. of Gen. Physiol. 30, 101—116.

PRESCOTT, D. M., and M. A. BENDER, 1963: Synthèsis and behaviour of nuclear proteins during the cell life cycle. J. of Cell comp. Physiol. 62, Suppl. 1, 175—194.

RASCH, E., and J. W. WOODARD, 1959: Basic proteins of plant nuclei during normal and pathological cell growth. J. Biophys. Biochem. Cytol. 6, 263—276.

RASMUSSEN, K. E., and K. LINDERSTRØM LANG, 1934: Clupeinuntersuchungen — II. Elektrometrische Titration von Clupein. Z. Physiol. Chem. 227, 181—212.

RASMUSSEN, P. S., K. MURRAY, and J. M. LUCK, 1962: On the complexity of calf thymus histone. Biochemistry 1, 79—89.

RAUEN, H. M., W. STAMM, and K. FELIX, 1953: Countercurrent separation of protamines. Z. Physiol. Chem. Hoppe-Seyler's 292, 101—109.

REID, B. R., and R. D. COLE, 1964: Biosynthesis of a lysine-rich histone in isolated calf thymus nuclei. Proc. Nat. Acad. Sci. (Washington) 51, 1044—1050.

RICHARDS, B. M., 1964: X-ray diffraction and electron microscopic studies of nucleohistones in "The Nucleohistones", J. BONNER and P. Ts'o, ed., Holden-Day, inc., San Francisco, London, Amsterdam; 108—116.

RINGERTZ, N. R., 1963: The effect of X-radiation on nuclear histone content. Exper. Cell Res. 32, 401—404.

Rotherham, J., J. L. Irvin, E. M. Irvin, and D. S. Holbrook, 1957: Incorporation of glycine into protein fraction of nuclei of liver and hepatoma. Proc. Soc. Exper. Biol. and Med. **96**, 21—24.

Sandritter, W., H. Fischer, K. Sussenberger, and H. G. Schiemer, 1959: Quantitative histochemische Untersuchungen zum Nachweis der Protaminspeicherung von Mäuseascitestumorzellen. Exper. Cell Res. **17**, 197—204.

Sanger, F., 1945: The free amino groups of insulin. Biochem. J. **39**, 507—517.

Satake, K., P. S. Rasmussen, and J. M. Luck, 1960: Arginine peptides obtained from thymus histone fractions after partial hydrolysis with Streptomyces griseus proteinase. J. Biol. Chem. **235**, 2801—2809.

Sautiere, P., M. Dantrevaux, and G. Biserte, 1965: Fractionnement des histones riches en arginine du thymus de veau par chromatographie de gel-filtration. Bull. Soc. Chim. Biol. **47**, 821—827.

Scanes, F. S., and B. T. Tozer, 1956: Fractionation of basic proteins and polypeptides clupeine and salmine. Biochem. J. **63**, 565—576.

Schwimmer, S., and J. Bonner, 1965: Nucleohistone as template for the replication of DNA. Biochim. Biophys. Acta **108**, 67—72.

Shooter, K. V., P. F. Davison, and J. A. V. Butler, 1954: Physical properties of thymus nucleoprotein. Biochim. Biophys. Acta. **13**, 192—198.

Siebert, G., 1963: Enzymes of cancer nuclei. Exper. Cell Res., Suppl. **9**, 389—417.

Signer, R., and H. Schwander, 1949: Isolierung hochmolekularer Nukleinsäure aus Kalbsthymus. Helv. Chim. Acta. **32**, 853—859.

Sluyser, M., P. J. Thung, and P. Emmelot, 1965: Inhibition of deoxyribonucleic acid synthesis in regenerating rat liver by the administration of histones in vivo. Biochim. Biophys. Acta **108**, 249—258.

Smellie, L. B., G. C. Butler, and D. B. Smith, 1958: Properties of calf thymus histone isolated by three different methods. Canad. J. Biochem. Physiol. **36**, 1—13.

Smellie, R. M. S., W. M. McIndoe, and J. N. Davison, 1953: The incorporation of ^{15}N and ^{14}C into nucleic acids and proteins of rat liver. Biochem. Biophys. Acta. **11**, 559—565.

Sonnenberg, B. P., and G. Zubay, 1965: Nucleohistone as a primer for RNA synthesis. Proc. Nat. Acad. Sci. (U. S. A.) **54**, 415—420.

Stedman, E., and E. Stedman, 1943: Probable function of histone as a regulator of mitosis. Nature **152**, 556.

— — 1947: The chemical nature and functions of the components of cell nuclei. Cold Spring Harbor Symp. on Quant. Biol. **12**, 224—236.

— — 1950: Cell specificity of histones. Nature **166**, 780.

— — 1951: The basic proteins of cell nuclei. Phil. Trans. Roy. Soc. London, Ser. B. **235**, 565—596.

Steele, W. J., and H. Busch. 1963: Studies on acidic nuclear proteins of the Walker tumor and liver. Cancer Res. **23**, 1153—1163.

Steinert, M., 1965: L'absence d'histone dans Kinetonucleus des trypanosomes. Etude cytochimique. Exptl. Cell Res. **39**, 69—73.

Stern, K. G., 1949: Experiments on the size and shape of chromosomal nucleoproteins and their bearing on gene structure. Exper. Cell Res., Suppl. **1**, 97—99.

— and S. Davis, 1946: Studies on thymus nucleohistone Feder. Prodeed. **5**, Nr. 1, 156.

— G. Goldstein, J. Wagman, and J. Schryver, 1947: Studies on desoxyribonucleoproteins, isolation and properties of genoprotein. Fed. Proc. **6**, 296.

Steudel, H., 1914: Über das Nucleohiston. Z. Physiol. Chem. **90**, 291—300.

Szybalski, W., and E. H. Szybalska, 1962: Genetics of human cell lines. IV. DNA mediated heritable transformation of a biochemical trait. Proc. Nat. Acad. Sci. (U. S. A.) **48**, 2026.

Tristram, G. R., 1947: Constitution of salmine. Nature (London) **160**, 637.

Ts'o, P. O. P., and J. Bonner, 1964: "The Nucleohistones", Holden-Day, inc., San Francisco, London, Amsterdam; 367—379.

Tsumita, T., and E. Chargaff, 1958: Studies on nucleoproteins. VI. The deoxyribonucleoprotein and the deoxyribonucleic acid of Bovine Tubercle Bacilli (BCG). Biochem. Biophys. Acta. **29**, 568—578.

Ui, N., 1956 a: Effect of association on the electrophoretic pattern of salmine. J. Chem. Soc. Japan. Pure Chem., Sect. 77, 947—951.
— 1956 b: On the molecular weight of calf thymus histone. Biochim. Biophys. Acta. 22, 205—206.
— 1957: Preparations, fractionation and properties of calf thymus histone. Biochim. Biophys. Acta. 25, 493—502.
Umana, R., S. Updike, J. Randall, and A. L. Dounce, 1964: Histone metabolism. In "The Nucleohistones", J. Bonner and P. Ts'o, ed., Holden-Day, inc., San Francisco, London, Amsterdam; 200—229.

Vanderplancke, A., 1964: De l'importance de la fraction proteinique dans le complexe nucleoprotidique (Essai de fractionnement des nucleohistones). Arch. Biochim. cosmet. 7, Nr. 65, 10—19.
Velick, S. F., and S. Udenfriend, 1951: The amino-end-groups and the amino acid composition of salmine. J. Biol. Chem. 191, 233—238.
Vendrely, C., 1952: L'acide désoxyribonucléique du noyau des cellules animales. Son rôle possible dans la Biochimie de l'hérédité. Bull. Biol. de la France et de la Belgique 1, 1—88.
Vendrely, R., 1964: The enzymatic degradation of nucleohistones. In "The Nucleohistones", J. Bonner and P. Ts'o, ed., Holden-Day, inc., San Francisco, London, Amsterdam; 307—314.
— M. Alfert, H. Matsudaira, and A. Knobloch, 1958 a: The composition of nucleohistone from pycnotic nuclei. Exper. Cell Res. 14, 295—300.
— N. Genty, and Y. Coirault, 1965: Etude comparée d'histones d'erythrocytes chez diverses espèces animales. Communication à la Réunion Biochimique Franco-Belge. Liège May 1965. Arch. Intern. de Physiologie et de Biochimie 73, 555.
— A. Knobloch, H. Matsudaira, 1958 b: A comparative biochemical study of nucleohistones from different vertebrates. Nature 181, 343.
— A. Knobloch, and C. Vendrely, 1957: An attempt of using biochemical methods for cytochemical problems; the DNP of spermatogenetic cells of bull testis. Exper. Cell. Res., Suppl. 4, 279—283.
— A. Knobloch-Mazen, and C. Vendrely, 1959: A comparative biochemical study of nucleohistones and nucleoprotamines in the cell nucleus "The cell nucleus". Butterworth and Co. pub., 200—205.
— — 1960: Données biochimiques récentes sur la relation entre acide desoxyribonucleique et protéines basiques dans le noyau. Biochem. Pharmacol. 4, 19—28.
— and C. Vendrely, 1953: Arginine and deoxyribonucleic acid content of erythrocyte nuclei and sperms of some species of fishes. Nature 172, 30.
Vorobyev, V. I., and V. M. Bresler, 1963: Unfractioned preparations of histones from normal mammalian tissues as agents inhibitory growth of transplanting tumors. Nature 198, 545—547.

Welsh, R. S., 1960: Water-soluble, non fibrous deoxyribonucleoprotein from calf thymus nuclei. Nature 187, 943—945.
Wilczok, T., and J. Mendecki, 1963: The effect of protamines and histones on incorporation of donor DNA into neoplastic cells. Neoplasma 10, 113—119.
Wilkins, M. H. F., 1956: Physical studies of the molecular structure of desoxyribose nucleic acid and nucleoprotein. Cold Spring Harbor Symp. on Quant. Biol. 21, 75—90.
— and G. Zubay, 1959: The absence of histone in the bacterium Escherichia coli. II. X-ray diffraction of nucleoprotein extract. J. Biophys. Biochem. Cytol. 5, 55—58.
— — and H. R. Wilson, 1959 a: X-ray diffraction studies of the molecular structure of nucleohistone and chromosomes. Trans. Faraday Soc. 55, 497.
— — 1959 b: X-ray diffraction studies of the molecular structure of nucleohistone and chromosomes. J. Mol. Biol. 1, 179—185.

Zalta, J. P., R. Rozencwajg, N. Carasso, and P. Favard, 1962: Isolement d'une fraction de noyaux cellulaires dont la pureté est contrôlée au microscope electronique. C. R. Acad. Sci. 255, 412—414.
Zbarkii, I. B., and K. A. Perebotchikova, 1954: Action des histones sur le developpement du sarcome experimental de la souris. Bull. Biol. Exper. Med. 38, 61—64.

ZBARSKY, I. B., 1962: La composition proteique et nucleique de structures nucleaires. Ann. Histochem. 7, 89—93.
— and G. P. GEORGIEV, 1959: Nouvelles données sur le fractionnement des noyaux cellulaires du foie de rat et la composition chimique des structures nucleaires (en russe). Biokhimiya 24, 192—199.
— O. P. SAMARINA, and L. P. YERMOLAYEVA, 1964: Nuclear proteins and their biosynthesis in the tumour cells. Acta XX, 937—940.
ZIMMERMANN, E., 1959: Dissertation, Johann-Wolfgang-Goethe-Universität, Frankfurt am Main, Germany. — In K. FELIX, 1960: Protamines Adv. in protein Chem. 15, 1—56.
ZUBAY, G., 1964: Nucleohistone structure and function. — In "The Nucleohistones", J. BONNER and P. Ts'o, ed., Holden-Day, inc., San Francisco, London, Amsterdam; 95—107.
— and P. DOTY, 1959: The isolation and properties of deoxyribonucleoprotein particles containing single nucleic acid molecules. J. Mol. Biol. 1, 1—20.
— and M. R. WATSON, 1959: The absence of histones in the bacterium Escherichia coli. I. Preparation and analysis of nucleoprotein extract. J. Biophys. Biochem. Cytol. 5, 51—54.

Protoplasmatologia
V. Karyoplasma (Nucleus)
3. Chemistry and Cytochemistry of Nucleic Acids and Nuclear Proteins
d) Cytochemistry of the Histones

Cytochemistry of the Histones

By

DAVID P. BLOCH

Botany Department, The University of Texas, Austin, Texas

With 21 Figures

Contents

Introduction

The histones and protamines were discovered along with the nucleic acids late in the last century by FRIEDRICH MIESCHER [1]. They were part of the "nuclein", first extracted by MIESCHER in 1869 from pus cell nuclei obtained from old bandages gotten from the hospitals of Tübingen

---

[1] A brief and useful history of the discovery of nucleic acids and histones is given by LEVENE and BASS (1931).

(Miescher 1897). This hard won material was one of a class of biological substances, including casein and phosphoproteins of egg yolk, which were then considered to be unique because of their acidic properties and phosphorus content. Twenty years later however, as the means were devised by Altmann (1889) for resolving the nuclein into its component parts, and the attributes which had previously aroused the most curiosity were found to reside in the nucleic acid portion, most attention was directed toward this substance and the associated proteins were treated as an unwanted by-product obtained during its isolation.

This disparity in popularity was long lasting, as attested to by approximately 1,890 listings in the 1962 indices of Chemical Abstracts under the headings of "desoxyribo" and "ribo-nucleic acids." but only 100, under "histones" and "protamines". There exists but a single monograph on the basic proteins, Kossel's "The Protamines and Histones," written in 1928, which is widely recognized as a classic. There are few reviews (e.g. Felix 1960). Not surprisingly, the fundamental roles played by the nucleic acids in the heredity and physiology of the cell have been sharply defined, while the functions of the histones remain vague and the few pertinent ideas have until recently found little factual support.

A recent awakening of interest in the histones probably stems from a growing sentiment, first expressed by the Stedmans in 1950, that histones may act in an inhibitory way in the regulation of gene activity. The succeptibility of gene activity to modification is well known through the works of McClintock (1955) and Brink (1958). One mode of control has been explained recently by the work of Jacob and Monod on induced enzyme formation in bacteria (1961). The more recent demonstrations by James Bonner's group at the California Institute of Technology (1962) and Mirsky's group at the Rockefeller Institute (1963), of a seemingly nonspecific inhibitory effect of histones on nucleic acid directed processes in *in vitro* systems, has excited further interest in these proteins as regulatory agents. Early in 1963 Bonner and Ts'o organized the "First World Conference on Histone Biology" at Rancho Santa Fe, California. The proceedings of this conference, published in 1964, represents the first major compilation of histone data between hard covers in 36 years, and the first ever to be guided by a central biological consideration. As such, it is a valuable contribution to modern histone research. The "inhibition hypothesis" remains tentative however, and the relevance to histone function of the information, largely chemical, accumulated between the times of Miescher and of the Stedmans can not yet be fully assessed.

Definition of Histone

Chemical Definition

Since their definition by Miescher and Kossel, histones and protamines have been distinguished from other proteins by their basicity and properties derived therefrom. Kossel (1928) lists a number of attributes, the concurrence of several of these serving to identify a given protein as a histone. These are; a high degree of basicity, high arginine content, sparing solubility of its complexes with other proteins, precipitability by ammonium hydroxide solutions and by alkaloids, formation of complexes with nucleic acids, and hydrolysis with hydrochloric acid to give simpler histopeptones. Such a test tube definition is admittedly arbitrary, and bears no necessary relation to biological function. In time, this class of substances came to include such basic proteins as the blood globins as well as those extracted from the nucleus. More recently however, it has again become customary to consider source as well as chemical properties, and the term now usually applies only to the basic proteins of the chromosome (MIRSKY and POLLISTER 1946).

Cytochemical Definition

Cytochemical methods for detecting (defining) histones reinforces this scheme of classification. The alkaline fast green method of ALFERT and GESCHWIND (1953) takes advantage of the high isoelectric point of histones and their resultant ability to bind anionic dyes at pH ranges where most other proteins fail to stain (Fig. 1). Here, as in chemical procedures, confidence that the method can always be relied upon to single out a biologically significant class of substances may be misplaced. Cytochrome and lysozyme, both basic proteins, and also methylated albumin stain well with this method (ALFERT and GESCHWIND 1953). Such proteins usually are not present in sufficient amounts to result in visible "non-specific" staining of cytological preparations, and chromatin specificity is often used as a measure of success in the application of the technique. The basic granules in the cytoplasm of eosinophilic leucocytes stain strongly with fast green under the conditions used for histone staining (BLOCH, unpublished). There is little question here, that the substances responsible for staining are unrelated to histones. In the typical case, where only the nucleus stains, the histones may be defined as those substances responsible for the positive reaction gained under the specified conditions. However, when cytoplasmic staining occurs, as in log phase *Tetrahymena* (ALFERT and GOLDSTEIN 1955), or in bud primordia during floral induction (GIFFORD 1964), or in the *Drosophila* salivary gland (HORN and WARD 1957), or in sea urchin oocytes (TALEPOROS 1959) there is uncertainty whether lack of nuclear (or chromosomal) specificity reflects non-specific staining or the presence of cytoplasmic (or non-chromosomal nuclear) histones. The concession of the possibility of the existance of cytoplasmic histones is important to the resolution of histone function and origin and is considered in more detail later.

Equally important is the recognition of the inadequacy of both cytochemical and biochemical methods to account for all the basic proteins which would unequivocally come under the heading of histones. DALY and MIRSKY (1955) have found that 5 per cent trichloracetic acid (such as is often used in the hydrolysis of DNA preparatory to the staining of histones) is insufficient to precipitate some lysine rich histones. This also holds true of the protamines of many sperm (ALFERT 1956). Conversely,

cytochemical methods demonstrate histones which may be lost during biochemical procedures. Guinea pig sperm contain histones very rich in
arginine. These were not detected by Crampton, using biochemical extraction procedures (Crampton et al. 1957), presumably because these tena-
ciously bound proteins were thrown away in the sediment from which the
other histones were extracted.

The Protamines

The protamines have traditionally been considered as a second category
of nuclear basic proteins. These were first found in the sperm of the
Rhine salmon by Miescher and were distinguished from histones by their
relative simplicity. It was soon recognized that similar proteins could be
obtained from the sperm of many fishes, and both simplicity and source
were taken into consideration in the classification of the protein. Never-
thelesss, some proteins from fish sperm are more complex than salmine:
Kossel (1928) made a distinction between mono-, di-, and triprotamines,
containing arginine, arginine and either lysine or histidine, and all three
basic amino acids, respectively. Salmine, a monoprotamine, contains only
arginine as its basic amino acid. Some of the triprotamines surpass the
simplest histones in their complexity, and their germinal source provided
the main justification for including them among the protamines. Had
they been found in erythrocytes, they would no doubt have been called
histones.

To confuse the issue further, is the existence of "intermediate histones"
in the sperm of some echinoderms, whose properties are intermediate
between those of salmine and somatic histones (Hultin and Herne 1949).
Finally, the sperms of other species, notably the carp and frog, contain
proteins very similar if not identical to the histones of somatic cells and
have accordingly been designated histones (Vendrely 1957).

Of the simplest monoprotamines, salmine has been reported to contain
as few as four different amino-acids, and clupeine, five; arginine in each
case accounting for approximately 2/3 of the total number of residues
(Waldschmitt-Leitz and Gunther 1948). The reports of such analyses conflict, however this is not surprising in view of the heterogeneity of protamines (Rauen 1952) and the variety of means utilized for their preparation. The more complex triprotamines contain most of the 16 amino
acids commonly found in somatic histones (Kossel 1928). The simplest
known histone, Stedman's a 1 fraction, contains only seven different amino
acids, of which, lysine, the sole basic component, comprises around 40 %
(Cruft et al. 1957). The molecular weights of monoprotamines range from
approximately 4,000 (Felix et al. 1950, Phillips 1955, Bloch 1962) to 10,800
(Ando et al. 1952) and possibly over 30,000 (d'Alcontres 1953). In contrast,

---

Fig. 1. Comparison of Feulgen staining of DNA with Fast Green staining of histones. Left, DNA; right, histones.
Top, *Drosophila* salivary gland chromosomes. The non-specificity for chromatin staining in the case of the fast
green can be minimized by staining at slightly higher pH's (8.3) than are ordinarily used. Center, ascites tumor
cell nuclei. Bottom, HeLa cells infected with adenovirus. Note that the aggregates of virus in the nuclei stain
with the Feulgen procedure, but not the fast green, and that the two types of DNA, viral and chromatin, can be
distinguished on this basis.

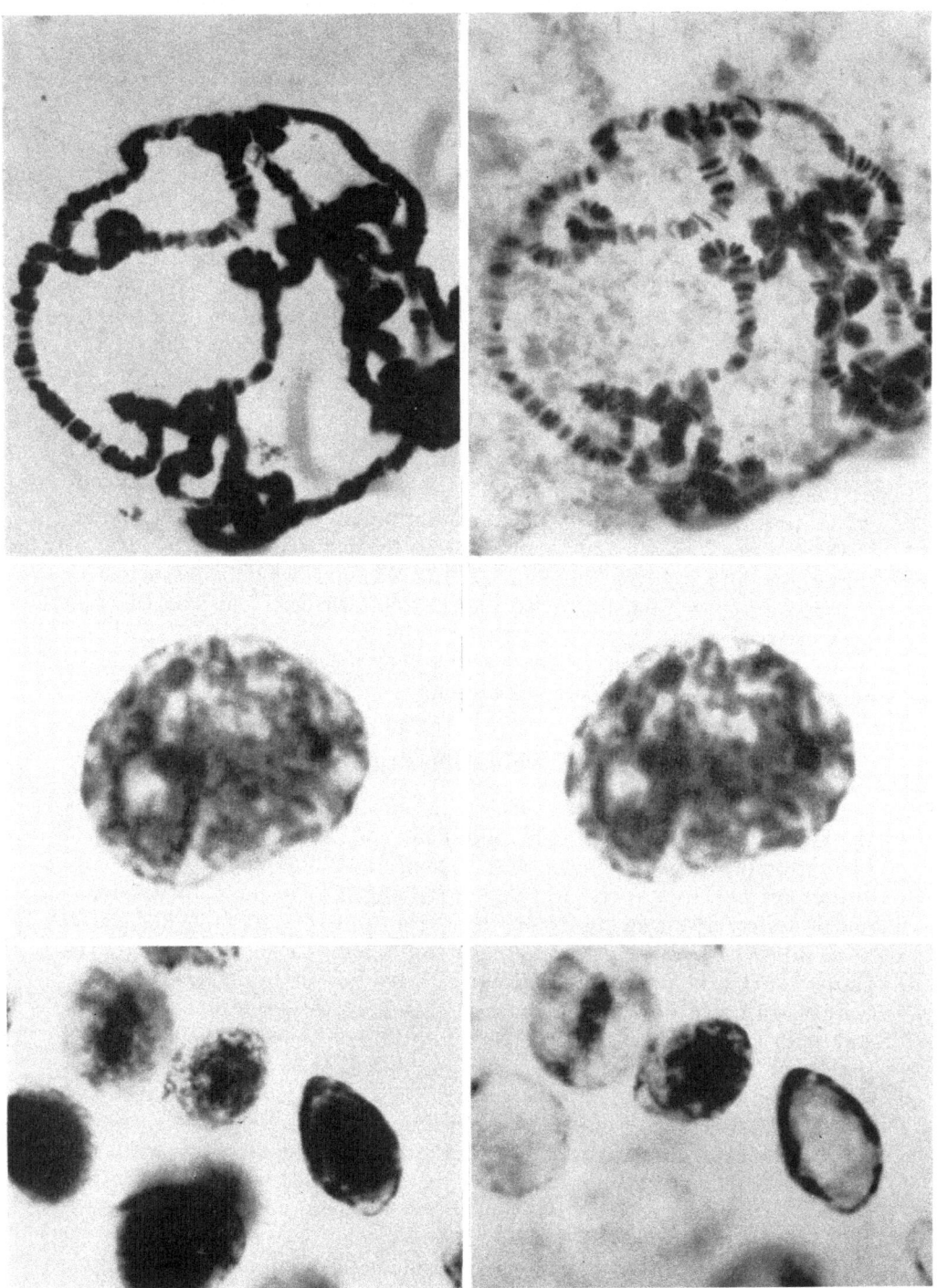

Fig. 1.

the molecular weights of somatic histones range from approximately 8,000 (UI 1956) to 74,000 (CRUFT *et al.* 1958). A protein from the immature spermatids of *Loligo* has been found to have a molecular weight of 8,000 (BLOCH 1962). This protein could be classed as either a histone or a triprotamine.

Thus no single criterion or set of criteria have been accepted which can be relied on to set histones apart from protamines, and the two classes merge indefinably in the developing sperm cells of various organisms. The term histone will be used in this review in a general way to include the broad spectra of basic proteins which have customarily come under the headings of both protamines and histones.

A tentative definition of histones agreed upon by an *ad hoc* "commitee" presided over by Dr. KENNETH MURRAY at the edge of a swimming pool at the Rancho Santa Fe conference is as follows: Histones are basic proteins, lacking in tryptophane, soluble in mercuric sulfate solutions, which are at some time during their existance complexed with DNA (MURRAY 1964). This is a useful, though tentative definition, which at this stage appropriately by-passes the question of function. It has hidden implications which were recognized and will become apparent in the following pages. It is, of course, an attempt to set apart a biologically significant class of substances with common function, without knowing what that function is, and is subject to periodic revision as our knowledge of histone function progresses.

It was also suggested at this meeting that the arginine lysine ratio be adopted as a means of describing histones.

The Localization Problem

Their biological functions uncertain, histones have been defined by the means used for their isolation and detection. They have been obtained by extracting isolated nuclei with strong salt solutions and then precipitating the DNA with alcohol, leaving the histone in solution (MIRSKY and POLLISTER 1946), or by extracting the histone directly from the nuclei with dilute acid, and precipitating it with picric acid (RASMUSSEN 1934, HAMER 1955). Histones in histological preparations can be selectively stained with acid dyes after removal of DNA which would otherwise interfere with staining (ALFERT and GESCHWIND 1953). Consecutive photomicrographs of identical cells, stained first for DNA, then for histone, show a similar distribution of both substances (BLOCH and GODMAN 1955, HORN and WARD 1957) (see also, Fig. 1). Thus cytochemical and biochemical methods conspire in providing a consistant set of criteria for the definition of histone which takes into consideration both the nuclear source and association with DNA, as well as taking advantage of its basic properties.

This happy accord between the biochemist and the cytochemist may not be as meaningful as it would seem. The biochemist obtained histones from nuclei after first removing the latter from the cytoplasm, consequently the cytochemist felt secure in his methods for detecting histones when he devised staining methods which would not stain the cytoplasm. In the latter instance, selection of the proper pH for specific nuclear staining is a matter of drawing a very thin line. When a "histone" is found in the cytoplasm, as in the cases cited above, it may be because the methods used are not sufficiently discriminating, or it may indicate a transit of histones from one side of the nuclear membrane to the other.

Basic proteins have been found in association with the RNA of ribosomes (Ts'o et al. 1958, Butler et al. 1960, Crampton and Petermann 1960, Setterfield et al. 1960, Zubay and Wilkins 1960). These proteins have been likened to histones and have even been called histones. The implication, intended or not, was that the basic proteins of ribosomes and those of chromatin may have a similar function, may be the same proteins, or may bear some sort of precursor product relationship. It would in fact be reasonable to suppose that histones, as proteins, do have an association with ribosomes, at least during their period of synthesis.

A restrictive definition of histones to exclude non-nuclear proteins may be as misleading in its implications as a relaxed definition. The custom established precedent of applying the term only to basic proteins of the chromatin might be extended to include also basic proteins which may have a chromosomal destination, whether or not they happen to be associated with DNA at any given time. Whether ribosomal proteins could be called histones in accordance with any useful definition remains to be determined, but their ultimate inclusion remains a possibility without sacrificing the biological significance of the class. This open door is particularly important in the case of presumed histones which are in the process of being synthesized prior to transfer to the chromatin.

It is also important to recognize the likelihood that the cytoplasmic histones obtained from the eggs of sea urchins (Taleporos 1959) and frog (Horn 1962, Bloch 1962) while not part of the chromatin proper, exist in association with the cytoplasmic DNA.

Proposed Functions of Histones

Histone function may be considered in the context of several somewhat divergent ideas. These are, in order of historical precedence: a) histones mediate nuclear controlled cytoplasmic processes, perhaps as chromosomal products which facilitate transfer of information from the gene to the cytoplasm (Caspersson 1950); b) histones are involved in the regulation of gene activity in differentiated cells (Stedman 1950); c) histones have a structural role in the chromosomes of higher organisms (Wilkins et al. 1959, Cole 1962) [2].

---

[2] Other functions have been attributed to histones. In 1943, when biochemists and geneticists were looking for the gene among the proteins, the fibrous nature of the nucleohistones and their nuclear origin suggested to some that these substances might be involved in the formation of spindle fibers (Stedman and Stedman 1942, 1947, Ryback 1948). Protamines had been thought to comprise the genetic substance itself (Wrinch 1934). Several papers in the 1930's dealt with protamines as a factor in

In 1941, CASPERSSON advanced a scheme for protein synthesis in which primary gene products described as "diamino-acid rich proteins," form at the chromocenter, migrate to the nucleolus then to the nuclear membrane, on the other side of which cytoplasmic RNA accumulates and subsequently cytoplasmic proteins are synthesized (CASPERSSON 1950). Implicit in this scheme is a direct or indirect role for these diamino-acid rich proteins, or histones, in the physical transfer of information from the nucleus to the cytoplasm which it controls. Later, doubt was cast upon the validity of these findings by the failure to find histones in nucleoli, either by methods of extraction applied to isolated nucleoli (VINCENT 1952) or by staining methods applied to histological preparations (ALFERT and GESCHWIND 1953). More recently however, basic proteins have been shown to exist in both nucleoli (BIRNSTIEL and CHIPCHASE 1963, FLAMM and BIRNSTIEL 1964), and in microsomes (see above). The implication of the nucleolus in the synthesis of ribosomes (BEERMAN 1960, 1961, BIRNSTIEL et al. 1963), suggests that CASPERSSON's findings may in many essentials be correct, but that the diamino-acid rich proteins from the nucleolus are ribosomal basic proteins and probably not histones.

DANIELLI later suggested that histones may combine with RNA and interfere with its function, thereby regulating rather than participating directly in gene expression (1953). Inhibition by complexing of histones with gene intermediates was presumed to be governed by laws of mass action, and it is interesting that some current views of histone action resemble DANIELLI's with the exception that the site of action of histone is with the DNA rather than the RNA (BUSCH et al. 1963). The basic tenor of the idea that histones control gene function, if not the mechanistic details, had already been advanced by STERN (1952), and the STEDMANS (1950), the latter supported by their findings of differences in the prevalence and compositions of histone fractions from different tissues. This variability would be compatible with either a catalytic (controlling) or a metabolic (transport) role for histone, but the impression of higher amounts of this substance and greater basicities in such synthetically inert cells as erythrocytes and sperm favored a regulatory role, particularly regulation by inhibition. Significantly also, this view arose during a time of mounting evidence for DNA constancy within the cells of an organism, and a consequent need for invoking controlling factors other than DNA to reconcile its constancy with variable gene expression. The recent findings of an inhibitory effect of histones on DNA primed RNA polymerase in vitro

---

melanogenesis (ROPSHAW 1933, GELLOWS 1935). Protamines had been considered as a reserve of arginine to be utilized during synthesis of nucleic acids (CLEMENTI 1921). Ansely has suggested that histones may control pairing during meiosis, on the basis of differences in stainability of histones of normal and asynaptic spermatocytes (ANSELY 1954, 1957). Current in the lore, are the ideas of uncertain origin that protamines are used for the "packaging" of nucleic acids in the sperm, and that histones provide a "coating" for DNA. That histones and DNA may act as templates for the synthesis of the respective partners, thereby obviating the need for self-replication, has also been proposed.

provides additional support for the "inhibition hypothesis" of the Stedmans (BONNER *et al.* 1962, ALLFREY *et al.* 1963).

Histones (or protamines) are ubiquitous in the nuclei of higher
organisms. They extend to the fungi in the plant kingdom (BLOCH, unpublished, D'ALCONTRES 1953), and to protozoans in the animal (ALFERT and
GOLDSTEIN 1955). They are apparently lacking in bacteria[3] (CHARGAFF and
SAIDEL 1949, ZUBAY and WATSON 1959, WILKINS and ZUBAY 1959). It would
appear then, that whatever their function, histones are unessential for the
basic phenomena of chromosome replication and regulation of cell syntheses, and must be accessory to the more complicated chromosomal functions of "higher" organisms. Perhaps a clue to their function lies in this
disparate distribution among the two major groups of organisms, the uni-
and the multicellular, or better, the "procaryotic" (containing no nuclear
membrane) and "eucaryotic" organisms (STANIER and VAN NIEL 1962).

Cytochemical Methods

Histones have no unique constituents, no known enzymatic activities,
(however see LESLIE 1961), and have yet to display detectable antigenicity,
hence their cytochemical identification has been a difficult problem. Their
lack of tryptophane may help provide the basis for identification of a
known protein as a histone if other proteins are known to be absent.
However, such a negative characteristic is of limited value. CASPERSSON in
1941 (See CASPERSSON 1950) used supposed differences in the ultra-violet
absorption spectra of ordinary and "diamino-acid" proteins (histones) to
characterize regions of the cell as rich or poor in the later. The validity
of this approach has been seriously challenged by MIRSKY and POLLISTER
(1946), who found no such differences between the spectra of histone and
non-histone proteins.

RIS and POLLISTER (1947) utilized differences in the solubilities of histone
and non-histone proteins in two cytochemical adaptations of the MILLON
test for protein. Histone is soluble in a sulfuric acid MILLON's reagent, and
precipitated along with most other proteins in a reagent made with trichloracetic acid. The difference in the results of the two tests on similar
preparations was used as a measure of histone protein. This approach
suffers from low sensitivity and lack of precision. Interestingly, CASPERSSON
found nucleoli to be rich in diamino-acid rich proteins, while RIS and
POLLISTER concluded the opposite to be true. As mentioned before, the
difference in the results may owe to the consideration by CASPERSSON of
both histones and nuclear ribosomal basic proteins, while the staining
methods of RIS and POLLISTER, and ALFERT and GESCHWIND, are apparently
sensitive only to the histone of chromatin.

---

[3] MIRSKY and POLLISTER (1946) had reported the presence of histones in pneumococcus type III, and PALMADE *et al.* (1958), in *E. coli,* but the latter finding has been
challenged by ZUBAY's more extensive analysis of the proteins in question, and may
reflect the difficulty in distinguishing between chromatin histone and ribosomal
basic protein when using acid extraction methods.

These methods were largely superceded by ALFERT and GESCHWIND's (1953) technique for selectively staining nuclear basic proteins. This method employs trichloracetic acid to remove nucleic acids which otherwise interfere with staining, then the anionic dye, fast green FCF, at a pH of 8.1 to stain the histones. This is an ingenous method, which, by using a pH delicately poised at an empirically determined value, results in the selective staining of proteins which have a "high enough" isoelectric point. Histones have higher isoelectric points than most proteins and the method usually results in the specific staining of the nucleus, the price of specificity being the loss of stoichiometry. As suggested by SINGER (1952) this type of staining would probably depend more on the overall charge of the protein than the number of individual basic amino acids.

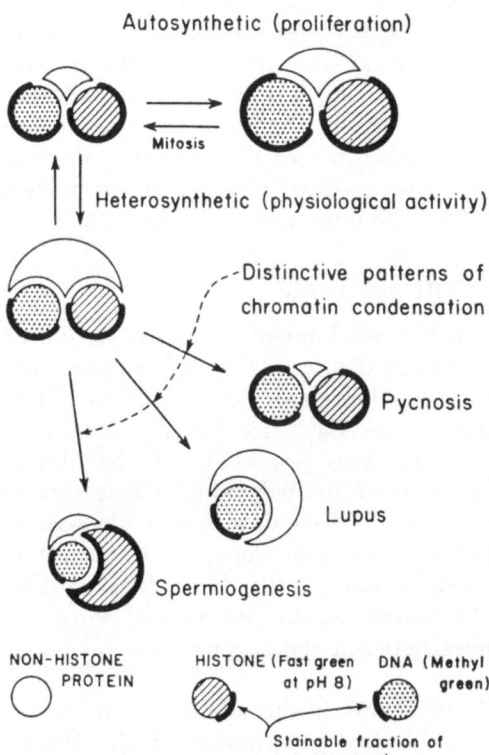

Fig. 2. A schematic representation of the effects of various activities on nuclear staining properties (ALFERT, 1958 b).

One might expect the staining under such conditions to be extremely sensitive to slight variations in both the nature of the histones and the nature of their environment within the cell. It is true that at pH 7.9 nuclear specificity is lost, and at 8.5 most nuclear staining is lost. It is also true that within this range variations in staining are seen which can be attributed to interference by competing substances, or masking. Examples are the decrease in histone stainability of many cells in the "heterosynthetic" or G-1 stages (BLOCH and GODMAN 1955), and also of nuclear material undergoing the degenerative changes associated with *lupus erythematosis* (GODMAN, DEITCH, and KLEMPERER 1958). On the other hand, increased staining is noted in some cases of nuclear pycnosis which is due to the opposite effect, an unmasking of protein (VENDRELY, ALFERT, KNOBLOCH, and MATSUDAIRA 1958). The schematic in Fig. 2 depicts the effects of changing interactions between histone, DNA, and associated proteins, in the staining of histone and DNA (ALFERT 1958 b).

Although subject to variation, the fast green method has proved valuable for detection, localization, and even measurement of histone within individual cells. The meaningfulness of the results is in many cases suggested by their straight-forwardness, for example, the proportionality

between histone and DNA during chromosome duplication (see Fig. 15).
Even so the interpretation given these findings is open to question. Does
the fast green method stain histones which are not complexed with DNA,
or does it only stain the DNA associated protein?

Lysine and arginine are the principal sources of histone's basicity. The
average histone complement contains more lysine than arginine although
the ratios of these two range, among different fractions of histones, from
exclusively lysine, to exclusively arginine. Deamination, or acetylation,
which alter the lysine residues, will lower the isoelectric point and the dye
binding capacity of these various proteins differently, depending upon the
relative contribution of lysine to the overall basicity of the protein (ALFERT
1956, BLOCH and HEW 1960 a). The staining of histone "rich in arginine" is
less effected by deamination. That of histones high in lysine, and in fact,
of most histones, is completely abolished by this treatment (see Fig. 3).
While methods such as these are too imprecise to tell much about the composition of most histones, they are very valuable in dealing with histones
extreme in composition, particularly when a change encompasses the entire
histone complement.

Protamines may be distinguished from histones by their inability to
stain with the fast green procedure, (ALFERT 1956) once the presence of
nuclear basic protein is made known by other methods. They are leached
out of the cell following or during removal of the nucleic acid with which
they are combined. Whether some other basic proteins, such as the high
lysine fraction of the STEDMANS, may be similarly lost is not known.
Protamines may be stained along with other histones by using eosin Y or
bromphenol blue after removal of nucleic acids by picric acid hydrolysis
(BLOCH and HEW 1960 a). These reagents form, in succession, insoluble complexes with protamines, and inhibit their loss during the staining procedure.

Histones found in some cleaving eggs, and basic proteins of nucleoli
remain unstained by the fast green and the eosin Y procedures. They may
stain quite well with the bromphenol blue method after picric acid hydrolysis (BLOCH and HEW 1960 b). The reasons for these differences in staining
behavior are unknown. They are probably due to differences in the
sensitivities of the various dyes to the effects of pH on their binding properties.

Histone arginine and lysine react differently to fast green and eosin,
and although both groups are involved in the binding of either of these
stains when used alone, when used consecutively or in combination, there is
a complex competition between the dyes and the protein such that the
slightly higher affinities of arginine and lysine for fast green and eosin
respectively result in a selective staining of arginine rich and lysine rich
histones. Typically, most nuclei of a tissue stain a bright green, a minority
stain a bright pink, and a still smaller minority display an ambivalence.
The staining of most nuclei is "pure," variations among tissues being
expressed as differences in frequencies of nuclei binding one or the other of
these dyes. Nevertheless differences among nuclei reflect intrinsic differences

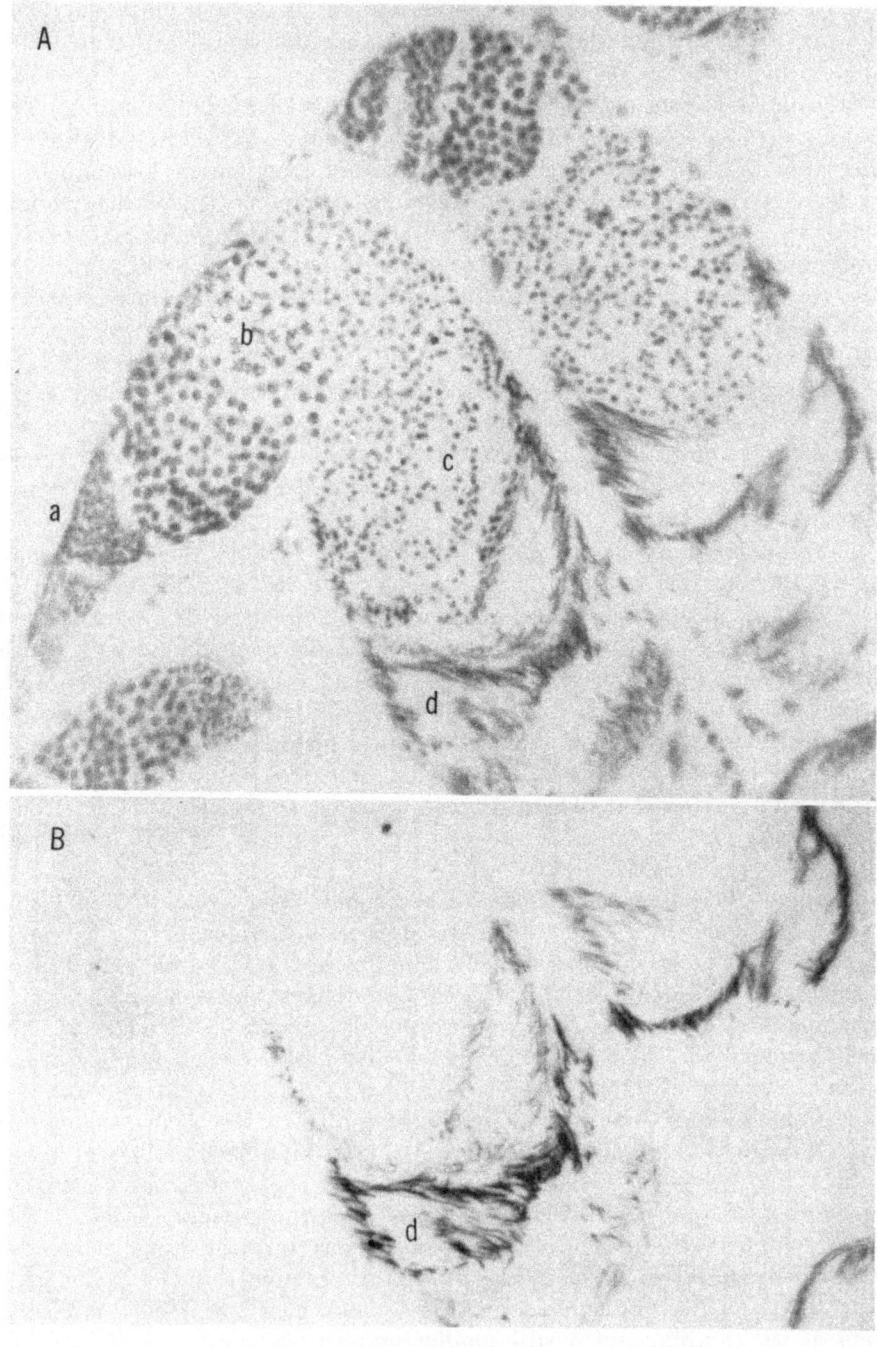

Fig. 3. Follicles of a testis of a species of *Scuderia* (katydid) stained with Feulgen (A) followed by fast green after deamination (B). The gonial cells (a) spermatocytes (b) and early spermatids (c) do not stain after deamination, while the maturing spermatids (d) stain strongly, indicating replacement of typical "somatic" histones by histones very rich in arginine during the late stages of spermiogenesis.

in the nuclei themselves. The percentage of nuclei in a tissue exhibiting a particular reaction is quite independent of the relative concentrations of the dyes over wide ranges. The biological significance of the staining differences is not known. The different color reactions do reflect different histone compositions where such differences are known to exist. However they are also manifested among nuclei with similar histone compositions, and here it is thought that they result from differential masking of lysine and arginine by other substances in the cell (BLOCH, unpublished).

Another staining reagent which may ultimately prove useful for selectively staining histones and perhaps also distinguishing among different histone fractions is an apurinic acid Schiff reagent (BLOCH, unpublished). This method stains histones beautifully and reacts somewhat differently than fast green as staining conditions alter. Its behavior has not been appreciably explored. The procedure is given below.

Modifications of the Sakaguchi reaction for arginine have also been used to study histones (DEITCH 1961, McLEISH et al. 1956, McLEISH and SHERRATT 1958). Most proteins contain arginine, and some of them a good deal more than many histones, so that the applicability of this test is limited. It has been used successfully in the study of sperm heads, (McLEISH et al. 1956, BLOCH and BRACK 1964) which contain a high ratio of histone to non-histone proteins, and whose histones, often simple protamines, are singularly high in arginine.

Fluorodinitrobenzene and other reagents which react with amino groups can be used to demonstrate lysine (DANIELLI 1953). Its utility in the study of histones is subject to the same limitations that apply to the Sakaguchi reaction. Most useful however, is the fact that the Sakaguchi, the FDNB, and the Feulgen reactions can be used in paired combinations, and the ratios of DNA/lysine, DNA/arginine (BLOCH and BRACK 1964) or arginine/lysine determined microspectrophotometrically. The latter is useful in determining at the intracellular level, a ratio which has been used by the biochemist to characterize isolated histones.

The following protocols describe cytological procedures which have been used for the study of histones. In those starred, slight modifications have been introduced into the cited procedure which this author has found to be of some advantage.

Fixation Procedures

No hard and fast rules may be made governing fixation of tissue for histone staining, however the following generalizations may serve as a guide. Formalin, most commonly used in a 10% solution (4% formaldehyde) buffered at pH 7.0 will immobilize most proteins, including histones, but excluding some protamines. This chemical reacts strongly with lysine amino groups, and should be used with consideration in any procedure in which the lysine amines must remain intact, or when they are being assayed. Formalin may be removed by hot acid treatment, such as the trichloracetic acid hydrolysis used in the fast green staining of histones. Carnoy fixation tends to result in loss of nuclear specificity, and most certainly results in actual loss of histone from the nucleus. Such acidic fixatives are to be avoided unless they contain strong protein precipitants, e.g. picric or trichloracetic acids. Under certain circumstances however, Carnoy fixation permits retention of basic proteins which are lost during the slower formaldehyde fixation (VAUGHN 1964). Many cytological practices have their empirically based origins, and except where these conflict with theory, the end result may be taken as the best test of their utility.

Alfert and Geschwind's alkaline fast green method. This procedure stains most histones, excepting simple protamines and some histones found during early embryonic development.

Fix in 10% neutral buffered formalin, 3 hours.

1. Hydrolyze in 5% trichloroacetic acid for 15 minutes at 90° C. (*or 3 hours at 60° C. in N TCA.)

2. Three 10 minutes rinses in 70% ethanol.

3. 30 minutes in a 0.1% solution of fast green FCF buffered at pH 8.1 with NaOH.

4. 5 minutes in distilled water (*or absolute methanol), dehydrate and mount.

Picric acid eosin method: Stains most histones, including protamines. Does not stain early cleavage histones (BLOCH and HEW 1960 a).

Fix in 10% neutral buffered formalin.

1. Hydrolyze for at least 6 hours in saturated picric acid solution at 60° C.

2. Rinse in water.

3. Overnight in a 0.1% solution of eosin Y buffered at pH 8.1–8.3.

4. Rinse in distilled water, 5 minutes, dehydrate and mount.

Picric acid bromphenol blue method: Stains most histones, including protamines and cleavage histones (BLOCH and HEW 1960 b).

Fix in 10% neutral buffered formalin.

1. Hydrolyze for at least 6 hours in saturated picric acid solution at 60° C.

2. Rinse in water.

3. Overnight in a 0.1% solution of bromphenol blue brought to a pH of 2.3.

4. Rinse for about 5 minutes in 95 per cent ethanol which had been made slightly alkaline by stirring with a glass rod previously wetted then held over the mouth of a bottle of concentrated ammonium hydroxide.

5. Bring without further rinsing to a fresh Coplin jar of 95% ethanol and let stand for a least 12 hours.

6. Dehydrate and mount.

Fast green-eosin method: Stains arginine histones green, lysine histones pink. This staining reaction is subject to masking and is very sensitive to the play of unknown factors. In a given instance, interpretation of observed differences among nuclei (whether these reflect inherent differences in the composition of their histones or differences in the reactivity of lysine and arginine of essentially similar histone complements) depends upon the application af ancillary techniques (BLOCH, unpublished).

Fix in 10% neutral buffered formalin.

1. Hydrolyze in 5% trichloracetic acid, 15 minutes, 90⁰ C.

2. Three 10 minute rinses in 70% ethanol (steps 2–7 in ice bath).

3. One half hour in 0.1% fast green FCF in 0.07 N tris buffer, pH 8.3.

4. One hour in a mixture of .05% fast green FCF and .05% eosin Y in the above buffer.

5. Ten minute rinse in buffer.

6. Ten minute rinse in 70% ethanol.

7. Ten minute rinse in 95% ethanol.

8. Complete dehydration and mount.

Apurinic acid Schiff reaction for histones:

Preparation of apurinic acid Schiff reagent.

1. Hydrolyze 250 mgs. purified DNA in 25 mls. .01 N. HCl, 37⁰ C. 24 hours. (TAMM et al. 1952.) Add acid if necessary to bring pH to 1.7.

2. Add 100 mls. Schiff's reagent. The solution should turn a dark magenta within an hour.

3. Dialyze for two days in several changes of .01 N HCl to remove unreacted leuco-Schiff. (In cold room.)

4. Dialyze for several days against several changes of distilled water.

5. Concentrate solution to about 10 mls. by dialyzing against 70% carbowax (polyethylene glycol, MW = 6,000).

6. Separate apurinic acid-Schiff from uncombined pararosaniline in a sephadex column. Washed sephadex G-75 should be used (Pharmacia Corp.). Several mls. of solution are layered on top of a column about 1 by 10 inches, then followed with water. The first band to come through is the high molecular weight apurinic acid Schiff complex. This should be used for staining within several days, as the free pararosaniline resulting from the slow but continuous decomposition of the product will eventually result in some non-specific staining. The three mls. which had been put into the column may be diluted to 10 to 20 mls. during the separation process. This dilution provides good staining.

Staining with apurinic acid-Schiff's reagent.

Fix tissue in 10% neutral buffered formalin.

1. Hydrolyze as for Alfert and Geschwind's alkaline fast green staining.

2. Stain for 30 minutes in APAS, pH 6.5.

3. Rinse for 30 minutes in buffer of same pH.

4. Dehydrate and mount.

The preparation looks like a crisply stained Feulgen. The procedure can be used with modifications comparable to those to which the alkaline FG, the picric acid eosin, and the bromphenol blue tests are subjected. The "cut off" pH's vary slightly from one batch of APAS reagent to the next, and the staining pH may have to be raised or lowered to give maximum staining compatible with nuclear specificity.

Millon's reaction for total protein and non-histone protein: (Pollister 1950).

Fix in alcohol. Avoid formalin, as it immobilizes histone.

Determination of histone by this method depends upon the micro-spectrophotometric determination of the differences in staining of similar objects, to which the following reactions have been applied.

Total protein.

Incubate in a solution containing 5% mercuric acetate in 30% trichloro-acetic for 5 to 10 minutes at 37⁰ C. Add one tenth volume of 1% sodium nitrite and incubate 25 minutes. Rinse in 70% ethanol, dehydrate, and mount.

Non-histone protein.

Incubate at 60⁰ C for 5 minutes in a solution containing 1 gm. mercuric sulfate to 9 mls. of .94 M H_2SO_4. Add one tenth volume of 1% sodium nitrite and incubate an additional 25 minutes. Rinse 15 minutes in three changes of 50% ethanol, dehydrate, and mount.

It is important to match the refractive index of the mounting medium with that of the tissue, since the Millon stain results in highly refractive tissue.

The tyrosine mercurial has an absorption peak at 480 mμ., and both tyrosine and tryptophane mercurials have peaks at around 350 mμ.

Methods for arginine rich spermatid and sperm histones:

Any of the above acid staining methods can be used after either deamination or acetylation, these steps following nucleic acid extraction.

*Deamination: (Van Slyke 1911). Three 15 minute rinses in freshly prepared solution of nitrous acid made by adding equal parts of 10% sodium nitrite and 10% acetic acid.

*Acetylation: (Monné and Slaughterbach 1957). Bring slides through alcohols to a solution of 1% glacial acetic acid in acetic anhydride. Set for one hour at 60⁰ C. Bring through alcohols to water.

These methods permit the specific staining of histones which have a very high ratio of arginine to lysine. The so-called arginine rich histones of somatic tissues, which contain an arginine to lysine ratio slightly over 1 do not stain. The arginine rich histones of the squid (*Loligo opalescens*) have approximately 6 times as much arginine as lysine and stain well after this treatment (Bloch 1962 b).

Sakaguchi reaction for protein bound arginine: (McLeish et al. 1958, Deitch 1961). Fix in 10% neutral buffered formation or any fixative which will not result in the loss of histone.

1. Place slide in a freshly prepared solution containing one part of 1.5% 2,4 dichloranapthol in 70% ethanol, 5 parts saturated Ba(OH)₂, and one part 1% NaOCl (Commercial Chlorox diluted 1 to 5 with water). These ingredients should be added in the above order when mixing the solution. The slide should be placed face down in a watch glass. Let stand 15 minutes.

2. Flood slide with a 5% urea in sat. Ba(OH)₂ so that the crust of $BaCO_3$ flows over the edge of the watch glass and does not come in contact with the section.

*3. Dehydrate in freshly distilled aniline. The tissue should vary from light to dark pink. Apparent fading is due to a shift of the spectral absorption toward the blue under less basic conditions. The pink color can

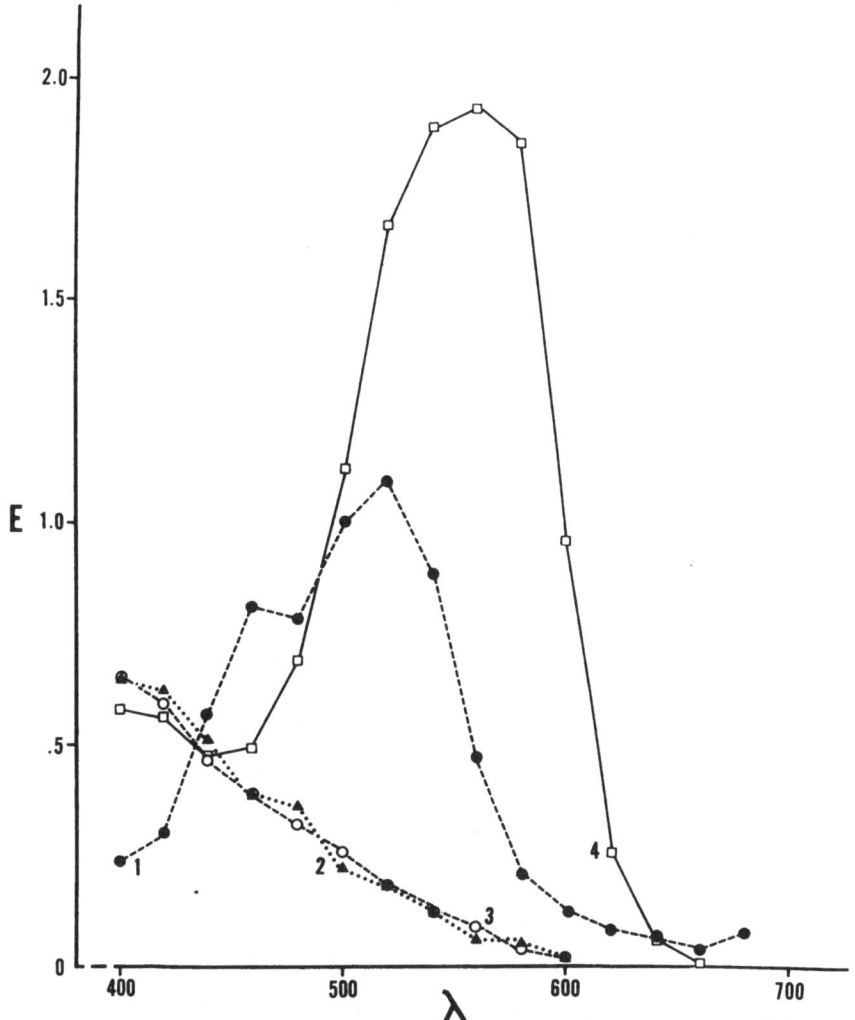

Fig. 4. Absorption spectra of nuclei stained with the Sakaguchi and Feulgen reactions. Curve 1: ●, Sakaguchi stain. Curve 2: ▲, Sakaguchi stain followed by treatment with cold acid. Curve 3: ○, Sakaguchi stain followed by acid hydrolysis used for the Feulgen reaction. Curve 4: □, Sakaguchi and Feulgen stain. A single grasshopper spermatid nucleus was used in obtaining all four curves. (BLOCH and BRACK, 1964.)

be recovered by retreating the slide with $Ba(OH)_2$ solution. The complex formed in this reaction is stable, even though the quality of the color may change.

Sakaguchi-Feulgen reaction for determination of arginine/DNA ratios: (BLOCH and BRACK 1964).

1. Carry out steps 1 and 2 in the Sakaguchi reaction.

2. Hydrolyze in N TCA, 60⁰ C., 12 minutes for alcohol fixed preparations, 25 minutes for formalin fixed preparations. The Sakaguchi complex turns yellow but is not lost.

3. Stain for one hour in TCA Schiff's or if the material was fixed in formalin, the usual HCl Schiff's reagent.

4. Three 5 minute rinses in sulfite bleach. (Freshly prepared solution containing 5 mls. each of 10 % potassium metasulfite and N TCA in 100 mls. water).

5. Rinse in water, dehydrate, and mount in refractive index oil 1.568 (or appropriate oil, depending upon the tissue used). Fig. 4 shows the absorption curves of the acidified Sakaguchi stain and the combined Sakaguchi and Feulgen stains. Measurements may be made at wavelengths corresponding to the respective peaks of the two stains and the relative ratios of arginine to DNA determined.

FDNB test for protein bound lysine:
Fix tissue in methanol. Formalin and acidic solutions should be avoided. If formalin has been used, it can be removed by hydrolysis in 5% TCA, 90⁰ C., 15 min.

1. Immerse slide in solution containing 1 ml. M $NaHCO_3$, 5 mls water, and 0.150 mls. fluorodinitrobenzene in 6.5 mls. absolute ethanol. Let stand one half hour.

2. Rinse well in 70% ethanol, followed by absolute ethanol, to remove excess FDNB.

3. Xylene, then mount.
Tyrosine contributes to a small degree to the staining. This can be avoided by blocking (see Danielli 1953).

FDNB-Sakaguchi reaction for determination of (lysine + tyrosine)/ arginine ratios:
Avoid formalin, or remove formalin.

1. Stain with FDNB as above.

2. Stain with the Sakaguchi reaction as above.

3. Measure extinctions at 520 and 400 mμ., correcting for absorption at 400 due to the Sakaguchi reaction.

Calculation, using the specific extinction coefficients of Sakaguchi and of DNFB stained protein, based upon the extinctions at the above wavelengths, (see figure 5), gives the ratio of (lysine + tyrosine)/arginine.

Fig. 6 shows the absorption curves of protamine containing sperm, and histone containing spermatids of squid, stained in this manner.

Trichloracetic acid Feulgen technique for staining of DNA when stain is to be followed by staining of histones (Bloch and Godman 1955 a).
Fix in 10 per cent neutral buffered formalin.

1. Hydrolyze 25 minutes in 1 N trichloracetic acid at 60⁰ C.

2. Stain for one hour in Schiff's reagent made with TCA.

3. Sulfite rinse made with TCA instead of HCl.

4. Rinse in water, dehydrate and mount.

5. Follow with fast green stain procedure for histones (see Fig. 1).

Differential staining of nuclear chromatin and viral inclusion bodies:

This method is applicable to certain viral infected cells such as adenovirus infected HeLa cells, in which the DNA is relatively resistant to acid

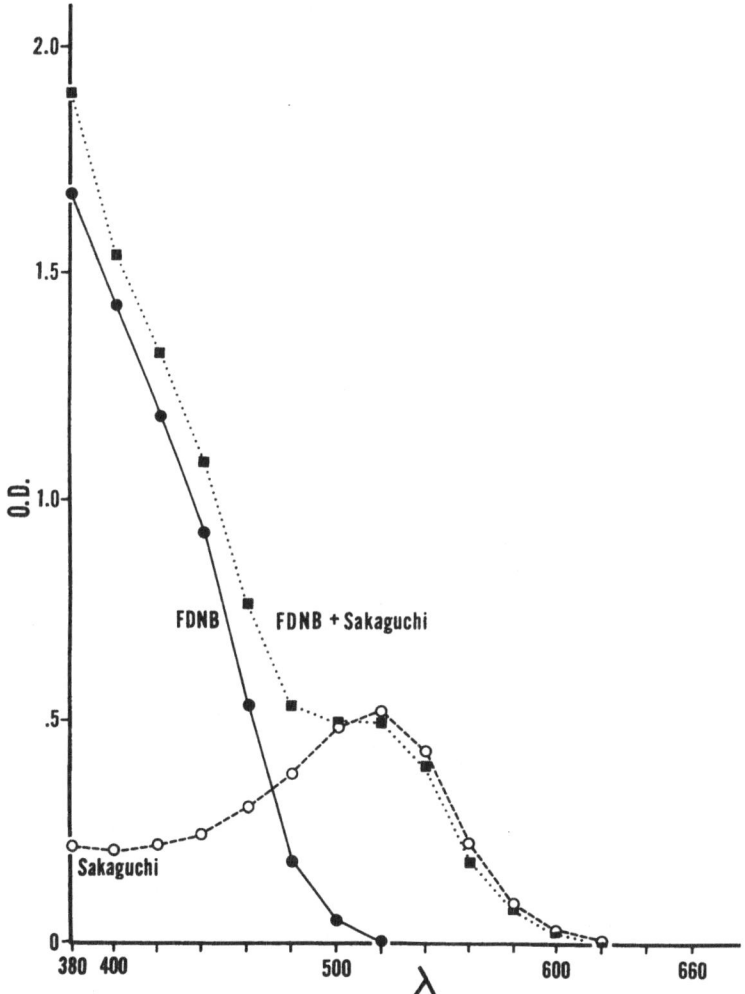

Fig. 5. Spectral absorption of three samples of bovine serum albumin stained with fluorodinitrobenzene (FDNB),
the Sakaguchi stain, and the combined stains, the Sakaguchi following the FDNB.

hydrolysis. The DNA of the chromatin can be sufficiently hydrolyzed to
permit histone staining, while the same hydrolysis of the more resistant viral
chromatin permits the DNA to be visualized with the Feulgen technique.

Fix smears in 10 per cent neutral buffered formalin.

1. Hydrolyze for 15 minutes in 5 per cent trichloracetic acid at 90° C,
or for one hour in N TCA at 60° C. (Hydrolysis conditions are best determined empirically).

2. Stain with TCA Schiff's reagent.

3. Three changes of sulfite rinse, 2 minutes each.

4. 0.1 per cent fast green FCF at pH 8.1 for one half hour.

5. Differentiate in distilled water, 5 minutes, dehydrate and mount.

Double staining of RNA and nuclear histones.

Fix in 10 per cent neutral buffered formalin.

1. Remove formalin by exposing slide briefly to boiling water.

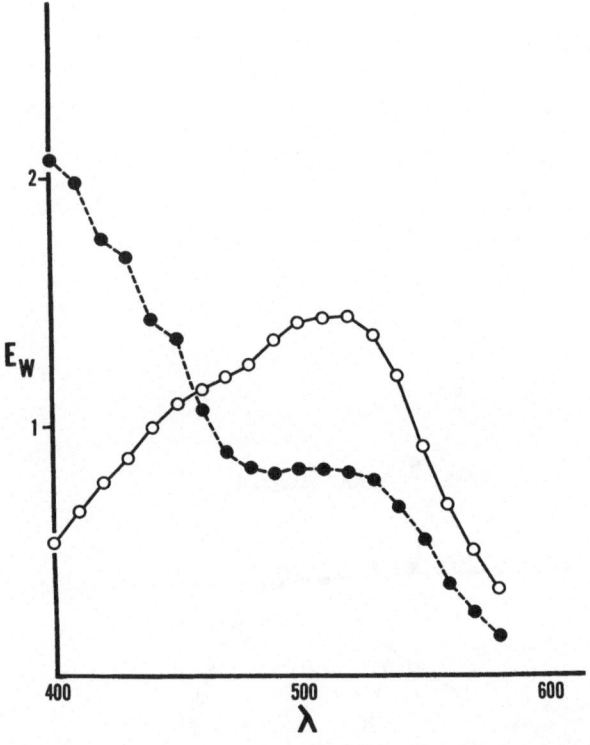

Fig. 6. Absorption spectrum of FDNB-Sakaguchi stained spermiogenic cells of the squid, *Loligo opalescens*. Solid circles: Histone bearing spermatids. Open circles: Protamine bearing sperms from the spermatophore. (Compare with Fig. 5.) (Bloch and Brack, unpublished.)

2. Digest with deoxyribonuclease. Treat preparation with a 1 mg. per ml. solution of crystalline deoxyribonuclease containing .003 M MgCl$_2$, pH 6.5, for 1 hour, 37° C. (Swift 1955).

3. Rinse in water.

4. Stain 1 hour with .1 % azure B buffered at pH 4.0 (Flax and Himes 1952).

5. Stain 30 minutes with eosin Y, as in eosin procedure for histones.

6. Differentiate 5 minutes in distilled water.

7. Dehydrate and mount.

Cytoplasmic RNA stains blue, nuclear histones stain orange. There is little evidence of nuclear basophilia.

Naphthol Yellow S stain for protein basic groups (DEITCH 1955).

Fix in alcoholic fixatives. If formalin is used, remove before staining.

1. Stain tissues for 15 minutes in 1 %Naphthol Yellow S in 1 % acetic acid.

2. Differentiate 15–24 hours in 1 % acetic acid. The differentiating bath should be made slightly yellow by the staining solution carried over from the staining bath.

3. Dehydrate in tertiary butanol.

4. Xylene.

5. Mount.

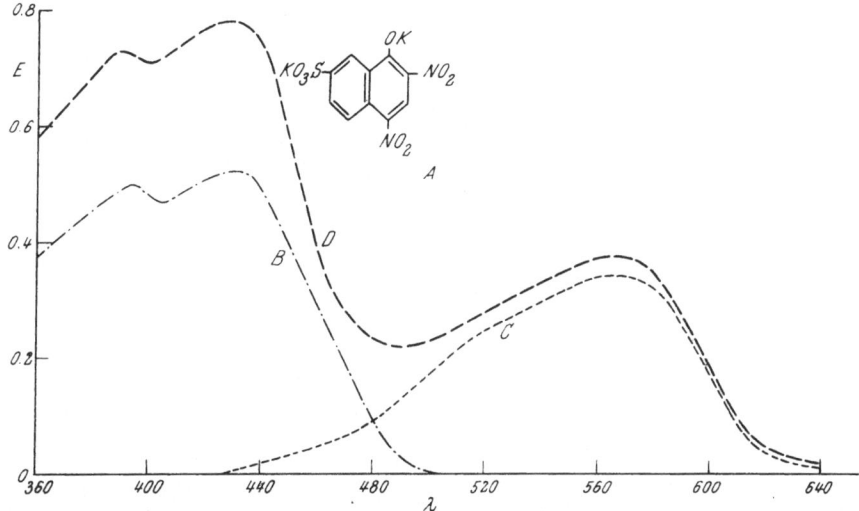

Fig. 7. Spectral absorption curves of Naphthol Yellow S, Feulgen, and the combined stains. *A*, structural formula of Naphthol Yellow S. *B*, Absorption curve of Carnoy fixed salamander (*Necturus*) liver cytoplasm stained with Naphthol Yellow S. *C*, Absorption curve of salamander liver nucleus stained with the Feulgen reaction. *D*, absorption curve of salamander liver nucleus stained with the Feulgen reaction and counterstained with Naphthol Yellow S. (DEITCH, 1955.)

This procedure stains proteins a deep yellow. It is quantitative under these conditions, binding stoichiometrically to the basic residues, lysine, arginine, and histidine. It can be used as a test for arginine (plus histidine where present in appreciable quantities) by preceding the staining with acetylation or deamination.

The Feulgen and naphthol yellow S stains can be combined by applying them successively, in that order, to the same slide. The ratios of the extinctions at appropriate wavelengths provide a measure of the ratio of DNA to protein (see Fig. 7).

Histone Metabolism

Histone Synthesis and Assemblage of deoxyribonucleohistone

Two aspects of the problem of histone synthesis deserve consideration. The first is the synthesis of the protein molecule, and the second, its assembly with DNA to form the DNH of chromatin. Until recently, little distinction was made between the two processes, and increases in staining of nuclear histone which may represent an assembly process were interpreted as indicating synthesis (Bloch and Godman 1955, Alfert 1956 a, Gall 1958, Rasch and Woodard 1959, Prescott and Kimball 1961). At present, little can be said of the true synthetic process other than that it is probably similar to that of most proteins, mediated by RNA which has its origin in the nucleus (Bloch and Brack 1964, see below).

There are two known orders of assembly of histone into the chromatin. One is a part of the process of chromosome replication, and the other, a process which occurs independently of chromosome replication, resulting in the qualitative alteration of the chromatin by replacement of one type of histone with another. These assemblages are seen as discontinuous processes. Other processes have been envisaged which would be in a sense more continuous. Caspersson's hypothesis, invoking basic protein synthesis at the chromosome level in the transport of material from the chromosome to the cytoplasm would serve as an example, although it appears unlikely that such synthesis involves histones. Busch (1963) has suggested that the DNA bound histone may be in equilibrium with free histone. The extent to which such phenomena may contribute to observed turnover of histones at the chromatin level is debatable and is considered in greater detail in the section on histone conservation.

Synthesis Leading to Chromatin Alteration

Because of the insensitivity of the methods applicable to the analysis of histones among different cell types, observed changes in the nuclear histones are based upon studies of cells in which the entire complement of histones is replaced by others whose compositions differ grossly from that of the average. Such changes have been seen to occur during spermiogenesis (Alfert 1956 b, Bloch and Hew 1960 a, Dass, Gay, and Kaufmann 1964 a), fertilization, and early embryonic development (Alfert 1958, Bloch and Hew 1960 b, Dass, Gay, and Kaufmann 1964 b). Early recognition of the change was a consequence of Miescher's and of Kossel's finding of changes in composition of basic proteins in the ripening testes of the Rhine salmon (see Kossel 1928). Kossel postulated conversion of the histones to protamines by way of intermediates. Comparable studies at the cellular level were not possible until more than a half century later, when means for the cytochemical detection of histones were devised. Initial cytochemical work on the transition was done by Alfert (1956 b) using salmon, the classical biochemical organism. His studies demonstrated several alterations in the staining properties of the developing spermatids: A typical staining of the early spermatid with alkaline fast green, which can be abolished by prior blocking of lysine amino groups by acetylation; an intense staining of the later spermatid which is indifferent to the effects of acetylation; and finally, loss of stainability of the mature sperm. Alfert's interpretation of this sequence was the following: The typical histone of the early spermatid is replaced by a protamine, devoid of lysine. The protamine is rendered insoluble, hence stainable, by a stabilizing protein; whose subsequent loss during further maturation results in loss of the protamine of the mature sperm during the staining procedure.

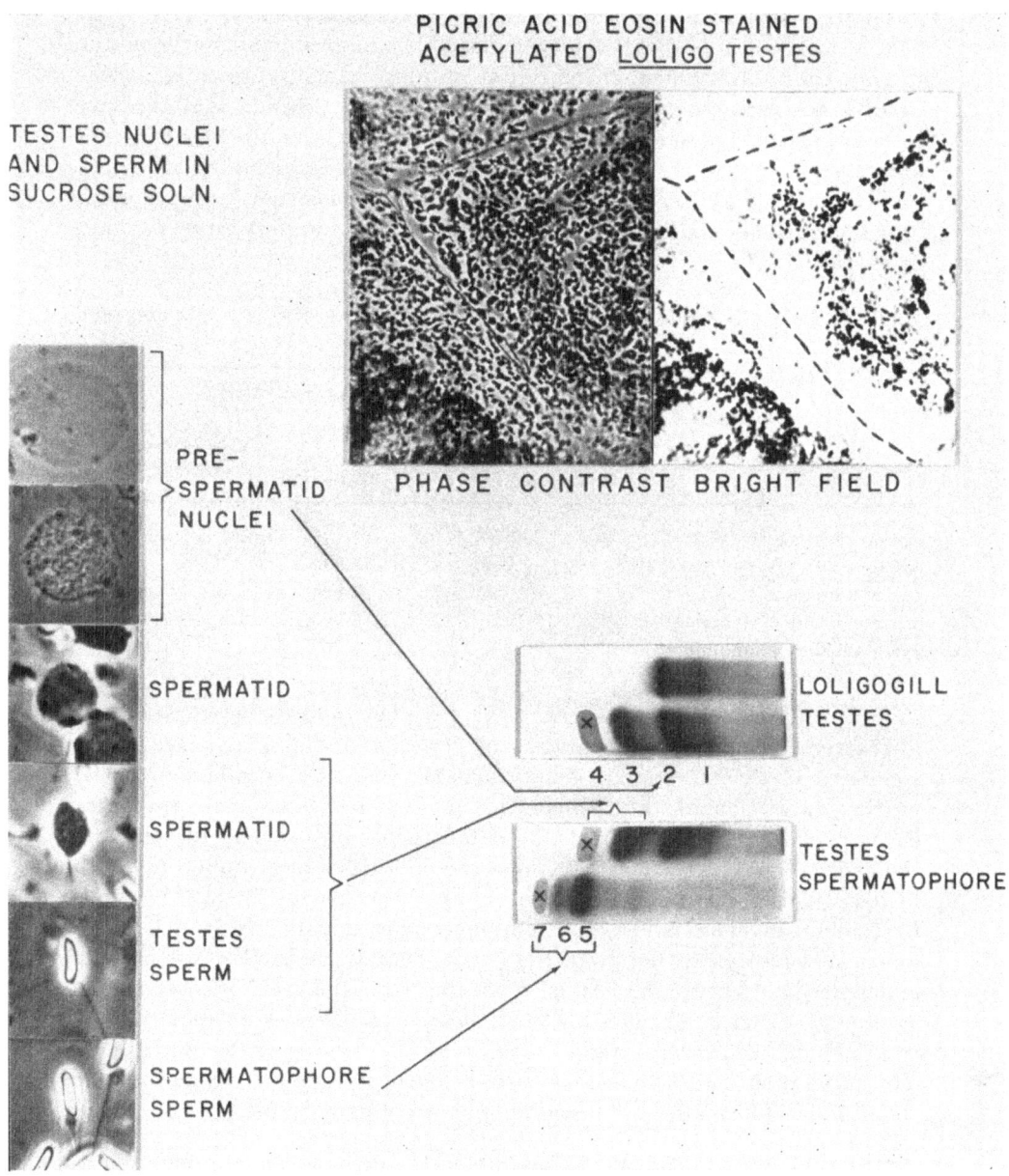

Fig. 8. Histone transition during spermiogenesis in *Loligo opalescens*.
The upper right hand photographs show the ability of the later spermatids and testes sperm to stain with acid dyes after deamination. The electropherograms compare the banding patterns of the basic proteins extracted from several organs and cells. The components marked "X" are present in variable amounts, and were not evident in these particular preparations. The arrows indicate the sources of the component proteins, when these are obtained from separated fractions of nuclei (from BLOCH, 1962b).

A similar sequence of staining changes during spermiogenesis has been observed in many other organisms [4] (see Fig. 3). The results of studies on the snail *Helix aspersa* were given a slightly different interpretation by Bloch and Hew (1960 a). The intense staining of the intermediate spermatid was attributed to the synthesis of a newly synthesized arginine rich, lysine poor histone which is subsequently replaced by the protamine in the mature sperm. The supposition that the initial staining change represented synthesis of a new protein was substantiated by the observation that the same cells displaying the acidophilia which withstands deamination, also incorporate arginine into the nucleus. However, apparent absence of incorporation into more mature spermatids, presumed to be incorporating protamine into the nucleus would appear on the surface to support Alfert's interpretation.

Table 1. *Molecular weights of squid nuclear proteins during spermatogenesis* (Bloch 1962 b).

Protein fraction	Molecular weight		
	Ultracentrifugation		End group analysis (Phillips, 1955)
	(Archibald, 1947)	(Yphantis, 1960)	
Somatic histone (band 2)	46,500		
Arginine-rich histone (band 3)	7,900	8,200	
Testes protamine (band 4)		4,600	3,800
Spermatophore protamine (band 5)		5,900	5,800
Spermatophore protamine (band 6)		5,000	5,200

The squid, *Loligo opalescens,* provided the opportunity to test these alternatives. The staining patterns of *Helix* and the salmon are closely simulated in *Loligo.* Spermiogenic tissue is present in large amounts, permitting extraction of the proteins in question. The maturing sperm are encased in spermatophores outside of the testes, aiding in identification of the stages from which the proteins were obtained. Electrophoresis of the histones from the testes showed two groups of proteins (Fig. 8). One group has mobilities identical to those obtained from somatic tissue. These histones were also extracted from the early spermatogenic cells (gonia, spermatocytes, and early spermatids) separated from the later cells by centrifugation in a sucrose gradient. The second group of testes histones contains two fractions which have higher mobilities. The slower of the two is a low molecular weight (8,000) histone (see Table 1) having all the amino acids found in most histones,[5] but containing an arginine to lysine ratio of about 6,

---

[4] It is of interest that the micronuclei of Tetrahymena shows a more intense staining with fast green, relative to Feulgen staining of DNA, than do the physiological macronuclei, suggesting a higher basicity of the gametic nuclear protein (Alfert and Goldstein 1955).

[5] The author is indebted to Dr. A. W. Schroeder for the amino acid analyses of the squid proteins.

(see Table 2) the arginine comprising almost 50 % of the amino acids on a molar basis. The faster fraction is a monoprotamine of molecular weight 4,200 containing approximately 70 % arginine. These two proteins are found in the later spermatids and immature sperm of the testes. The fast green staining of those spermatids which are indifferent to the effect of deamination is attributed to the presence of the arginine rich histone. The

Table 2. *Amino acid compositions of some histones found during spermatogenesis in the squid (Loligo opalescens). The amounts are expressed in mole %.*

Amino acid	"somatic" (band 2)	testes spermatid arginine rich (band 3)	testes protamine (band 4)	spermatophore slow (band 5)
lysine	27.0	8.5	4.3	1.8
histidine	1.2	.9	1.6	1.5
NH$_3$	—	—	—	—
arginine	5.2	47.5	70.3	77.5
aspartic acid	4.3	3.1	.7	.4
threonine	2.8	3.1	1.7	1.4
serine	5.7	6.5	8.8	9.6
glutamic acid	6.7	3.0	.9	.5
proline	6.9	2.5	1.9	2.1
glycine	7.6	5.3	2.2	.4
alanine	16.3	6.0	.7	.3
1/2 cystine				
valine	5.0	2.8	.4	
methionine	.5	1.3	.1	
isoleucine	3.3	1.0	.2	.2
leucine	5.5	5.0	1.2	.2
tyrosine	1.2	3.4	4.8	4.2
phenylalanine	1.0			
	100.2	99.9	99.8	100.1
Mol. wt.	46,500	8,000	4,200	5,800

spermatophore proteins show one minor and two major bands, the slowest corresponding in mobility to the fastest band of the testes, the faster bands having no counterpart in the testes. These proteins are fairly typical protamines. Insofar as these findings can be generalized to include the salmon, and snail, it may be presumed that the staining changes common to this class of organisms reflect replacement of a typical spectrum of histones by histones which are very rich in arginine, and subsequent replacement of these by protamines.

The Pacific coast mussel, *Mytilus edulis,* differs from the Salmon, *Loligo, Helix* group. During spermiogenesis, some of the late stages do stain but only faintly with fast green after deamination, in striking contrast to the stark staining of intermediate spermatids of the other organisms. Electrophoresis of the testicular histones shows, as in the squid, three major bands whose mobilities are higher than those of the general run of somatic histones, although somatic tissues were not studied in this organism. These three bands designated β, γ, and δ in order of increasing mobility, become sequentially labelled after a single injection of C^{14} labelled amino acids (Fig. 9–11). The order of labelling is β, δ, and finally γ. The β protein is labelled one day after injection. Twelve days after injection, β and δ are both labelled, but no trace is seen in the γ fraction. After several weeks,

Table 3. *Amino acid compositions of some histones found during spermatogenesis in the Pacific coast mussel, Mytilus edulis. (Amount expressed in mole per cent.)*

Amino acid	α band	γ band	δ band
lysine	22.0	23.1	29.0
histidine	0.9	0.1	0.9
NH$_3$	—	—	—
arginine	8.3	29.4	7.9
aspartic acid	5.8	0.6	4.7
threonine	5.1	2.8	4.1
serine	12.7	16.2	9.7
glutamic acid	3.5	0.6	4.3
proline	6.8	4.7	6.8
glycine	7.2	6.6	5.9
alanine	12.3	12.9	12.5
1/2 cystine	0.5	0.4	
valine	4.0	1.2	4.0
methionine	0.8		0.7
isoleucine	3.5	0.3	2.5
leucine	5.0	0.4	3.8
tyrosine	0.8	0.2	1.7
phenylalanine	1.4	0.3	1.6
	100.6	99.8	100.1

most of the label is in the γ fraction. This would suggest that the γ fraction is the protein contained in the mature sperm, and accords also with the high amounts of this protein as compared with the β and δ proteins. The γ protein is also very basic, its amino acid composition given in Table 2. [6] Its relatively low mobility, as compared with the δ protein is due to its high molecular weight. The composition of the β fraction is as yet unknown.

Several alternative interpretations can be given these data. The most obvious is a precursor product relationship, the β being converted to the δ, and this in turn to the γ of the mature sperm. The conversion may involve complete breakdown, the products of hydrolysis utilized for *de novo* synthesis of the successive proteins, or may represent conversion of large

---

[6] Dr. LUBOMIR HNILICA generously provided the data on the analysis of these mussel proteins.

molecules, such as has been observed in the case of acid phosphatase
(KECK and CHARLES 1963). If the former, one would have to postulate complete inaccessibility of the pool containing the breakdown products to

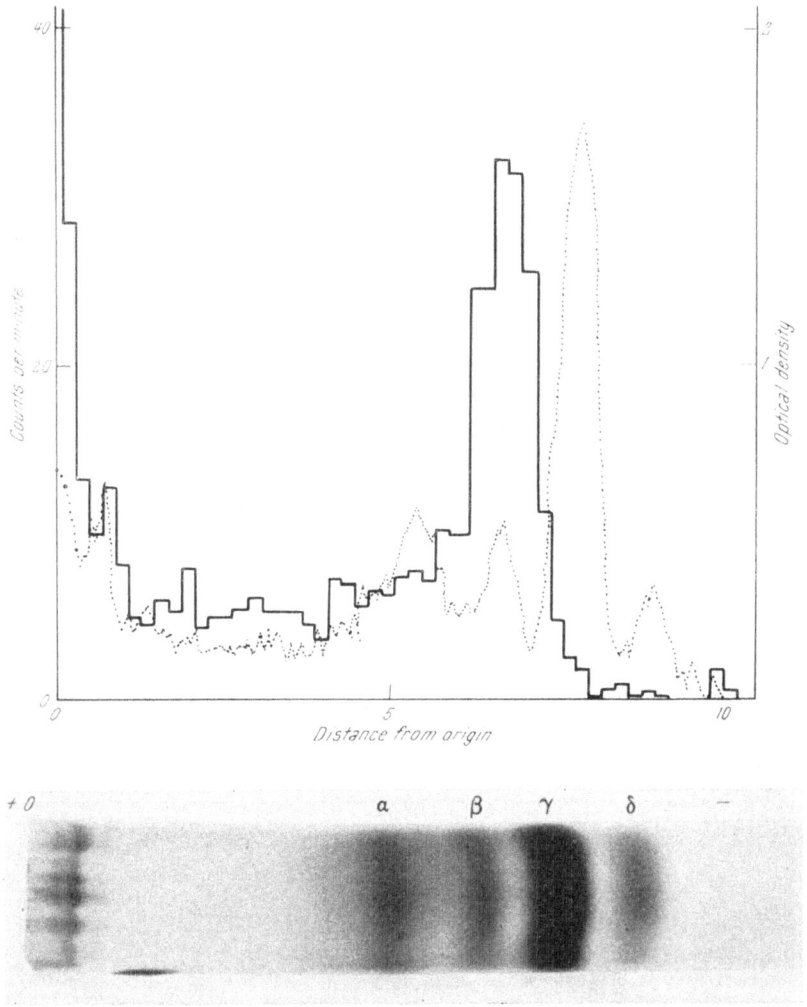

Fig. 9-11. Incorporation of C[14] amino acids into the different spermiogenic histones of *Mytilus*. The electropherograms show the major bands obtained from histones extracted from the testes. The β, γ, and δ bands have
mobilities typical of spermiogenic histones and protamines. The solid bar shows the radioactivity of samples
taken along the gel, and the dotted lines, the optical density of the stained gel. The histone samples shown in
figures 8, 9 and 10 were taken from animals sacrificed 1, 12 and 20 days respectively, after injection of labelled
amino acids. (BLOCH and BALLATINE, unpublished.)

exogenous amino acids else the newly synthesized proteins would show
immediate incorporation. A third possibility, which is favored, is that the
three proteins are synthesized more or less at the same time in the early
spermatid, but that they are successively incorporated into the chromatin
during development, and the method of isolation only detects the histone
associated with DNA.

This conclusion is arrived at via a rather tortuous route. Its advantages are that it would offer an explanation why the earliest of the protamine bearing spermatids of *Helix* do not exhibit incorporation; the protamine

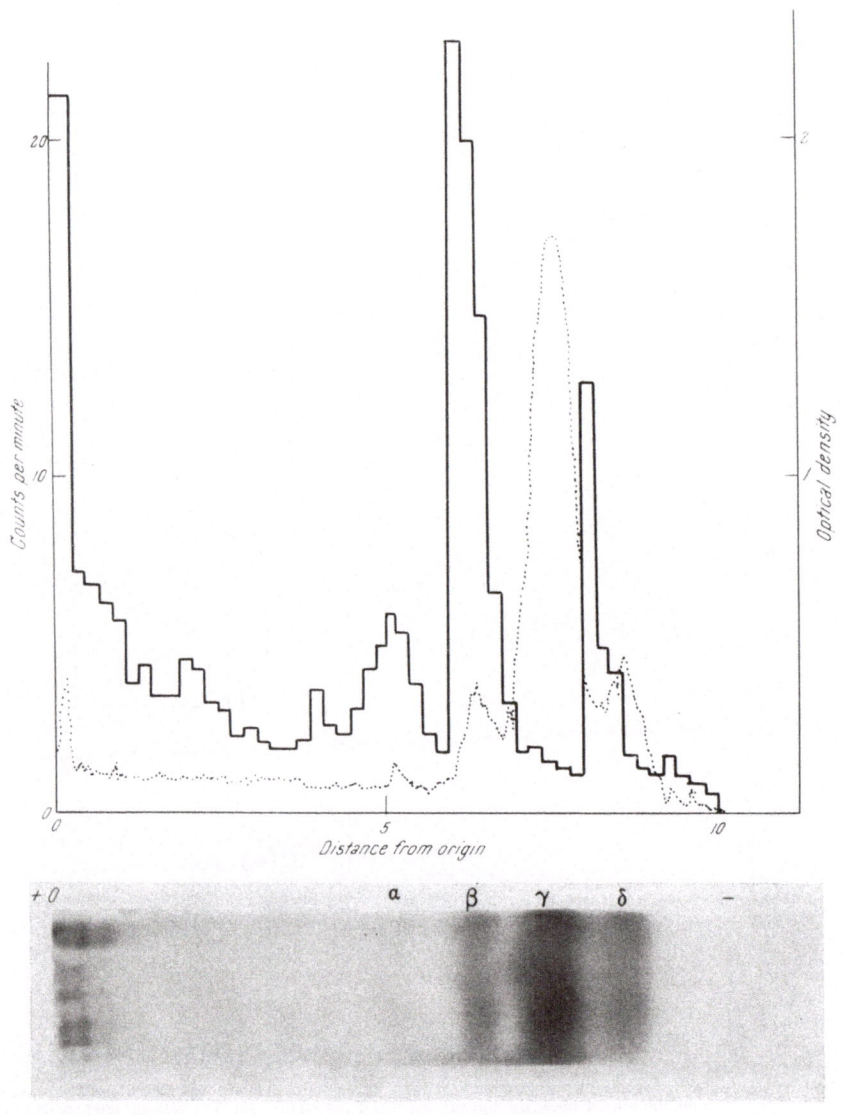

Fig. 10.

being incorporated into the nucleus had actually been synthesized at an earlier stage. It is also compatible with the succession of replacements seen in *Loligo*. The chromosomal histones *only* are isolated, and the hypothesized presursors of the chromosomal protein, perhaps in the cytoplasm, or perhaps in fact in the nucleus, but not yet associated with the chromatin, escape detection.

That this explanation is plausible is borne out by the results of studies on spermiogenesis in the grasshopper, *Chortophaga viridifasciata,* which clearly indicate that the nuclear histones involved in replacement can be

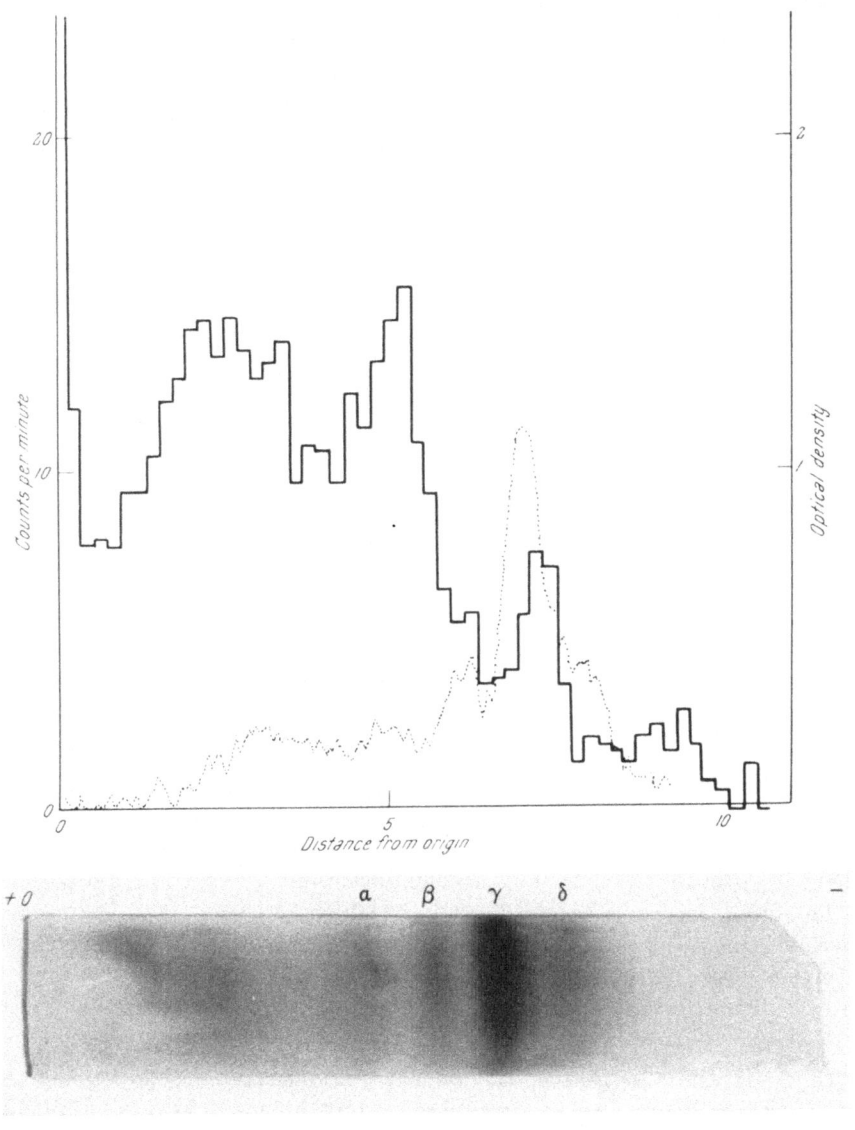

Fig. 11

synthesized in the cytoplasm (BLOCH and BRACK 1964). In this organism, as in many mammals studied, the transformation is "incomplete" in that the mature sperm resembles in its staining, the intermediate spermatid of *Loligo* and *Helix,* indicating that the final protein is an arginine rich histone rather than a protamine. The high refractoriness of the protein to extraction procedures further attests to this.

Cell elongation during the process provides a useful parameter in the study of the metabolism of the spermatid. It occurs in two phases. The initial phase involves the cytoplasm and terminates in a spermatid characterized by a large spherical nucleus and a long axial filament, both covered by a thin layer of cytoplasm. Some residual cytoplasm is still retained in the form of beads intermittently strung along the tail, but most of the cytoplasm contained in the earlier stages has been sloughed. During the second phase, the nucleus changes from a spherical to a highly elongated form. The progress during the second phase can be measured in terms of the increasing axial ratio of the nucleus.

The transformation of the nuclear protein occurs during the second phase of elongation, and is perhaps somewhat unique in this respect, for it begins well after sloughing of the cytoplasm. The alterations in the staining properties appear to be most rapid as the nuclear length to width ratio changes from around 5 to 10. The rate of synthesis as determined by incorporation of arginine into the spermatid also appears to be most rapid at this stage.

While the postmeiotic spermatid is very rich in RNA, this substance being detectable both in the nucleus and the cytoplasm by staining and by incorporation of cytidine, by the end of the first stage of elongation, the synthesis of RNA has practically ceased, all detectable RNA has been lost from the still spherical nucleus, and most of the RNA of the cell has been sloughed. As the histone synthesis begins, the only detectable RNA exists in the form of a few scattered granules in the cortical cytoplasmic layer surrounding the nucleus, and along the axial filament. Autoradiography of thin sections (1 micron) of plastic embedded tissue taken one hour after injection of the animals with tritium labelled arginine shows the silver grains to overlie primarily the thin cortical cytoplasmic layer immediately surrounding the nucleus (Fig. 12). Electron microscopy of these cells shows this area to be rich in ribosome like granules. At later times, most of the label is seen in the nucleus. It was concluded that the histone is synthesized in the cytoplasm by ribosomal machinery which had itself been synthesized a week or so earlier. The histone then migrates into the nucleus and combines with the DNA. Surprisingly, the histone in the cytoplasm appears not to stain with the techniques used for histone staining. This lack may be due to masking, dilution, or to ease of extraction during staining.

If cytoplasmic synthesis of nuclear histones during spermiogenesis occurs generally during the relatively early stages while RNA is still available for synthesis, one might expect to see a sequential labelling of histone fractions extracted f r o m n u c l e i, which reflects the succession of incorporations of previously synthesized histones into the chromatin during sperm development.

This generalization is depicted schematically in Fig. 13. A consequence of these alterations which may be important in development is the changing relationship between the genes which synthesize histones and the genes which associate with their products.

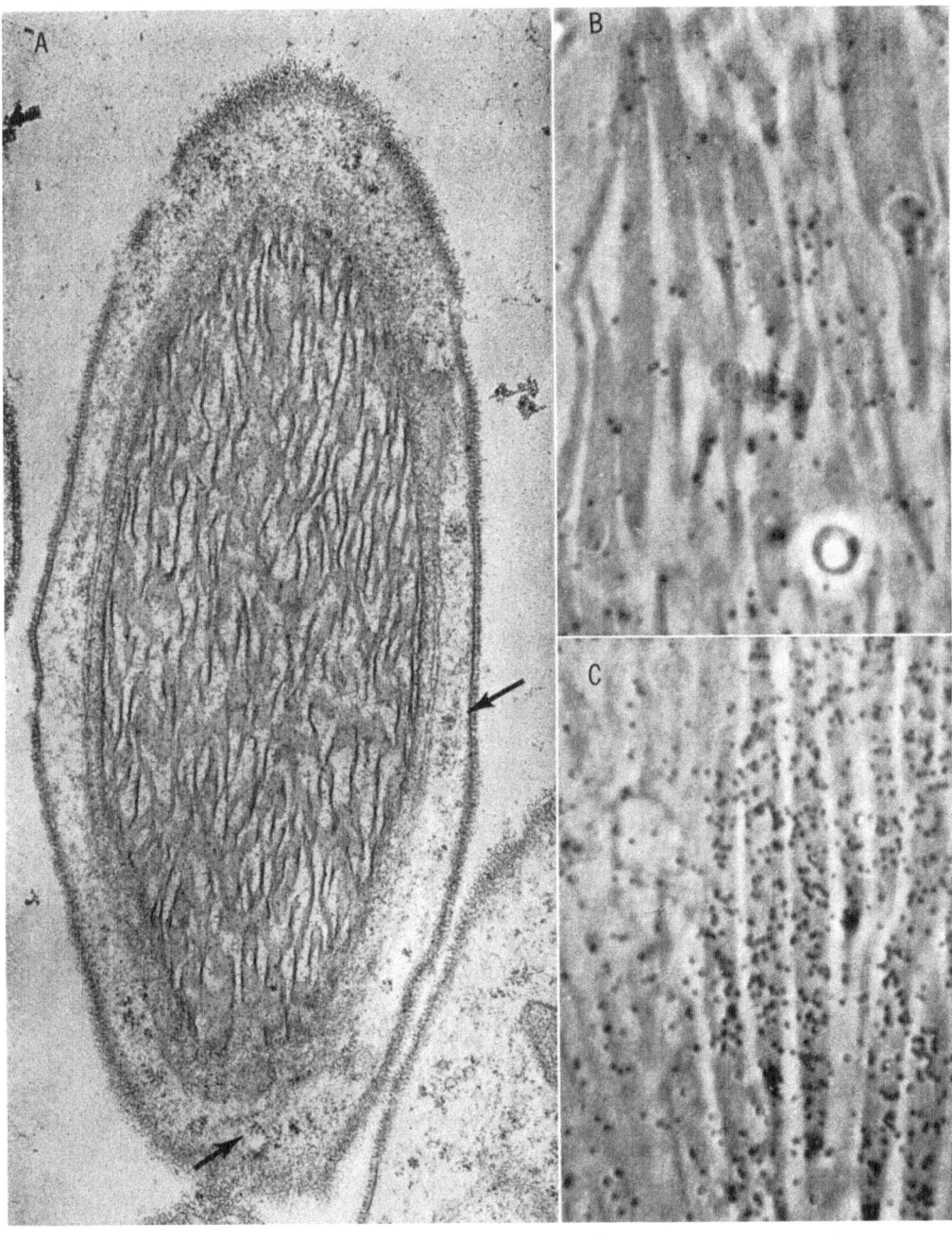

Fig. 12. Incorporation of arginine into spermatids of the grasshopper *Chortophaga viridifasciata*. 12 *A*; Electron micrograph of a spermatid head in oblique section. Magnification, 21,000. Note presence of ribosomes in cytoplasm. Figures 12 *B* and *C*: Autoradiographs of 1 micron plastic embedded sections of testes taken from animals 1 hour and 1 day respectively after injection with tritium labelled arginine. Note initial peripheral labeling followed by labeling over the entire nucleus. Magnification, 2,000. (BLOCH and BRACK, 1964.)

The fate of the old histone is of interest, insofar as it pertains to the question of convertibility of proteins. Vaughn (1965) has found that concomitant with the formation of mature sperm histone in the rat, is the appearance of numerous flecks of acidophilic substance in the nucleus. These move out into the cytoplasm and eventually aggregate into fairly dense acidophilic granules which are finally sloughed or disappear. These behave as one might expect a residual nuclear substance which must be disposed of to behave. They are basic proteins, but interestingly, are fixed by Carnoy, not formalin.

Other evidence for cytoplasmic accumulation of basic proteins which may be histones has been reported for the grass snake (Sud 1961) and has been seen in the Mole crab *Emerita analoga.* The sperm of this latter creature is fairly typical of crustacean sperm in its bizarre morphology (Fig. 14). The mature sperm contains a "capsule," which consists of a thick walled hollow cylinder capped at one end by a structure resembling a vikings headgear, complete with "horns". At the other end of the capsule is a collar from which project radial spines. At the other side of the collar is the nucleus. The nucleus is rather flaccid, phase light, and loses its spherical form during fixation and staining, becoming highly irregular in shape. Nevertheless it stains fairly well with the Feulgen procedure. It does not stain well with fast green for histones, nor with the picric acid eosin method for protamines. The presence of a faintly basic protein is indicated by a faint positive reaction with bromphenol blue as used for staining nuclear basic proteins. The capsule stains strongly with fast green as used for staining typical histones. As might be expected, this staining does not require prior removal of DNA. Deamination abolishes staining of most of the capsule, however the horns stain brightly after deamination, reacting as typical arginine rich spermatid histones. They also stand out a bright pink after Sakaguchi staining. Application of the FDNB Sakaguchi combination shows the cap to contain little if any lysine, the body of the capsule to contain primarily lysine. The nuclear protein contains approximately equal amounts of lysine and arginine. When these cells are stained with a mixture of fast green and eosin, the capsule body stains pink, the cap a blue green, and the nucleus a bright green.

These basic proteins first arise in a vacuole in the early spermatid, and are presumed to be synthesized there, although this is not certain. The histones of the nucleus during this stage stain typically, but fast green stainability is soon lost. It is thought that the main difference between the sperm of the mole crab and that of the vast majority of organisms is that in the former, the newly arising sperm histones accumulate and are stored before entering the nucleus. Whether the capsular proteins are indeed histones, i.e. whether they combine with DNA, perhaps during fertilization or later, remains to be determined.

Chromatin alteration has also been seen to occur during fertilization and during early ontogenetic development in several organisms. Little is known of the synthesis leading to these events however, and they are considered in somewhat more detail in the section on histone variation.

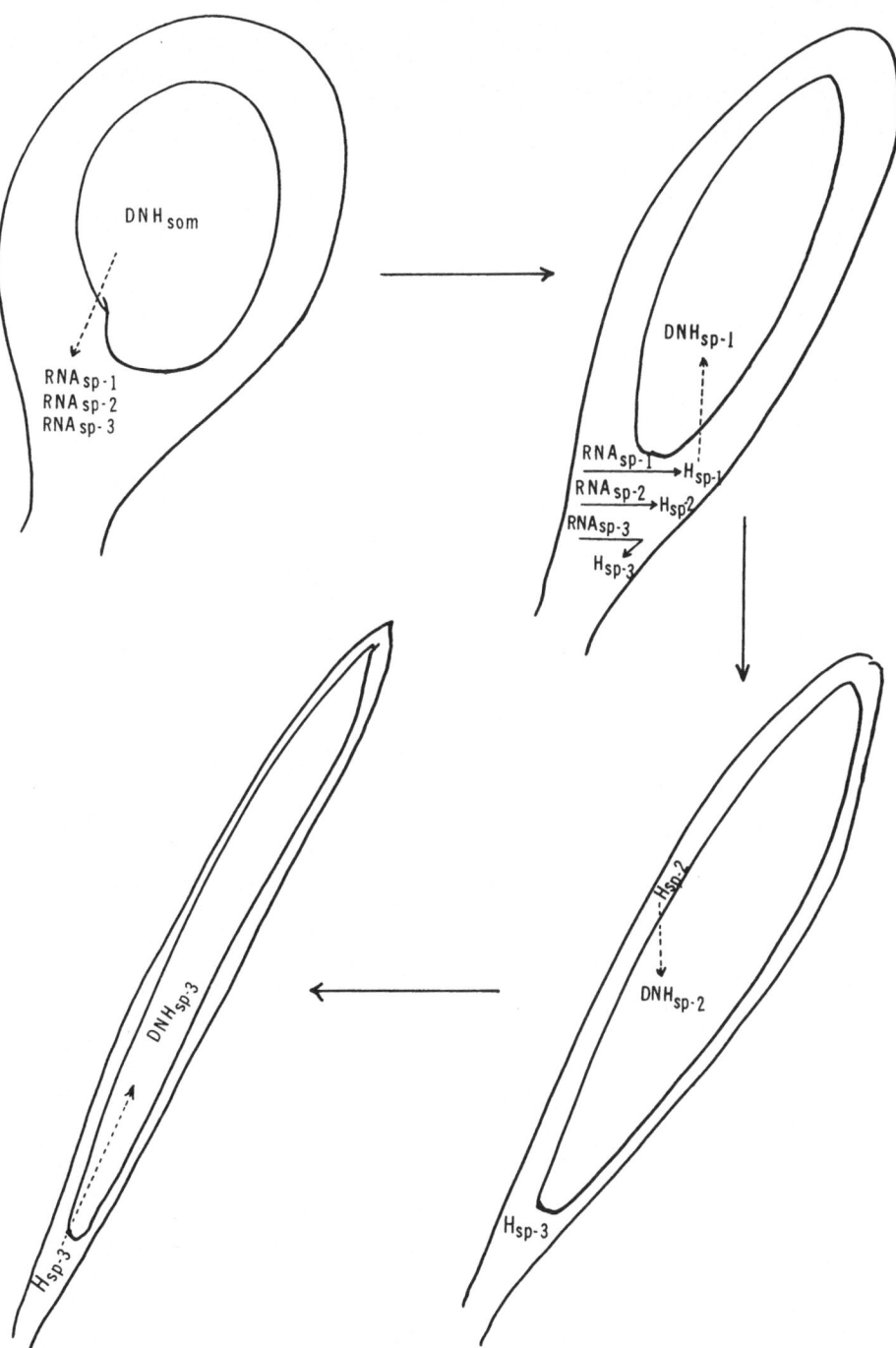

Fig. 13. Summary of histone synthesis and replacement during spermiogenesis. Spermiogenic histones appear to be synthesized during a definite period during spermiogenesis, and are incorporated consecutively into the chromatin during later stages. Alternatively, the proteins of the earlier stages may be converted to those of the later stages.

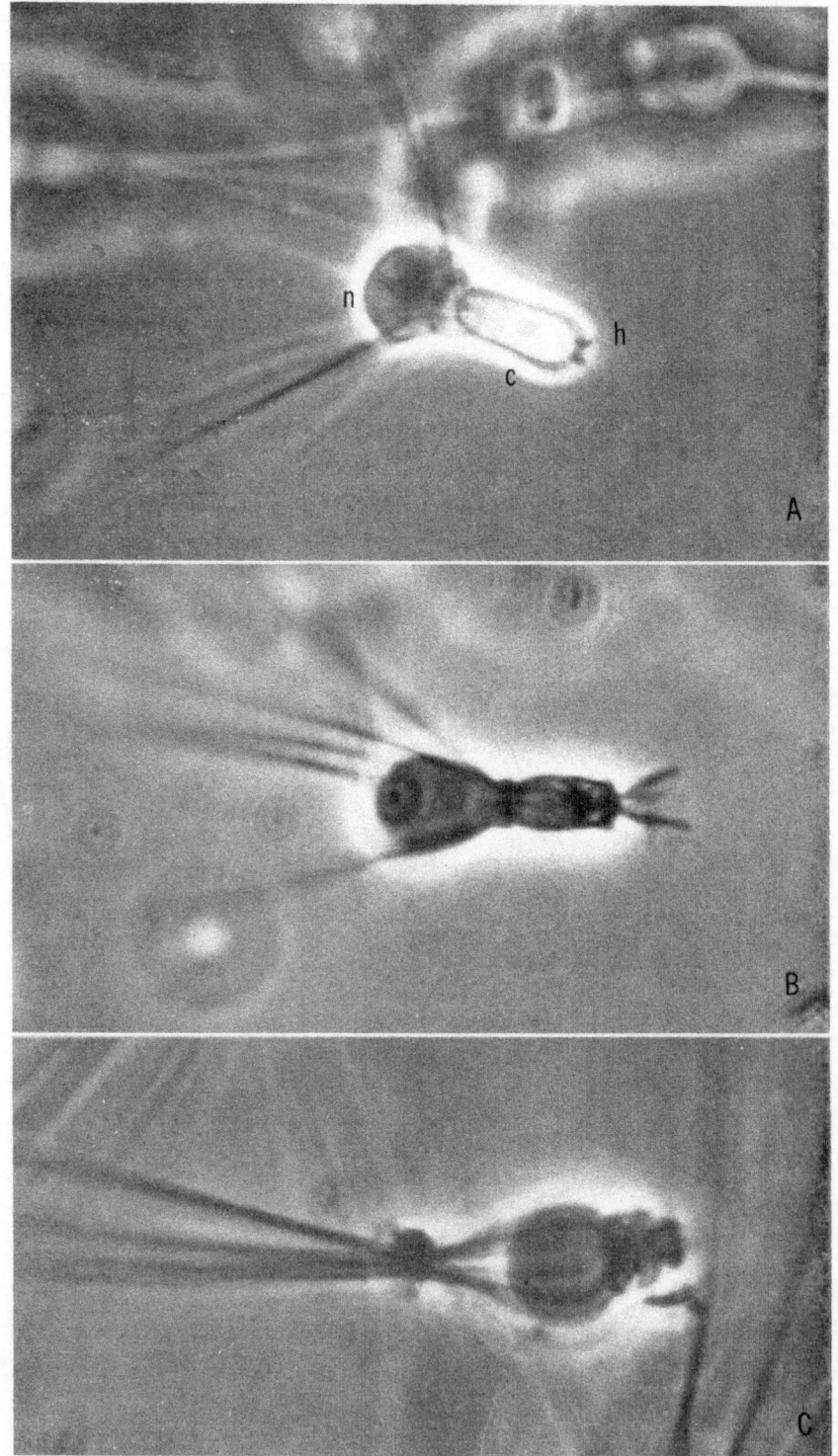

Fig. 14.

Histones during Chromosome Replication

The first attempts to correlate histone synthesis with chromosome replication were carried out in 1955, at a time when the mechanics of protein
synthesis were unknown. The results seemed clear cut and readily lent
themselves to an interpretation which seemed reasonable at the time;
namely, that DNA and histones are synthesized together during what has

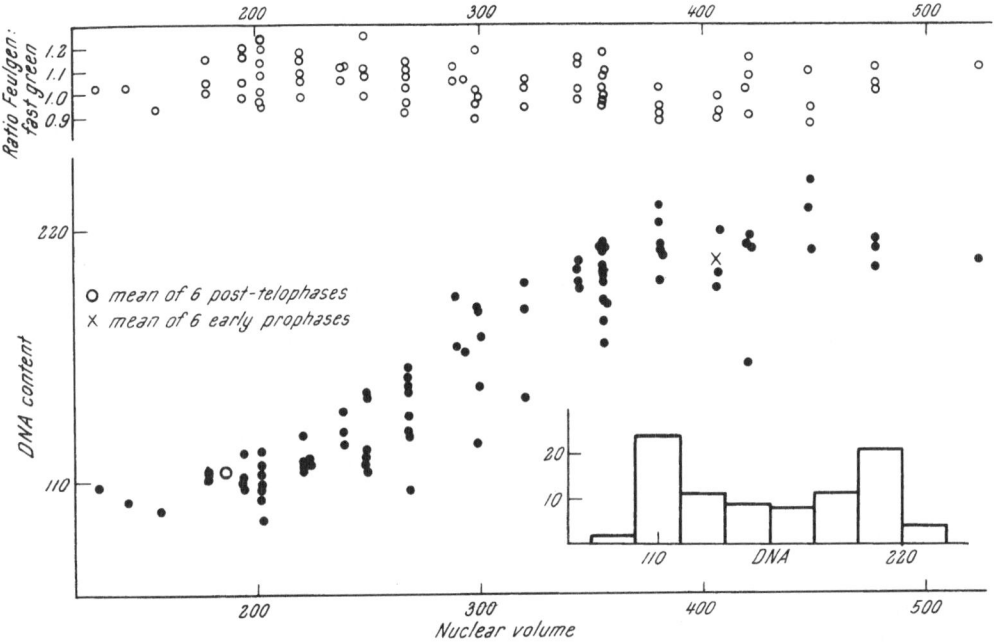

Fig. 15. The relation between volume, DNA, and histone content in intermitotic nuclei of growing onion root
tips. Lower right: A frequency distribution of DNA values typical of a mitotically active tissue. Middle: Individual measurements of DNA content plotted against nuclear volumes. Upper: Individual DNA histone ratios
of the same nuclei, plotted against nuclear volume. (From ALFERT, 1956a.)

since been designated the "S" period of interphase. This conclusion was
the result of two essentially similar experiments. BLOCH and GODMAN (1955)
measured consecutively, using the cytophotometric technique of POLLISTER,
the Feulgen stained DNA and the fast green stained histones of the same
individual cells. Proliferating animal cells showed an excellent correlation
between the amounts of both stains; post-telophase nuclei showing the
diploid amounts, pre-prophase nuclei, twice these amounts, and intermediate
cells, intermediate amounts of both substances. ALFERT (1956 a, see Fig. 15)
RASCH and WOODARD (1959), and WOODARD et al. (1961) using plant material,

Fig. 14. Phase photomicrographs of sperm of the mole crab, *Emerita analoga*. *n*, is the nucleus; *c*, the capsule,
and *h*, the horns. The histone of the nucleus contains approximately equal amounts of arginine and lysine. The
capsule contains basic protein which stains as do typical histones and has a lysine to arginine ratio of about 8.
The horns contain a basic protein which stains as do arginine rich histones of sperms of other species and has
little if any lysine. *A*, shows the sperm in sea water, and *B* and *C*, various stages of inversion of the nucleus
into the space previously occupied by the capsule, when the sperms are put into diluted sea water. Magnification:
2,750 times.

showed a similar dependence of Feulgen staining and fast green staining on nuclear size, the latter parameter being used as an indication of the progress of the cell through the cell cycle. A study by Gall (1959) on the events leading to division of *Euplotes* provided a more precise localization of the replication phenomenon. This protozoan has a large elongated macronucleus. Prior to division darkly staining "reorganization bands" originate at both ends and these procede as waves toward the middle. Gall showed parallel increases in both Feulgen staining and fast green staining at the band. Prescott and Kimball later demonstrated that thymidine and amino acids are indeed actively incorporated in a dilute phase light region immediately behind the band [7] (1961).

These results were interpreted as indicating parallel syntheses at the nuclear level, and in *Euplotes,* at a subnuclear level, and accorded with the view that the DNH complex may somehow be synthesized as a unit, parallel syntheses occurring at and possibly even below the gene level. Some reservations have been voiced. Swift (personal communication), and also Umaña et al. (1962), have suggested that the fast green staining procedure may be sensitive only to histone which is in association with DNA at the time the tissue is fixed, rather than to histone *per se,* and that histone staining may reflect its physical state. In retrospect, this seems reasonable. Histones, as proteins, owe their origin to specific genes rather than to the genes with which they associate (Bloch 1962 b, Bloch and Brack 1964). While a correlated control of the two syntheses might be expected, much as the synthesis of DNA and spindle material are correlated yet are independently initiated (Bucher and Mazia 1960) a closer relationship would be fortuitous.

There are in fact a number of recent biochemical investigations which suggest that histone and DNA syntheses may not be associated. Lindner and Kutkam (1962), have shown that the ratio of histone to DNA of Ehrlich ascites cells, as measured microspectrophotometrically, increases after treatment with 5 fluorouracil which inhibits DNA synthesis. Similarly, 5 fluorodeoxyuridine, a more potent and specific inhibitor of DNA synthesis, appears not to affect histone synthesis appreciably at least in short term experiments (Zeevaart 1964, Flamm and Birnstiel 1964). Evans et al. (1962) found that during liver regeneration, the incorporation of amino acids into histones precedes that of nucleotides into DNA, occurs over a longer range, and accordingly, the specific activity increases to a greater extent in the DNA when its synthesis does occur. Umaña et al. (1962) found higher ratios of histones to DNA in whole cells than in isolated chromosomes, and suggested that the histones are synthesized shortly after mitosis and that these histones are unstainable with standard cytological procedures. According to this view, the parallel increases in Feulgen and fast green staining noted during interphase would reflect association of previously synthesized

---

[7] It may be significant that while thymidine was shown to be incorporated immediately in this region, the incorporation of an amino acid was shown 45 minutes after administration of isotope.

histone with DNA immediately following DNA replication, the complexing rendering the histone stainable with fast green by aiding in its fixation. It would not be unreasonable to look for the actual synthesis of this protein in the cytoplasm.

It is interesting in this connection that a reexamination of the temporal relationships between histone and DNA synthesis reinforce the original

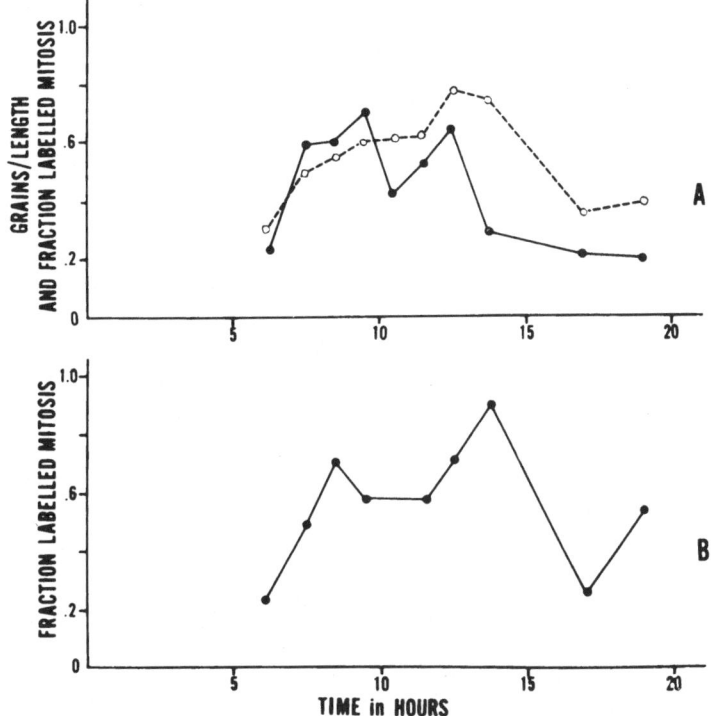

Fig. 16. Comparison of the times needed for appearance of labelled amino acids (*A*), and labelled thymidine (*B*), into mitotic figures of onion root tips. The solid curve in the upper figure is for the density of grains over the chromosomes (grains/chromosome length) and the broken curve, the fraction of "heavily labelled" division figures as subjectively estimated.

view of simultaneity. Comparison of the times needed for labeled thymidine and labeled amino acids to become incorporated into metaphase chromosomes, and the times necessary for the reappearance of isotope free metaphases after pulse labeling mark the durations of the G-2 and S periods for synthesis of DNA and chromosomal proteins respectively, and such studies are not subject to the vagaries of staining. The results show that DNA and histone are indeed synthesized at the same time during interphase (BLOCH and BRACK, unpublished, see Fig. 16).

Whether both substances are synthesized at the same time at subnuclear levels, that is whether the DNA and histone which will be associated are synthesized together is not known, but seems unlikely. Independent syntheses would simplify the replication picture, but of course would beg the question of the specificity of complexing during chromosome replication. This problem is considered more fully in the section "General Considerations".

Histone Turnover

The problem of histone turnover is resolved into separate questions as attention is directed toward different sites within the cell. Turnover doubtlessly occurs at the immediate synthetic site, probably ribosomal templates, and here the problem reduces to one of protein synthesis. Limited exchange also occurs at the level of the chromosome during replacement of one histone type by another, as during spermiogenesis. The main issue is the extent to which histone turnover at the chromosome occurs in non-differentiating cells, and it is here that two current views may be contrasted. The slightly older one, perhaps more prevalent among cytologists, regards histones as a fairly constant and stable constituent of the chromosome, varying not at all, or only by small fixed degress as chromosomal function changes. This view is the result of the apparent similarities in the behavior of histone and DNA and has evolved from the concept of chromosome constancy. The second view, typically biochemical, pictures histone in a more dynamic state. Constant turnover would be predicted if the DNA-histone association is in equilibrium with free histone and DNA as pictured by Busch et al. (1963), or if histone acts in some manner as a conveyor of information between the nucleus and cytoplasm as suggested in Caspersson's early scheme of protein synthesis (1950). If the assumpton of gene activity depends upon the unmasking of DNA by removal of histone (Bonner et al. 1962) or if other modification of activity requires small localized replacements of histones, one might predict a turnover which is more limited, in accordance with a catalytic function.

A variety of approaches to the problem has been attempted. One is the study of incorporation into histones in non-dividing non-differentiating cells, where incorporation can be attributed to neither replication, nor gross change of function, and another, the study of histone conservation between cell divisions, to determine whether all incorporation into histones can be accounted for by replication. The results, while not unequivocal, do suggest that some turnover may occur.

In 1952 Brunish and Luck, and independently, Daly, Allfrey and Mirsky, found that histones obtained from tissues of animals injected with radioactive amino acids are labeled, but that the incorporation is lower than that into other proteins. For example; among the histone, residual protein, ribosomal, and supernatant proteins, of liver, kidney and pancreas, all non-proliferating tissues, the histones showed the least incorporation (Allfrey et al. 1955). These findings were extended in 1957 in their studies on protein synthesis in isolated nuclei (Allfrey, Mirsky and Osawa 1957). Relative incorporation into histone, a pH 7.1 soluble nuclear fraction (probably nuclear ribosomes), and DNA associated non-histone protein (probably their "residual protein") occurred in a ratio of 36 : 179 : 243.

Such studies supported the early view that histones are r e l a t i v e l y inert, and later work suggested that much of the incorporation into histones might indeed be attributed to chromosome replication leading to division. For example, comparisons of incorporation into histones of tumor and normal tissues (BUSCH, DAVIS and ANDERSON 1958, HOLBROOK *et al.* 1960) have generally shown that of the tumor to be higher, and have been interpreted as indicating a higher rate of proliferation in the tumor tissues.

However not all incorporation can be accounted for by replication. It was well recognized that incorporation into histones of non-dividing cells is considerably higher than zero, and that in those studies in which incorporation into histones and DNA are directly comparable, that into the histone is generally higher (DALY, ALLFREY, and MIRSKY 1952). These findings are supported by the demonstration of independent syntheses of histones and DNA in proliferating cells, and the suggestion that synthesis of the former may show less periodicity and occurs over a greater span of the division cycle (EVANS *et al.* 1962).

The more recent work of BUSCH *et al.* and HOLBROOK *et al.* stresses the activity of the histones rather than the lack of it. BUSCH *et al.* (1958) found a histone which incorporates labelled amino acids more rapidly than non-histone proteins, and HOLBROOK *et al.* (1960) reported a citrate soluble "histone" in hepatoma whose kinetics of incorporation are consistent with the view that this protein may serve as a precursor of the more typical histones. Whether this protein or BUSCH's are indeed histones, that is, whether they are, or are ever in association with DNA, remains to be determined. In view of the methods used to obtain them (in HOLBROOK's case, extraction with citrate, precipitation with cold perchloric acid, then extraction with dilute HCl) it seems likely that they may be derived from nuclear ribosomes, also found by MIRSKY and his coworkers to be metabolically active. The similarity between ribosomal basic proteins and histones, as stated before, could easily lead to their confusion unless precautions are taken to obtain the "histones" from their complex with DNA rather than simply from nuclei. While an actively incorporating basic protein fraction obtained from ribosomes may indeed be histones caught in the act of being synthesized, such turnover still does not indicate turnover at the chromatin level.

PRESCOTT and BENDER have sought to study the problem of histone conservation and partitioning among progeny chromosomes during division (1963) in the manner of TAYLOR WOODS and HUGHES in their studies on DNA (1957). One of the major obstacles to this approach to histone metabolism is the necessity of obtaining chromosomes free of both labelled cytoplasm, and non-chromosomal nuclear protein which would otherwise obscure chromosomal label. This requirement prompted the use of a preparative treatment by PRESCOTT (dispersal of cytoplasm by acetic acid or acetic acid and alcohol) which unfortunately also results in removal of most of the histones. PRESCOTT found very light labelling in metaphase chromosomes in spite of his attempts to load the system with radioactivity and to stabilize the histone with light formalin treatment. He concluded that most of the label was lost between the initial incorporation and metaphase, suggesting turnover. Further decrease in labelling of the metaphases of subsequent divisions above that expected by dilution also led to the suggestion that histones are not conserved from one division to the next. However, in view of the loss of histones during the preparation of the chromosomes for autoradiography, and also, of the possible delaying effect on division of even much lower doses of radioactive material than were used by Prescott (see Wimber and Quastler 1963), Prescott's interpretation of his observations must be considered tentative, and the question of histone turnover in the chromatin of non-differentiating cells remains unanswered.

One of the most interesting instances of protein turnover in the nucleus has come to light in the work of Goldstein (1963) and of Prescott and Bender (1963) on what is called "Goldstein's protein" by the latter. There seems to be no reason for calling this protein a histone. The work is cited here because it illustrates the behavior of a labile nuclear protein which has some peculiar hereditary consequences, and which might serve as a model which could be used in reconciling ideas of a materially labile yet qualitatively constant association between DNA and histone.

Prescott has found that on administration of labelled amino acids to an amoeba, both nucleus and cytoplasm become labelled. By repeated excision of a portion of the cytoplasm, the cytoplasm is depleted of its label much more rapidly than is the nucleus, whose label leaks relatively slowly into the cytoplasm. After 20 to 30 excisions, the cytoplasm is almost devoid of label, while the nucleus still exhibits a high degree of radioactivity. The amoeba has been prevented from dividing by the surgical treatment, but resumes division as soon as it is permitted to reach a critical size. Autoradiography of the dividing cells shows that during division, the label goes into the cytoplasm, and is evenly distributed to all regions of the cell. Another amoeba, let divide, then autoradiographed, showed that the dispersed label immediately re-enters the reconstituting nuclei. This protein is apparently synthesized in the cytoplasm (Goldstein 1963), yet may properly be considered a nuclear protein because of its affinity for the nucleus. Its removal to the cytoplasm during divisions indicate that it is not a constant chromosomal protein, at least, not associated with these structures during the stage of division when the chromosomes are usually held accountable for nuclear hereditary phenomena. Its interest here lies in its ability to maintain a specific association with a particular cellular structure which, although interrupted, is capable of being recovered.

Histone Variation

Histone Variation: Within Cells

A number of factors could conceivably contribute to histone variability. Histone composition may reflect the composition, the secondary or higher order structure, or the nature of the activity of its associated DNA. A compositionally dependent variability is suggested by the findings of Lucy and BUTLER of nucleo-histone fractions in which GC DNA's tend to be associated with lysine rich histones and AT DNA's with arginine rich histones (1956). COLE in a highly speculative treatment, has postulated that histone may be involved in modifying secondary structure, as in chromosome coiling during mitosis (1962). An activity dependent variability is implied by the current theories of ALLFREY, LITTAU, and MIRSKY (1963) in which different degrees of inhibition of DNA primed RNA polymerase are attributed to different varieties of histones. Interestingly, a compositionally dependent histone variability, while providing for intracellular histone

Table 4. *Ratios of (lysine + tyrosine)/arginine of the meiotic chromosomes of the male grasshopper (see Fig. 17).* (BLOCH and BRACK 1965.)

Chromosomes	(lysine + tyrosine) / arginine
1	4.7
2	4.6
3	5.2
4	4.0
5	5.2
6	5.1
7	4.9
8	5.6
9	5.0
10	4.8
11	5.3
12	4.7

variation, would contribute little toward variation among cells (granted DNA constancy among cells). Activity dependent variation would presume both inter- and intracellular histone variation.

An important requirement for histone variability, histone heterogeneity, has been firmly established by the application of biochemical methods. Heterogeneity has been demonstrated by chemical fractionation (KOSSEL 1928), chromatography (CRAMPTON *et al.* 1955), electrophoresis (HNILICA *et al.* 1962), sedimentation (RAUEN 1952), and end group analysis (HNILICA *et al.* 1962). Heterogeneity is seen even when the analytical methods are applied to homogeneous populations of nucleated erythrocytes (NEELIN and CONNELL 1959), and sperm cells (RAUEN 1952), indicating a variation among histones within a given cell type. Furthermore, heterogeneity is very extensive, and the fact that a single fraction obtained by one method (say chromatography), is proven to be heterogeneous when subjected to further analysis (end group) shows the difficulties inherent in the preparation of useful quantities of a single histone fraction.

The demonstration of heterogeneity *per se* does not constitute proof of variation among genes, as long as a random or indeterminate association of histones with DNA remains a possibility. As yet, biochemical methods have not been sufficiently refined to permit the assignment of any of the classes of histones (arginine rich, lysine rich) to specific locations within the genome, or with a few exceptions (sperm cells) even to specific cells. Cytochemical methods have only recently been applicable to the characterization of histones, but then only providing information of the "average" histone, and this only when the histones exist apart from other proteins.

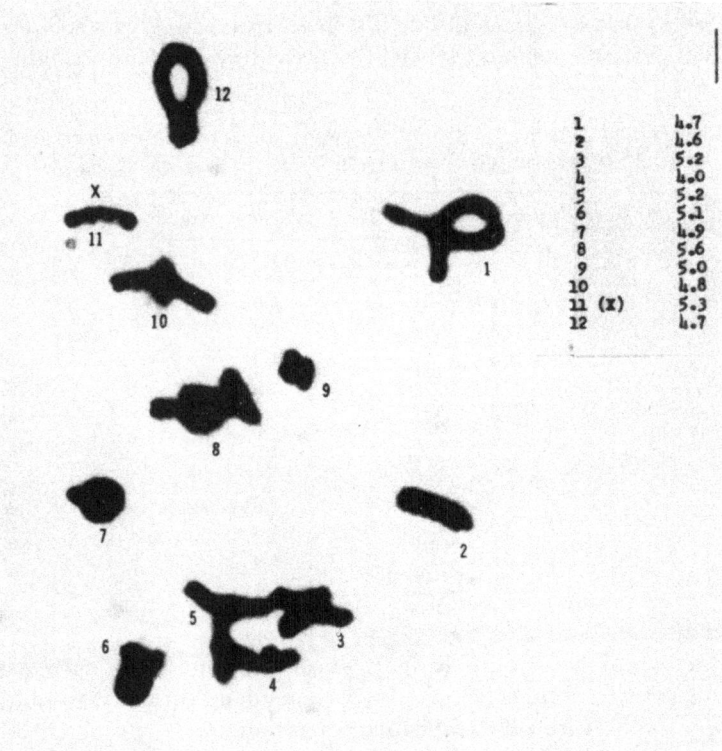

1	4.7
2	4.6
3	5.2
4	4.0
5	5.2
6	5.1
7	4.9
8	5.6
9	5.0
10	4.8
11 (X)	5.3
12	4.7

Fig. 17. Methyl green stained grasshopper spermatocyte chromosomes. The chromosomes were used for the determinations of the lysine + tyrosine/arginine ratios given in Table IV. (Bloch and Brack, 1965.)

Table 4 gives the ratios of (lysine plus tyrosine)/arginine in the protein of the grasshopper spermatocyte chromosomes seen in Fig. 17 (Bloch and Brack 1965). These chromosomes do not differ significantly, all having a ratio of about 2–3. Proximal and distal regions are similarly alike. Finer analysis has not yet been attempted.

Some inferences may be made of the extent of histone variability at the molecular level, and the factors contributing thereto. These are theoretically based, but rest only on the premise that histones are proteins whose synthesis is not radically different from that of other proteins (Bloch 1962 a). The DNA histone complex consists of approximately equal weights of both moieties, and roughly 3 4 amino acid residues per nucleotide. Since at least three nucleotides are needed for the specification of a single amino acid (six if only one strand codes) there is insufficient infor-

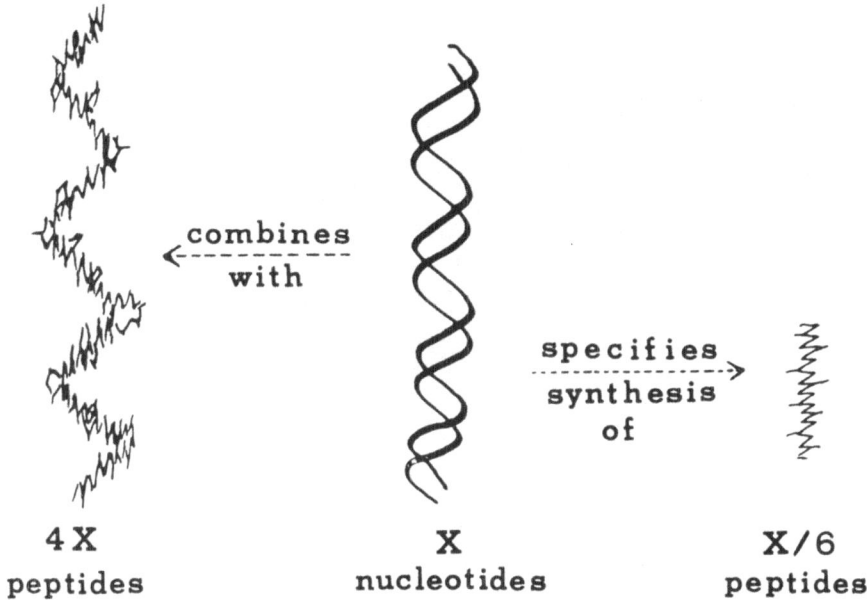

combines with

specifies synthesis of

4 X peptides

X nucleotides

X/6 peptides

Fig. 18. Schematic comparision of the sizes of proteins with which a stretch of DNA can combine, and whose synthesis the DNA can code. (BLOCH, 1962a.)

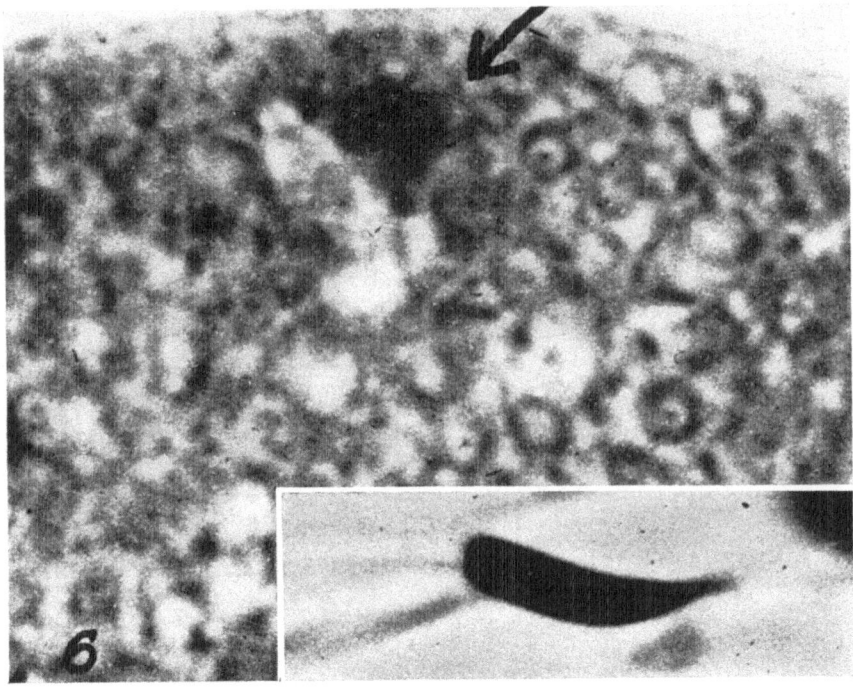

Fig. 19. Alteration in staining of the sperm nucleus during fertilization in *Helix aspersa*. Insert shows a sperm head, stained with bromphenol blue. The arrow shows the male pronucleus after fertilization. The sperm head and pronucleus are the only objects stained. The remainder of the photographic image is due to refractility. (BLOCH and HEW, 1960b.)

mation in a stretch of DNA to code the synthesis of its associated histone, and at the level of the nucleus, insufficient information in the entire DNA complement to code a unique histone for each DNA (Fig. 18). As a con-

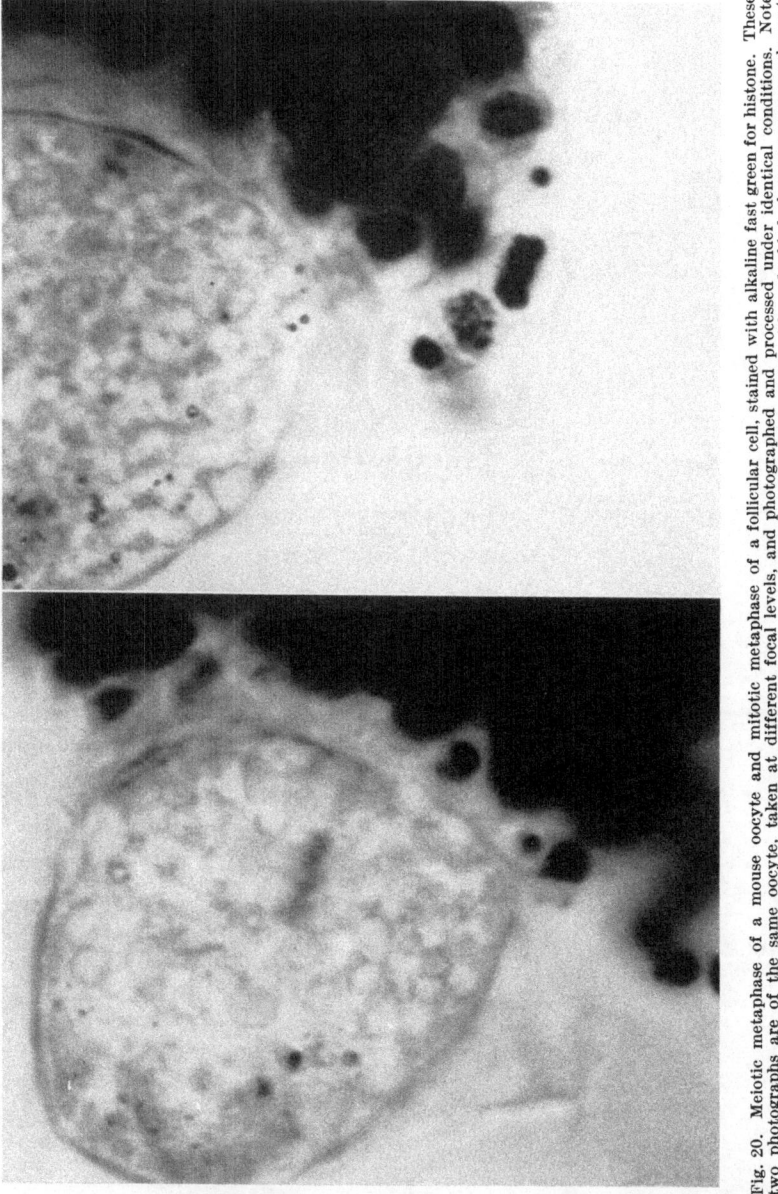

Fig. 20. Meiotic metaphase of a mouse oocyte and mitotic metaphase of a follicular cell, stained with alkaline fast green for histone. These two photographs are of the same oocyte, taken at different focal levels, and processed under identical conditions. Note the difference in staining of the two metaphase figures. Both are presumed to be diploid nuclei with a doubled chromosome complement.

sequence, the histone associated with a given stretch of DNA must have its origin elsewhere, in a limited fraction of DNA whose function is to provide for the synthesis of histone for the entire complement. Since most of the DNA is covered (Davision and Butler 1956, Felix *et al.* 1951, Vendrely *et al.*

1959, Feughelman *et al.* 1955, Zubay and Doty 1959), some different DNA's must associate with common histones. Histone variability, therefore, while perhaps extensive, is limited and relatively low compared with the heterogeneity of its associated DNA.

If the genome is envisaged as consisting of units, the unit defined as the stretch of DNA associated with a single histone molecule, the genome of a given cell must be composed of a number of different units which may be categorized on the basis of common or uncommon histone associations (Fig. 21). A small fraction of the total DNA is involved in the synthesis of histones for the entire complement and may be presumed to enjoy some special status in the hierarchy of gene activity.

Variation: Among Cells

Histone variation among cells is attested to by the observations of changes during spermiogenesis, fertilization, and early embryonic development. The case of spermiogenesis has already been extensively treated. Little is known of subsequent changes, other than the fact of their occurrance. During fertilization in the snail, *Helix aspersa,* the protamine is lost from the sperm and is replaced in the male pronucleus by a protein of rather weak basicity, as suggested by the requirement of a particularly sensitive bromphenol blue test for its detection (Fig. 19) (Bloch and Hew 1960 b). This change was originally thought to indicate the presence of a unique "cleavage" histone, but may reflect combination of the total DNA with a protein which is unique only in the extent of its coverage of the genome. During later stages of development, approximating gastrulation, the weakly basic complement is replaced by a spectrum similar to that of the adult organism. Comparable changes have been indicated in *Drosophila* (Das, Gay, and Kaufmann 1964 a, b) the mouse (Alfert 1958), and perhaps the frog (Horn 1962), although the latter case is not clear cut (Bloch 1962 b). Fig. 20 shows differences in staining of meiotic and mitotic chromosomes in the mouse ovary. These alterations vary widely among organisms, in the degree of the change, and its timing. In every case studied the typical complement of histones is recovered early in development, at or before gastrulation. Whether similar changes extend further into development is problematical. It would appear likely, since it is only because of the extremity of the changes that cytochemical techniques have been able to discern them at all. The detection of large changes with crude methods suggests the occurrence of smaller ones which are undetectable [8].

---

[8] However it is true that most cytological studies made on cells other than germinal or of early developmental stages suggest a quantitative and possibly a qualitative constancy of histone. A constant ratio of histone to DNA staining has been found among a variety of plant tissues including germinal cells (Rasch and Woodard 1959), among cells in different physiological stages, e.g. different size classes of nuclei brought about through hormonal control (Alfert, Bern, and Kahn 1955), and in different pathological states, e.g. *Vicia* crown gall tumor (Rasch and Woodard 1959). Although Alfert (1955) found an increase in the ratio of fast green staining of histone' to Feulgen staining of DNA during pycnosis in the

It is interesting that even biochemists have been quite unable to agree on whether or not histone differences exist among the different cell types of the adult organism, even though the current interest in the inhibition hypothesis gained its initial impetus through the STEDMAN's report of small differences.

The few changes which are readily demonstrable indicate that a stretch of DNA does combine with alternative histones in different cells (Fig. 21). This, and the inference that different DNA's associate with identical histones within a cell, indicate that a given combination of DNA and histone is not the result of specific affinities between the two substances. Presumably, any DNA in the cell can combine with any histone. The factors determining a specific set of associations must then lie outside of the components themselves. Whatever these factors are, they permit changes in the relationship of the histone producing genes and the histone associating genes during at least a limited span of the life cycle of an organism.

HISTONE——▶

DNA	CELL	A	B	C	D....
1		α	δ	ε	γ
2		α	δ	ε	γ
3		β	δ	ε	γ
4		β	δ	ε	γ
5		β	δ	ε	γ
6		α	δ	ε	γ
7		α	δ	ε	γ
8		α	δ	ε	β
9		γ	δ	ε	β
10		γ	δ	ε	β
⋮					

Fig. 21. Schematic representation of the variability of the DNA histone association among different DNA units within a given cell (vertical columns), and among homologous units of different cells (horizontal rows). Cell A represents any particular somatic cell. Cells B and C represent spermatids or sperms, and early cleavage cells. Cell D is an extrapolation representing a hypothetical cell, showing by comparison with A, that a given DNA is potentially capable of associating with any of the several histones found within a cell.

General Considerations

There have been numerous examples in biology where knowledge of phenomena without visible cause, represented by a set of formalisms, and study of material substances with no obvious biological functions to justify their existance, suddenly dovetail, giving the phenomenon credibility and the substance importance. The anchoring of abstract hereditary phenomena to the very material chromosome serves as a classic example. The anchor pin was SUTTON's (1903) recognition of the identity of the patterns of inheritance of traits and of chromosomal arms.

In seeking a function for the histones, one encounters a parallel which is striking in its precision. The biological phenomenon is a class of hereditary variations in gene expression which are attributable to "chromosome

guinea pig ovarian follicular cells, this increase was thought to stem from removal of substances during the process of pycnosis which are capable of interfering with the electrostatic binding of dyes to chromatin. This unmasking results in a relative increase in methyl green staining of DNA as well as fast green staining of histone. No evidence was found for change in amino acid composition of histones during comparable pycnosis in the kidney (VENDRELY, ALFERT, MATSUDAIRA, and KNOBLOCH 1958). As mentioned above, other variations in ratios of histone to DNA staining have been seen which were similarly attributed to masking effects, rather than to actual changes in the amounts or quality of the histone.

differentiation" (changes in the properties of the chromosome). The material substance is histone. A connection is made plausible through the formulation by BRINK, in 1960, of a set of rules governing the behavior of a hypothetical chromosomally based determinant, designated "parachromatin". The behavior of parachromatin, as predicted to account for its biological effect, a variable chromosome, closely parallels the behavior of histones as observed and independently inferred from chemical data, thereby circumstantially attaching to histones parachromatin's function.

Chromosome differentiation — Definition

"Chromosome differentiation" may be defined as the process resulting in differences, among isologous or homologous chromosomes, *which are capable of being maintained when these chromosomes are brought into similar intracellular environments.* If a chromosomal locus is capable of existing in alternative metastable states.

$$\text{e.g. if locus }_a \xleftrightarrow{} \text{locus }_{a1}$$

its chromosome may be said to differentiate. Differentiated states will be most readily detected by changes in phenotypic expression, although such changes do not by themselves suffice in establishing chromosomal differentiation, since expression may also reflect environment. A simple criterion which may ultimately serve as a test, but presently serves to illustrate the point, would be the preservation of a given state of expression on transplantation of a chromosome region from one cell to another. If the region responsible for the production of visual purple remained inactive on transplantation from a muscle cell to a retinal cell, it would exhibit differentiation. If on the other hand, it adapted to the environment of the new host cell and behaved as its homolog in the retinal cell it would not exhibit differentiation. In the latter case any differences in expression between the two cell types could be attributed to the flexibility of the undifferentiated chromosome in the different cytoplasms. Accordingly, the beta-galactosidase region of *E. coli* does not differentiate in this sense for transfer of this segment by mating from a cell in one state of induction to a cell in the other state has shown the chromosome to act in accordance with its new environment (JACOB and MONOD 1961). Perhaps the term "modulation" may be applied here to the chromosome, as it is by embryologists to the cell, to denote easily reversed changes in function which are under environmental control.

Another important aspect of chromosome differentiation which distinguishes it from more transitory chromosome variations is that while induction of a change from one state to another may be the result of an extrinsic stimulus, the maintenance of this state rests upon an intrinsic change in the locus itself and does not depend upon continued application of the stimulus. Any stability displayed after β-galactosidase induction depends upon continued presence of inducer or of persistance of the enzyme induced.

15*

An important implication of the concept of a differentiating chromosome is the heritability of the alternative states during cell proliferation. If a locus can exist in alternative metastable states, these states should be propagable and their physiological consequences inherited during cell division. In other words,

$$\text{locus }_a \rightleftarrows \text{locus }_a1$$
$$\downarrow \qquad\qquad \downarrow$$
$$2 \text{ locus }_a \qquad 2 \text{ locus }_a1$$

The concept of a differentiating chromosome presupposes an actual material alteration of the chromosome. One might envisage several bases for such differentiation. Gross changes in the genome resulting from chromosome elimination (Metz 1938), diminution (Boveri 1887), and differential DNA synthesis (Stich and Naylor 1957, Bayreuther 1956, Ficq and Pavan 1959), do play a limited role in development in many organisms, although these are not of sufficiently general occurrence to account for most differentiation even in those organisms in which they occur. Alterations in the primary structure of DNA by nucleotide substitution, if such could be brought under biological control, might provide a means. Changes in secondary or higher order structure, as in heterochromatization, or changes in the association of DNA with accessory substances, such as histones, might produce the required result if these structures or associations could be replicated. At this point however, examples of chromosome differentiation would be more instructive than conjecture on how they might be effected.

Evidence for Chromosome Differentiation

"Paramutation" demonstrates the ability of a locus to exist in alternative states. This phenomenon was first clearly recognized by Brink (1958), and the term used to describe a change brought about in a gene in *Zea mays* under the influence of an allele introduced from a strain of diverse origin. The normal dominant R allele results in red pigmentation of the aleurone. The R' obtained from a heterozygous RRst individual (the latter allele results in a stippled color pattern) is seen to have lost almost all of its ability to produce color. The change in R is apparently induced by the "paramutagenic" Rst allele. Like a mutation, the change can be transmitted through successive generations even after removal of the inducing factor by segregation. Unlike mutation however, the change is under biological control, is predictable, occurring 100% of the time, and can be partially reversed by bringing the affected R' allele into the homozygous condition. Possible arguments attributing the change to cytoplasmic factors were countered by appropriate crosses by Brink, and the phenomenon stands as an example of the gene's ability to exist in alternative states. Interestingly, the R and R' states can coexist in the same cell, and replicate as such through generations of cells and of organisms, thereby obviating a simple play of the intracellular milieu upon the gene as an explanation of its variable expression.

Another example is seen in "phase variation" in *Salmonella* (Lederberg and Iino 1956). Phase variation describes the ability of different cells of a strain of *Salmonella* to exhibit either one of two flagellar antigens. For example, cells of *S. typhimurium* exhibit either the i or the 1.2 antigen, those of *S. abony,* either b or enx, the pair of potentialities defining the strain. The ability to form one of the pair of antigens is clonally inherited, occasional cells, (approximately one in several thousand) reverting to the other form, which is then similarly inherited. Two distinct loci, the H-1 and H-2 are responsible for the production of their respective antigens (e.g., i and 1.2). The H-1 locus is hypostatic to the H-2, being expressed when the H-2 is inactive, and suppressed when the H-2 is active. The control of the H-2 unknown. Perhaps, as suggested by Lederberg, fluctuations are governed by stochastic behavior (are indeterminate). At any rate transduction of the H-2 locus apparently does not effect the state of its activity, for its expression in recipient cells resembles that in the donor cells. This stability contrasts with the change of activity of the beta-galactosidase region of *E. coli* after transfer by mating or by "sexduction". It is difficult to envisage how a gene whose activity is governed solely by its environment can be transduced in a given state, unless the critical "environment" remains sufficiently closely applied to the locus to be included within the bacteriophage particle and is replicated along with the locus after transduction. In any case, it is conceptually meaningful to consider the H-2 locus, like the paramutable R locus in maize, as a genetic unit which exhibits alternative metastable and hereditary states.

The heteropycnotic condition of the X-chromosome of the females of some mammals provides another example of chromosome differentiation. Here, homologous blocks of chromatin exhibit differences in morphology (Ohno *et al.* 1959), metabolism (Taylor 1960), and the expression of the genes contained therein (Lyon 1961, Beutler *et al.* 1962). The heteropycnotic state arises early during embryonic development (Austin 1962), occurring independently in the individual cells of the blastocyst and randomly involving one of the two homologs. The alternative states are mitotically transmissible, and as in the case of paramutation, the two states coexist within the same cell thereby precluding as a basis a simple repressor mechanism such as operates in enzyme induction in bacteria. The maintenance of the two states indicates the existence of intrinsic differences in the chromosomes themselves.

Parachromatin and Histone

In 1960 Brink advanced a major new hypothesis to account for a number of "quasi-genetic" phenomena which implicate the genome in the direct control of its activity (Brink 1960). These phenomena include the well known "position effect" describing the influence of the position of a gene in the genome on its activity (Schultz 1936, Lewis 1950), "activation" or "modulation," in which the expression of a gene of usual rank is modified by controlling genes (McClintock 1955, Brink and Nilan 1952), and a

"historical effect" in which the activity of a gene at various levels is conditioned by its past as well as its present circumstances, the most dramatic example being paramutation. The genome was depicted by Brink as a dual entity consisting of "orthochromatin" or the classical gene bridging generations of organisms, and "parachromatin" an accessory substance through which the gene in somatic cells interacts with its environment.

This system is described in the following terms. The ortho- and parachromatin form an association which is coextensive throughout the genome. The orthochromatin, probably DNA, determines genetic potential and is constant. In contrast, the parachromatin, whose nature is unknown, plays an epigenic role and varies among cells, imparting a variable quality to the genome. Parachromatin is relatively homogeneous as compared with orthochromatin, and is accordingly ambivalent, i.e. a unit of parachromatin is capable of variable association. This ambivalence is a consequence of the origin of parachromatin in a limited fraction of the orthochromatin, and is shared by the orthochromatin, as indicated by its own variable association during development.

The orthochromatin-parachromatin association is capable of replication, in the sense that the pattern is reproduced during chromosome replication. Changes in the association occur in response to specific stimuli during development. The association reaches a ground state during gamete formation, permitting repetition of the pattern of changes during subsequent generations.

The equation of orthochromatin-parachromatin with DNA-histone is suggested by the parallel behavior of the two systems. The DNA-histone association, consisting of approximately equal weights of the two components, and exhibiting a near 1 : 1 ratio of protein basic groups and nucleic acid phosphate groups, appears to be coextensive throughout the genome or nearly so. The DNA, which determines genetic potential, is at least qualitatively constant among cells while the histone, like the hypothetical parachromatin is known to vary at least during spermiogenesis, fertilization, and early cleavage. The histones, comprising but few among multitudes of proteins within the cell, and synthesized in a conventional manner, must owe their origin to a limited fraction of the DNA. Even if all DNA synthesized histone, the coding ratio would require that the histone be relatively homogeneous. As a consequence the histone exhibits an ambivalence in its association with DNA. A similar ambivalence on the part of DNA is indicated by the changes in association with histone undergone during spermiogenesis, *inter alia.*

Since DNA replicates itself in a milieu containing a variety of histones, one may ask what determines the nature of the association after replication. Since a given DNA combines with a number of histones during development, and a given histone combines with a variety of DNAs within a cell, maintenance of a given association cannot be the result of an inherent affinity of a DNA for an opposite member, but must reflect an inherent tendency to the *status quo* or a continued extrinsic stimulus conditioning the association (unless of course, the association is indeterminate and

random). If the association which exists prior to replication determines the association after replication, the association is capable of replication and is heritable. (The fact that histone is synthesized elsewhere no more precludes replication of the association than does the fact that nucleotides are synthesized elsewhere prevent one from speaking of replication of DNA.) The predicted "ground state" attained during gamete formation may be effected by the atypical basic proteins found in many sperms and during early embryonic development. It is interesting that nowhere but in these presumably crucial stages in development are seen such extremes in histones as to make the changes apparent.

These close parallels between parachromatin and histone provide grounds for speculation, and the predictions made by BRINK may serve as a guide to determine the extent to which predictions of the behavior of histones, made on this basis, are fulfilled in future experiments.

Bibliography

D'ALCONTRES, G. S., 1953: Acerca de la presencia de protaminas en el polen. Acta cient. Venezolana **4**, 23—24.

ALFERT, M., 1955: Changes in the staining capacity of nuclear components during cell degeneration. Biol. Bull. **109**, 1—12.

— 1956 a: Quantitative cytochemical studies on patterns of nuclear growth. In: Fine Structure of Cells, pp. 137—163, Interscience, New York.

— 1956 b: Chemical differentiation of nuclear proteins during spermatogenesis in the salmon. J. Biophys. Biochem. Cytol. **2**, 109—114.

— 1958 a: Cytochemische Untersuchung an basischen Kernproteinen während der Gametenbildung, Befruchtung und Entwicklung. Ges. physiol. Chem., Colloq. 9, 73—84.

— 1958 b: Variations in cytochemical properties of cell nuclei. Exper. Cell Res. Suppl., **6**, 227—235.

— H. A. BERN, and R. H. KAHN, 1955: Hormonal influence on nuclear synthesis II. Karyometric and microspectrophotometric studies of rat thyroid nuclei in different functional states. Acta anatomica **23**, 185—205.

— and I. I. GESCHWIND, 1953: A selective staining method for the basic proteins of cell nuclei. Proc. Nat. Acad. Sci., U. S. **39**, 991—999.

— and N. O. GOLDSTEIN, 1955: Cytochemical properties of nucleoproteins in *Tetrahymena pyriformis;* a difference in protein composition between macro- and micronuclei. J. Exper. Zool. **130**, 403—422.

ALLFREY. V. G., M. M. DALY, and A. E. MIRSKY, 1955: Some observations on protein metabolism in chromosomes of non-dividing cells. J. Gen. Physiol. **38**, 415—424.

— V. C. LITTAU, and A. E. MIRSKY, 1963: On the role of histones in regulating RNA synthesis in the cell nucleus. Proc. Nat. Acad. Sci., U. S. **49**, 414—421.

— A. E. MIRSKY, and S. OSAWA, 1957: Protein synthesis in isolated cell nuclei. J. Gen. Physiol. **40**, 451—490.

ALTMANN, R., 1889: Über Nucleinsäure. Arch. Anat. u. Physiol., Physiol. Abt., 409—411.

ANDO, T., I. SHINICHI, G. HASHIMOTO, M. YAMASAKI, and K. IWAI, 1952: Constituent amino acids, N-terminal residues, and the molecular weights of protamines. Bull. Chem. Soc. (Japan) **25**, 132.

ANSELY, H. R., 1954: A cytological and cytophotometric study of alternative pathways of meiosis in the house centipede (*Scutigera forceps,* Rafinesque) Chromosoma **6**, 656—695.

— 1957: A cytophotometric study of chromosome pairing. Chromosoma **8**, 380—395.

ARCHIBALD, W. J., 1947: A demonstration of some new methods of determining molecular weights from the data of the ultracentrifuge. J. Phys. Colloid Chem. **51**, 1204—1214.

AUSTIN, C. R., 1962: Sex chromatin in embryonic and fetal tissue. Acta Cytologica **6**, 61—68.

Bayreuther, K., 1956: Die Oogenese der Tipuliden. Chromosoma 7, 508—557.
Beermann, W., 1960: Der Nukleolus als lebenswichtiger Bestandteil des Zellkernes. Chromosoma 11, 263—296.
— 1961: Genaktivität und Genaktivierung in Riesenchromosomen. Verh. Deutsch. Zool. Ges. Akademische Verlagsgesellschaft, Leipzig, pp. 44—75.
Beutler, E., M. Yeh, and V. F. Fairbanks, 1962: The normal human female as a mosaic of x-chromosome activity: Studies using the gene for G-6-PD deficiency as a marker. Proc. Nat. Acad. Sci., U. S. 48, 9—16.
Birnstiel, M. L., M. I. H. Chipchase, 1963: The chemical and physical fractionation of nucleoli. Fed. Proc. 22, 473.
— — and B. B. Hyde: The nucleolus, a source of ribosomes. Biophys. Biochem. Acta. (In press.)
Bloch, D. P., 1958: Changes in the desoxyribonucleoprotein complex during the cell cycle. In: Frontiers in Cytology. Ed. by S. L. Palay, Yale Univ. Press, pp. 113—166.
— 1962: On the derivation of histone specificity. Proc. Nat. Acad. Sci., U. S. 48, 324—326.
— 1962 b: Histone synthesis in non-replication chromosomes. J. Histochem. Cytochem. 10, 137—144.
— 1963: The histones: syntheses, functions, transitions. In: The Cell in Mitosis. Ed. by L. Levine, Academic Press, New York, pp. 205—224.
— and S. D. Brack, 1964: Evidence for cytoplasmic synthesis of nuclear histone during spermatogenesis in the grasshopper Chortophaga viridifasciata (de Geer). J. Cell. Biol. 22, 327—340.
— — 1965 a: Comparison of the timing of DNA and histone synthesis during the cell cycle. (In preparation.)
— — 1965 b: A cytochemical method for the estimation of the ratio of protein bound (lysine + tyrosine) to ardinine, as applied in the characterization of histones. (In preparation.)
— and G. C. Godman, 1955 a: A microspectrophotometric study of the synthesis of desoxyribonucleic acid and nuclear histone. J. Biophys. Biochem. Cytol. 1, 17—28.
— — 1955 b: Evidence of differences in the desoxyribonucleoprotein complex of rapidly dividing and non-proliferating cells. J. Biophys. Biochem. Cytol. 1, 531—550.
— C. Morgan, G. C. Godman, C. Howe, and H. M. Rose, 1958: A correlated histochemical and electron microscopic study of the intranuclear crystalline aggregates of adenovirus (RI-APC virus) in HeLa cells. J. Biophys. Biochem. Cytol. 3, 1—8.
— and H. Y. C. Hew, 1960 b: Changes in nuclear histones during fertilization, and early embryonic development in the pulmonate snail, Helix aspersa. J. Biophys. Biochem. Cytol. 8, 69—81.
— — 1960 a: Schedule of spermatogenesis in the pulmonate snail Helix aspersa, with special reference to histone transition. J. Biophys. Biochem. Cytol. 7, 515—532.
Bonner, J., and P. O. P. Ts'o, 1964: Then Nucleohistones. Holden-Day, San Francisco.
Boveri, T., 1887: Über Differenzierung der Zellkerne während der Furchung des Eies von Ascaris megalocephala. Anat. Anz. 2, 688—693.
Brink, R. A., 1958: Paramutation at the R locus in maize. Cold Spring Harbor Symp. Quant. Biol., 23, 379—391.
— 1960: Paramutation and chromosome organization. Quart. Rev. Biol. 35, 120—135.
— and Nilan, 1952: The relationship between light variegated and medium variegated pericarp in Maize. Genetics 37, 519—544.
Brunish, R., and J. M. Luck, 1952: Amino acid incorporation in vivo into liver proteins. J. Biol. Chem. 198, 621—628.
Bucher, N. L. R., and D. Mazia, 1960: Deoxyribonucleic acid synthesis in relation to duplication of centers in dividing eggs of the sea urchin, Strongylocentrotus purpuratus. J. Biophys. Biochem. Cytol. 7, 651—655.
Busch, H., J. R. Davis, and D. C. Anderson, 1958: Labelling of histones and other nuclear proteins with L-lysine C¹⁴ in tissues of tumor bearing rats. Cancer Res. 18, 916—926.
— W. J. Steele, L. S. Hnilica, C. W. Taylor, and H. Mavioglu, 1963: Biochemistry of histones and the cell cycle. J. Comp. Cellular Physiol. 62, Suppl. 95—110.
Butler, J. A. V., P. Simson, and P. Cohn, 1960: The presence of basic proteins in microsomes. Biochem. Biophys. Acta 38, 386—388.

CASPERSSON, T. O., 1950: Cell Growth and Cell Function. New York, W. W. Norton and Co.

CHARGAFF, E., and H. F. SAIDEL, 1949: On the nucleoproteins of avian tubercule bacilli. J. Biol. Chem. **177**, 417—428.

CLEMENTI, A., 1921: Une novelle hypothese de travail sur la signification physiologique des protamines et des histones par rapport au metabolisme nucleaire. Arch. Intern. Physiol. **16**, 100—118.

COLE, A., 1962: A molecular model for biological contractility: Implications in chromosome structure and function. Nature **196**, 211—214.

CRAMPTON, C. F., S. MOORE, and W. H. STEIN, 1955: Chromatographic fractionation of calf thymus histone. J. Biol. Chem. **215**, 787—801.

— and M. L. PETERMANN, 1960: The amino acid composition of proteins isolated from the ribonucleoprotein particles of rat liver. J. Biol. Chem. **234**, 1642—1644.

— W. H. STEIN, and S. MOORE, 1957: Comparative studies on chromatographically purified histones J. Biol. Chem. **225**, 363—386.

CRUFT, H. J., J. HINDLEY, C. M. MAURITZEN, and E. STEDMAN, 1957: Amino acid compositions of the six histones of calf-thymocytes. Nature **180**, 1107—1109.

— C. M. MAURITZEN, and E. STEDMAN, 1958: The isolation and properties of 1.6 Sγ-histone from calf thymocytes. Proc. Roy. Soc. (London) B, **149**, 36—41.

DALY, M. M., V. ALLFREY, and A. E. MIRSKY, 1952: Uptake of Glycine-N^{15} by components of cell nuclei. J. Gen. Physiol. **36**, 173—179.

— and A. E. MIRSKY, 1955: Histones with high lysine content. J. Gen. Physiol **38**, 405—414.

DANIELLI, J. F., 1953: Cytochemistry, John Wiley and Sons, Inc., New York.

DAS., C. C., H. GAY, and B. P. KAUFMANN, 1964 a: Histone-protein transition in *Drosophila melanogaster*. I. Changes during spermatogenesis. Exper. Cell Res. **35**, 507—514.

— B. P. KAUFMANN, and H. GAY, 1964 b: Histone protein transition in *Drosophila melanogaster*. II. Changes during early embryonic development. J. Cell Biol. **23**, 423—430.

DASS, S., and H. RIS, 1959: Submicroscopic organization of the nuclei during spermiogenesis in the grasshopper. J. Biophys. Biochem. Cytol. **4**, 129—132.

DAVISON, P. F., and J. A. V. BUTLER, 1956: Chemical composition of calf thymus nucleoprotein. Biochim. Biophys. Acta **21**, 568—573.

DEITCH, A. D., 1955: Microspectrophotometric study of the binding of the anionic dye, napthol yellow S, by tissue sections and by purified proteins. Lab. Invest. **4**, 324—351.

— 1961: An improved Sakaguchi reaction for cytophotometric use. J. Histochem. Cytochem. **9**, 477—483.

EVANS, J. H., D. J. HOLBROOK Jr., and J. L. IRVIN, 1962: Incorporation of labelled precursors into protein and nucleic acid of nuclei of regenerating liver. Exper. Cell Res. **28**, 120—125.

FELIX, K., 1960: Protamines, Adv. Protein Chem. **15**, 1—56.

— H. FISCHER, A. KREKELS, and R. MOHR, 1951: Nucleoprotamin I. Z. physiol. Chem. **287**, 224—234.

— — — and M. RAUEN, 1950: Clupein IX. Z. physiol. Chem. **286**, 67—78.

— and A. MAGER, 1937: Clupeine VIII. Z. physiol. Chem. **249**, 111—123.

FEUGHELMAN, M., R. LANGRIDGE, W. E. SEEDS, A. R. STOKES, H. R. WILSON, C. W. HOOPER, M. H. F. WILKINS, R. K. BARCLAY, and L. D. HAMILTON, 1955: Molecular structure of deoxyribose nucleic acid and nucleoprotein. Nature **175**, 834—838.

FICQ, A., and C. PAVAN, 1957: Autoradiography of polytene chromosomes of *Rhynchosciara angelae* at different stages of larval development. Nature **180**, 983—984.

FLAMM, W. G., and M. L. BIRNSTEIL, 1964: Recent studies on the metabolism of nuclear basic proteins. In: The Nucleohistones, edited by J. Bonner and P. O. P. Ts'o, Holden-Day, San Francisco.

GALL, J., 1959: Macronuclear duplication in the ciliated protozoan *Euplotes*. J. Biophys. Biochem. Cytol. **5**, 295—308.

— and L. B. BJORK, 1958: The spermatid nucleus of two species of grasshopper. J. Biophys. Biochem. Cytol. **4**, 479—484.

GELLOWS, J. G., 1935: The biochemistry of the lens. IV. The origin of the pigment in the lens. Arch. Ophthalmol. **14**, 99—107.

Gifford, E., 1964: Variation in histone concentration during floral induction in *Chenopodium album* and *Xanthium*. The Nucleohistones, ed. by J. Bonner and P. O. P. Ts'o, Holden-Day, San Francisco.

Godman, G. C., A. D. Deitch, and P. Klemperer, 1958: The composition of the LE and hematoxylin bodies of systemic lupus erythematosus. Amer. J. Pathol. 34, 1—23.

Goldstein, L., 1963: RNA and protein in nucleocytoplasmic interactions. Cell Growth and Cell Division, ed. R. J. C. Harris, Academic Press, New York.

Grassé, P. P., N. Carasso, and P. Favard, 1956: Les ultrastructures cellulaires au cours de la spermiogenèse de l'escargot (*Helix pomatia*, L.). Ann. Sci. Nat. Zool. 18, 339—380.

Hamer, D., 1955: The composition of the basic proteins of echinoderm sperm. Biol. Bull. 108, 35—39.

Hnilica, L., E. W. Johns, and J. A. V. Butler, 1962: Observation of the species and tissue specificity of histones. Biochem. J. 82, 123—124.

Holbrook, D. J. Jr., J. L. Irvin, E. M. Irvin, and J. Rotherham, 1960: Incorporation of glycine into protein and nucleic acid fractions of nuclei of liver and hepatoma. Cancer Res. 20, 1329—1337.

Horn, E. C., 1962: Extranuclear histone in the amphibian oocyte. Proc. Nat. Acad. Sci., U. S. 48, 257—265.

— and C. L. Ward, 1957: The localization of the basic proteins in the nuclei of larval *Drosophila* salivary glands. Proc. Nat. Acad. Sci., U. S. 43, 776—779.

Hsu, T. C., 1962: Differential rate in RNA synthesis between euchromatin and heterochromatin. Exper. Cell Res. 27, 332—334.

Huang, R. C., and J. Bonner, 1962: Histone, a suppressor of chromosomal RNA synthesis. Proc. Nat. Acad. Sci., U. S. 48, 1216—1222.

Hultin, T., and R. Herne, 1949: Amino acid analysis of a basic protein fraction from sperm nuclei of some different invertebrates. Ark. Kemi. Minerol. Geol. 26 A. #20, 1—8.

Jacob, F., and J. Monod, 1961: Genetic regulatory mechanisms in the synthesis of proteins. J. Mol. Biol. 3, 318—356.

Kaye, J., 1958: Changes in the fine structure of the nuclei during spermiogenesis. J. Morph. 103, 311—330.

Keck, K., and E. A. Choules, 1963: An analysis of cellular and subcellular systems which transform the species character of acid phosphatase in *Acetabularia*. J. Cell Biol. 18, 459—469.

Kossel, A., 1928: The Protamines and Histones. Longmans Green and Co., London and New York.

Lederberg, J., and T. Iino, 1956: Phase variation in *Salmonella*, Genetics 41, 743—757.

Leslie, I., 1961: Biochemistry of heredity: A general hypothesis. Nature 189, 260—268.

Levene, P. A., and L. W. Bass, 1931: Nucleic Acids. The Chemical Catalog Co., New York.

Lewis, E. S., 1950: The phenomenon of position effect. Adv. Genetics 3, 73—115.

Lindner, A., and T. Kutkam, 1962: Histone, DNA, and protein changes after treatment of tumor cells with 5-fluorouracil (5-FU). Abst. 2nd Ann. Meeting Amer. Soc. Cell Biol. p. 105.

Lucy, J. A., and J. A. V. Butler, 1955: Fractionation of desoxyribonucleoprotein. Biochim. Biophys. Acta 16, 431—432.

Lyon, M. F., 1961: Gene action in the x-chromosome of the mouse (*Mus musculus*, L.). Nature 190, 732—733.

McClintock, B., 1955: Intranuclear systems controlling gene action and mutation. Brookhaven Symp. Biol. 8, 58—74.

— 1961: Some parallels between gene controlled systems in maize and bacteria. Amer. Nat. 95, 265—277.

McLeish, J., L. G. E. Bell, L. F. LaCour, and J. Chayen, 1956: The quantitative cytochemical estimation of arginine. Exper. Cell Res. 12, 120—125.

— and H. S. A. Sherratt, 1958: The use of the Sakaguchi reaction for the cytochemical determination of combined arginine. Exper. Cell Res. 14, 625—628.

Metz, C. W., 1938: Chromosome behavior, inheritance, and sex determination in *Sciara*. Amer. Nat. 72, 485—520.

Miescher, F., 1897: Die histochemischen und physiologischen Arbeiten. Leipzig.

Mirsky, A. E., and A. W. Pollister, 1946: Chromosin, a desoxyribose nucleoprotein complex of the cell nucleus. J. Gen. Physiol. 30, 117—148.
— and H. Ris, 1948: The chemical composition of isolated chromosomes. J. Gen. Physiol. 31, 7—18.
Monné, L., and D. P. Slaughterback, 1951: The disappearence of protoplasmic acidophilia upon deamination. Ark. Zool. Ser. 2,1, 455—462.
Murray, K., 1964: In: The Nucleohistones. Ed. J. Bonner and P. O. P. Ts'o, Holden-Day, San Francisco.
Neelin, J. M., and G. E. Connell, 1959: Zone electrophoresis of chicken erythrocyte histone in starch gel. Biochim. Biophys. Acta 31, 539—541.

Ohno, S., W. D. Kaplan, and R. Kinosita, 1959: Formation of the sex chromatin by a single x-chromosome in liver cells of *Rattus norvegicus*. Exper. Cell Res. 18, 415—418.
Palmade, C., M. R. Chevallier, A. Knobloch, and R. Vendrely, 1958: Isolement d'une deoxyribonucleohistone a partir *d'Escherichia coli*. C. r. Acad. Sci. 246, 2534—2537.
Phillips, D. M. P., 1955: N-terminal groups in salmine. Biochem. J., 60, 403—409.
Pollister, A. W., 1950: Quelques méthodes de cytologie chimique quantitative. Rev. d'Hematologie 5, 527—554.
Prescott, D. M., 1962: Nucleic acid and protein metabolism in the macronuclei of two ciliated protozoa. J. Histochem. Cytochem. 10, 145—153.
— and M. A. Bender, 1963: Synthesis and behavior of nuclear proteins during the cell cycle. J. Cell Comp. Physiol. 62, Suppl. 1, 175—194.
— and R. F. Kimball, 1961: Relation between RNA, DNA, and protein synthesis in the replicating nucleus of *Euplotes*. Proc. Nat. Acad. Sci., U. S. 47, 686—693.

Rasch, E., and J. W. Woodard, 1959: Basic proteins of plant nuclei during normal and pathological cell growth. J. Biophys. Biochem. Cytol. 6, 263—276.
Rasmussen, K. E., 1934: Clupeinuntersuchungen. I. Darstellung und Fraktionierung von Clupein. Z. physiol. Chem. 224, 97—114.
Rauen, H. M., 1952: On the knowledge of clupeine and related proteins. Résumés des Communications, Deuxième Congrès International de Biochimie, Paris, Masson et Cie., p. 190.
Rebhun, L., 1957: Nuclear changes during spermiogenesis in a pulmonate snail. J. Biophys. Biochem. Cytol. 3, 509—524.
Ris, H., and A. W. Pollister, 1947: Nucleoprotein determination in cytological preparations. Cold Spring Harbor Symp. Quant. Biol. 12, 147—157.
Ropshaw, H. J., 1933: Melanogenesis with special reference to sufhydryls and protamines. Amer. J. Physiol. 103, 535—552.
Rybak, B., 1948: La rectification de la fecondations et de la mitose. C. r. Acad. Sci. 226, 1145—1146.

Schultz, J., 1936: Variegation in *Drosophila* and the inert chromosome regions. Proc. Nat. Acad. Sci., U. S. 22, 27—33.
Setterfield, G., J. M. Neelin, E. M. Neelin, and S. T. Bayley, 1960: Studies of basic proteins from ribosomes from buds of pea seedlings. J. Mol. Biol. 2, 416—424.
Singer, M., 1952: Factors which control the staining of tissue sections with acid and basic dyes. Int. Rev. Cytology 1, 211—255.
Stanier, R. Y., and C. B. Van Neil, 1962: The concept of a bacterium. Arch. für Mikrobiol. 42, 17—35.
Stedman, E., and E. Stedman, 1943: Probable function of histone as a regulator of mitosis. Nature 152, 556—557.
— — 1947: The chemical nature and functions of the components of cell nuclei. Cold Spring Harbor Symp. Quant. Biol. 12, 224—236.
— — 1950: Cell specificity of histones. Nature 166, 780—781.
Stern, K., 1952: Problems in nuclear chemistry and biology. Exper. Cell Res. Suppl. 2, 1—15.
Stich, H. V., and J. K. Naylor, 1957: Variation of desoxyribonucleic acid content of specific chromosome regions. Exper. Cell Res. 14, 442—445.
Sud, B. N., 1961: Histochemistry and significance of the chromatoid body in spermatogenesis of the grass snake *Natrix natrix*. Biochem. J. 78, 16 P.
Sutton, W. S., 1903: The chromosomes in heredity. Biol. Bull 4, 231—251.

Taleporos, P., 1959: Cytoplasmic "histones" and "protamines" in the egg of the sea urchin *Strongylocentrotus purpuratus*. J. Histochem. Cytochem. 7, 322.

Tamm, C., M. E. Hodes, and E. Chargaff, 1952: The formation of apurinic acid from the desoxyribonucleic acid of calf thymus. J. Biol. Chem. 195, 49—63.

Taylor, J. H., 1960: Asynchronous duplication of chromosomes in cultured cells of chinese hamster. J. Biophys. Biochem. Cytol. 7, 455—464.

Taylor, J. H., P. S. Woods, and W. L. Hughes, 1957: Organization and duplication of chromosomes as revealed by autoradiographic studies using tritium labelled thymidine. Proc. Nat. Acad. Sci. (U. S.) 43, 122—128.

Ts'o, P. O. P., J. Bonner, and H. Dintzis, 1958: On the similarity of amino acid composition of microsomal nucleoprotein particles. Arch. Biochem. Biophys. 76, 225—227.

Ui, N., 1956: Molecular weights of calf thymus histones. Biochem. Biophys. Acta 22, 205—206.

— 1957: Preparation, fractionation, and properties of calf thymus histone. Biophys. Biochim. Acta 25, 493—502.

Umaña, R.. S. Updike. and A. L. Dounce, 1962: Ratio of total histone to DNA in dividing and interphase cells. Fed. Proc. Soc. Exper. Biol. 21, 156.

Van Slyke, D. D., 1911: A method for the quantitative determination of aliphatic amino groups. J. Biol. Chem. 9, 185—207.

Vaughn, J. C., 1964: Personal communication.

Vendrely, R.: 1957: Données récentes sur le chimie de l'ADN et des desoxyribonucleoproteines. Arch. Julius-Klaus-Stiftung Vererbungsforschung, Sozialanthropologie, Rassenhygiene 32, 538—553.

— M. Alfert, A. Knobloch, and H. Matsudaira, 1958: The composition of histone from pycnotic nuclei. Exper. Cell Res. 2, 295—300.

— A. Knobloch, and H. Matsudaira, 1959: A comparative biochemical study of nucleohistone from different vertebrates. Nature 181, 343.

— A. Knobloch-Mazen, and C. Vendrely, 1959: A comparative biochemical study of nucleohistones and nucleoprotamines in the cell nucleus. In: The Cell Nucleus. Ed. J. S. Mitchell, London, Butterworths, pp. 200—205.

— and C. Vendrely, 1953: Arginine and deoxyribonucleic acid content of erythrocyte nuclei and sperms of some species of fishes. Nature 172, 30—31.

Vincent, W. A., 1952: The isolation and chemical properties of the nucleoli of starfish oocytes. Proc. Nat. Acad. Sci., U. S. 38, 139—145.

Waldschmidt-Leitz, E., and E. Gunther, 1948: Bausteinanalyse des Clupeins und Salmines. Makromol. Chem. 2, 120—126.

Wilkins, M. H. F., and G. Zubay, 1959: The absence of histone in the bacterium Escherischia coli. II. X-ray diffraction of the nucleoprotein extract. J. Biophys. Biochem. Cytol. 5, 55—58.

— — and H. R. Wilson, 1959: X-ray diffraction studies of the molecular structure of nucleohistone and chromosomes. J. Mol. Biol. 1, 179—185.

Wimber, D. E., and H. Quastler, 1963: A ^{14}C- and ^{3}H-thymidine double labelling technique in the study of cell proliferation in Tradescantia root rips. Exper. Cell Res. 30, 8—22.

Woodard, J. W., E. Rasch, and H. Swift, 1961: Nucleic acid and protein metabolism during the mitotic cycle in Vicia faba. J. Biophys. Biochem. Cytol. 9, 445—462.

Wrinch, D. M., 1934: Chromosome behavior in terms of protein pattern. Nature 134, 978—979.

Yasuzumi, G., and K. H. Ishida, 1957: Spermatogenesis in animals as revealed by electron microscopy. II. Submicroscopic structure of developing spermatid nuclei of grasshopper. J. Biophys. Biochem. Cytol. 3, 633—668.

Yphantis, D., 1960: Rapid determination of molecular weights of peptides and proteins. 88, 586—601.

Zbarskii, I. B., and P. A. Perevoshchikova, 1951: Makromol. Chem. 2, 120—126.

Zeevaart, J. A. D., 1964: Chemical Basic of induction. In: The Nucleohistones. Ed. by J. Bonner and P. O. P. Ts'o.

Zubay, G., and P. Doty, 1959: The isolation and properties of deoxyribonucleoprotein particles containing single nucleic acid molecules. J. Mol. Biol. 1, 1—20.

— and M. R. Watson, 1959: The absence of histone in the bacterium Escherichia coli. I. Preparation and analysis of the nucleoprotein extract. J. Biophys. Biochem. Cytol. 5, 51—54.

— and M. H. F. Wilkins, 1960: X-ray diffraction studies of the structure of ribosomes from Escherichia coli. J. Mol. Biol. 2, 105—112.